Joseph L. Awange and Erik W. Grafarend
Solving Algebraic Computational Problems in Geodesy and Geoinformatics
The Answer to Modern Challenges

Joseph L. Awange
Erik W. Grafarend

Solving Algebraic Computational Problems in Geodesy and Geoinformatics

The Answer to Modern Challenges

With 79 Figures

 Springer

DR.-ING. JOSEPH L. AWANGE
DEPARTMENT OF GEOPHYSICS
KYOTO UNIVERSITY
KITASHIRAKAWA OIWAKE-CHO
SAKYO-KU KYOTO-SHI
606-8502
JAPAN

E-mail: jawange@yahoo.co.uk

PROF. DR.-ING. HABIL. ERIK W. GRAFAREND
DEPARTMENT OF GEODESY
STUTTGART UNIVERSITY
GESCHWISTER-SCHOLL-STRAßE 24D
70174 STUTTGART
GERMANY

E-mail: grafarend@gis.uni-stuttgart.de

Library of Congress Control Number: 2004114234

ISBN 3-540-23425-X **Springer Berlin Heidelberg New York**

Springer is a part of Springer Science+Business Media
springeronline.com
© Springer-Verlag Berlin Heidelberg 2005
Printed in Germany

The use of general descriptive names, registered names, trademarks, etc. in this publication does not imply, even in the absence of
a specific statement, that such names are exempt from the relevant protective laws and regulations and therefore free for general use.

Cover design: E. Kirchner, Heidelberg
Production: A. Oelschläger
Typesetting: Camera-ready by the Authors
Printing: Mercedes-Druck, Berlin
Binding: Stein + Lehmann, Berlin

Printed on acid-free paper   32/2132/AO   5 4 3 2 1 0

This book is dedicated to my co-author, Ph.D. supervisor and mentor Professor Dr.-Ing. habil. Dr.h.c. mult. Erik W. Grafarend.

Joseph L. Awange
September 2004

Preface

While preparing and teaching 'Introduction to Geodesy I and II' to undergraduate students at Stuttgart University, we noticed a gap which motivated the writing of the present book: Almost every topic that we taught required some skills in algebra, and in particular, computer algebra! From positioning to transformation problems inherent in geodesy and geoinformatics, knowledge of algebra and application of computer algebra software were required. In preparing this book therefore, we have attempted to put together basic concepts of *abstract algebra* which underpin the techniques for solving algebraic problems. Algebraic computational algorithms useful for solving problems which require exact solutions to nonlinear systems of equations are presented and tested on various problems. Though the present book focuses mainly on the two fields, the concepts and techniques presented herein are nonetheless applicable to other fields where algebraic computational problems might be encountered. In Engineering for example, network densification and robotics apply resection and intersection techniques which require algebraic solutions.

Solution of nonlinear systems of equations is an indispensable task in almost all geosciences such as geodesy, geoinformatics, geophysics (just to mention but a few) as well as robotics. These equations which require exact solutions underpin the operations of ranging, resection, intersection and other techniques that are normally used. Examples of problems that require exact solutions include;

- three-dimensional resection problem for determining positions and orientation of sensors, e.g., camera, theodolites, robots, scanners etc.,

- coordinate transformation to match shapes and sizes of points in different systems,
- mapping from topography to reference ellipsoid and,
- analytical determination of refraction angles in GPS meteorology.

The difficulty in solving explicitly these nonlinear systems of equations has led practitioners and researchers to adopt approximate numerical procedures; which often have to do with linearization, approximate starting values, iterations and sometimes require alot of computational time. In-order to offer solutions to the challenges posed by nonlinear systems of equations, this book provides in a pioneering work, the application of *ring* and *polynomial theories, Groebner basis, polynomial resultants, Gauss-Jacobi combinatorial* and *Procrustes algorithms*. Users faced with algebraic computational problems are thus provided with algebraic tools that are not only a MUST, but essential and have been out of reach. For these users, most algebraic books at their disposal have unfortunately been written in mathematical formulations suitable to mathematicians. We strive to simplify the algebraic notions and provide examples where necessary to enhance easier understanding.

For those in mathematical fields such as applied algebra, symbolic computations and application of mathematics to geosciences etc., the book provides some practical examples of application of mathematical concepts. Several geodetic and geoinformatics problems are solved in the book using methods of abstract algebra and multidimensional scaling. These examples might be of interest to some mathematicians.

Chapter 1 introduces the book and provides a general outlook on the main challenges that call for algebraic computational approaches. It is a motivation for those who would wish to perform analytical solutions. Chapter 2 presents the basic concepts of ring theory relevant for those readers who are unfamiliar with abstract algebra and therefore prepare them for latter chapters which require knowledge of ring axioms. Number concept from operational point of view is presented. It is illustrated how the various sets of natural numbers $\mathbb{N}$, integers $\mathbb{Z}$, quotients $\mathbb{Q}$, real numbers $\mathbb{R}$, complex numbers $\mathbb{C}$ and quaternions $\mathbb{H}$ are vital for daily operations. The chapter then presents the concept of ring theory. Chapter 3 looks at the basics of polynomial theory; the main object used by the algebraic algorithms that will be discussed in the book. The basics of polynomials are recaptured for readers who wish to refreshen their memory on the basics of algebraic operations. Starting with the definition of polynomials, Chap. 3 expounds on the concept of polynomial rings thus linking it to the number ring theory presented in Chap. 2. Indeed, the theorem developed in the chapter enables the solution of nonlinear systems of equations that can be converted into (algebraic) polynomials.

Having presented the basics in Chaps. 2 and 3, Chaps. 4, 5, 6 and 7 present algorithms which offer algebraic solutions to nonlinear systems of equations. They present theories of the procedures starting with the basic concepts and showing how they are developed to algorithms for solving different problems. Chapters 4, 5 and 6 are based on *polynomial ring theory* and offer an in-depth look at the basics of *Groebner basis, polynomial resultants* and *Gauss-Jacobi combinatorial algorithms*. Using these algorithms, users can develop their own codes to solve problems requiring exact solutions.

In Chap. 7, the Global Positioning System (GPS) and the Local Positioning Systems (LPS) that form the operational basis are presented. The concepts of local datum choice of types $\mathbb{E}^*$ and $\mathbb{F}^*$ are elaborated and the relationship between Local Reference Frame $\mathbb{F}^*$ and the global reference frame $\mathbb{F}^\bullet$, together with the resulting observational equations are presented. The test network "Stuttgart Central" in Germany that we use to test the algorithms of Chaps. 4, 5 and 6 is also presented in this chapter. Chapter 8 deviates from the polynomial approaches to present a linear algebraic (analytical) approach of Procrustes that has found application in fields such as *medicine* for gene recognition and *sociology* for crime mapping. The chapter presents only the partial Procrustes algorithm. The technique is presented as an efficient tool for solving algebraically *the three-dimensional orientation problem* and the *determination of vertical deflection*.

From Chaps. 9 to 15, various computational problems of algebraic nature are solved. Chapter 9 looks at the ranging problem and considers both the GPS pseudo-range observations and ranging within the LPS systems, e.g., using EDMs. The chapter presents a complete algebraic solution starting with the simple planar case to the three-dimensional ranging in both closed and overdetermined forms. Critical conditions where the solutions fail are also presented. Chapter 10 considers the Gauss ellipsoidal coordinates and applies the algebraic technique of Groebner basis to map topographic points onto the reference ellipsoid. The example based on the baltic sea level project is presented. Chapters 11 and 12 consider the problems of resection and intersection respectively.

Chapter 13 discusses a modern and relatively new area in geodesy; the GPS meteorology. The chapter presents the theory of GPS meteorology and discusses both the space borne and ground based types of GPS meteorology. The ability of applying the algebraic techniques to derive refraction angles from GPS signals is presented. Chapter 14 presents an algebraic deterministic version to outlier problem thus deviating from the statistical approaches that have been widely publicized. Chapter 15 introduces the 7-parameter datum transformation problem commonly encountered in practice and presents the general Procrustes algorithm. Since this is an extension of the partial Procrustes algorithm presented in Chap. 8, it is referred to as Procrustes algorithm II. The chapter further presents an algebraic solution of the transformation problem using Groebner basis and Gauss-Jacobi combinatorial algorithms. The book is completed in Chap. 16 by presenting an overview of modern computer algebra systems that may be of use to geodesists and geoinformatists.

Many thanks to Prof. B. Buchberger for his positive comments on our Groebner basis solutions, Prof. D. Manocha who discussed the resultant approach, Prof. D. Cox who also provided much insight in his two books on rings, fields and algebraic geometry and Prof. W. Keller of Stuttgart University Germany, whose door was always open for discussions. We sincerely thank Dr. J. Skidmore for granting us permission to use the Procrustes 'magic bed' and related materials from Mythweb.com. Thanks to Dr. J. Smith (editor of Survey Review), Dr. S. J. Gordon and Dr. D. D. Lichti for granting us permission to use the scanner resection figures appearing in Chap. 12. We are also grateful to Chapman and Hall Press for granting us permission to use Fig. 8.2 where malarial parasites are identified using Procrustes. Special thanks to Prof. I. L. Dryden for permitting us to refer to his work and all the help. Many thanks to Ms F. Wild for preparing Figs. 11.7 and 12.7. We acknowledge your efforts and valuable time. Special thanks to Prof. A. Kleusberg of Stuttgart University Germany, Prof. T. Tsuda of Radio Center for Space and Atmosphere, Kyoto University Japan, Dr. J. Wickert of GeoForschungsZentrum Potsdam (GFZ) Germany and Dr. A. Steiner of the Institute of Meteorology and Geophysics, University of Graz, Austria for the support in terms of literature and discussions on Chap. 13. The data used in Chap. 13 were provided by GeoForschungsZentrum Potsdam (GFZ). For these, the authors express their utmost appreciation.

The first author also wishes to express his utmost sincere thanks to Prof. S. Takemoto and Prof. Y. Fukuda of Department of Geophysics, Kyoto University Japan for hosting him during the period of September 2002 to September 2004. In particular Chap. 13 was prepared under the supervision and guidance of Prof. Y. Fukuda: Your ideas, suggestions and motivation enriched the book. For these, we say "*arigato gozaimashita*" – Japanese equivalent to *thank you very much*. The first author's stay at Kyoto University was supported by Japan Society of Promotion of Science (JSPS): The author is very grateful for this support. The first author is grateful to his wife Mrs. *Naomi Awange* and his two daughters *Lucy* and *Ruth* who always brightened him up with their cheerful faces. Your support, especially family time that I denied you in-order to prepare this book is greatly acknowledged. Naomi, thanks for carefully reading the book and correcting typographical errors. However, the authors take full responsibility of any typographical error. Last but not least, the second author wants to thank his wife *Ulrike Grafarend*, his daughter *Birgit* and his son *Jens* for all support over these many years they were following him at various places around the Globe.

Kyoto (Japan) and Stuttgart (Germany)   *Joseph L. Awange*
September 2004   *Erik W. Grafarend*

Contents

1

Introduction

Since the advent of the Global Navigation Satellite System (GNSS), and particularly Global Positioning System (GPS), the fields within geosciences such as geodesy, geoinformatics, geophysics, hydrology etc., have undergone tremendous changes. GPS satellites have revolutionized operations in these fields and the entire world in ways that its inventors never fathomed. The initial goal of GPS satellites was to provide the capability for the US army to position from space. This way, they could be able to know the positions of their submarines without necessarily relying on fixed ground targets which were liable to enemy attack. Slowly, but surely, the civilian community, led by geodesists, begun to device methods of exploiting the potentials of this system. The initial focus of research was on the improvement of positioning accuracies since civilians were only accessible to the coarse acquisition C/A-code of the GPS signal. This code is less precise as compared to the P-code used by the US army and its allies. The other source of error in positioning was the Selective Availability (SA). However, in May 2000, the then United States of America's president Bill Clinton officially discontinued the selective availability.

As research on GPS progressed, so were new discoveries of its uses. For example, previous research focussed on modelling or eliminating atmospheric effects such as refraction and multipath on the transmitted signals. In the last decade, however, Melbourne et al. [248] suggested that this negative effect of the atmosphere on GPS signals could be inverted to remote sense the atmosphere for vertical profiles of temperature and pressure. This gave birth to the new field of GPS meteorology which is currently an active area of research. GPS meteorology has now propelled environmental and atmospheric studies with the enhancement of weather prediction and forecasting. This new technique is presented in Chap. 13, where the algebraic computations involved are solved. One would be forgiven to say that the world will soon be unable to operate without GPS satellites. This, however, will not be an understatement either. GPS satellites have influenced our lives such that almost every operation is increasingly becoming GPS dependent! From the use of mobile phones, fertilizer regulation in farming, fish tracking in fisheries, vehicle navigation etc., the word is GPS. These numerous advantages of GPS satellites have led the European countries to prepare GALILEO satellites which are the equivalent of GPS, scheduled to be operational around 2008 [335, p. 24]. The Russian based Globalnaya Navigationnaya Sputnikovaya Sistema (or simply Global Navigation Satellite System) GLONASS are still operational albeit with financial constraints.

The direct impact of using these satellites is the requirement that operations be almost entirely three-dimensional. The major challenge posed by this requirement is that of integrating the satellite system, which operates globally to the traditional techniques that operate locally. In geodesy and geoinformatics for example, satellite positioning has necessitated the transformation of coordinates from local systems to GPS global system (World Geodetic System WGS-84). This problem, and others involving GPS satellites such as those discussed in this book have one thing in common: *They require the solution of nonlinear equations that relate the unknowns to the measured values.*

In daily operations, nonlinear equations are encountered in several applications, thus necessitating the need for developing efficient and reliable computational tools. In cases where the number of observations n and the number of unknowns m are equal, i.e., $n = m$, the unknown parameters may be obtained by solving explicitly (in a closed form) nonlinear systems of equations. Because of the difficulty in obtaining reliable closed form procedures, approximate numerical methods have been adopted in practice. Such procedures depend on some approximate starting values, linearization and iterations. In some cases, the numerical methods used are unstable or the iterations fail to converge depending on the initial "guess" [269, pp. 340–342]. The other shortcoming of the approximate numerical procedures has been pointed out by Cox et al. [95, pp. 28–32]; who in their book have illustrated that systems of equations with exact solutions become vulnerable to small errors introduced during the process of establishing the roots. In case of extending the partial solution to the complete solutions of the system, errors may accumulate and thus become so large. If the partial solution was derived by iterative procedures then the errors incurred during the root-finding may blow up during the extension of the partial solution to the complete solution (back substitution). There exists therefore a strong need for unified procedures that can be applied in general to offer exact solutions to nonlinear systems of equations.

In some applications, explicit formulae rather than numerical solutions are desirable. In such cases, explicit procedures are usually employed. The resulting explicit formulae often consists of univariate polynomials relating the unknown parameters (unknown variables) to the known variables (observations). By inserting numeric values into these explicit formulae, solutions can immediately be computed for the unknown variables. In-order to understand the foregoing discussion, let us consider a case where students have been asked to integrate the function $f(x) = x^5$ with respect to x. In this case, the power of x, i.e., 5 is definite and the integration can easily be performed. Assume now that for a specific purpose, the power of x can be varied taking on different values say $n = 1, 2, 3, ...$ In such a case, it is not prudent to integrate x raised to each power, but to seek a general explicit formula by integrating

$$\int x^n \mathrm{d}x, \tag{1.1}$$

to give

$$\frac{x^{n+1}}{n+1}. \tag{1.2}$$

One thereafter needs only to insert a given value of n in (1.2) to obtain a solution. In practice, several problems require explicit formulae as they are performed repeatedly.

Besides the requirement of exact solutions by some applications, there also exist the problem of exact solutions of overdetermined systems (i.e., where many observations than unknown exist). In reality, field observations often result in more data being collected than is required to determine the unknown parameters. Exact solution to the overdetermined problems is just but one of the challenges. In some applications, such as the 7-parameter datum transformation discussed in Chap. 15, where coordinates have to be transformed from local (national) systems to GPS (global) system and vice versa, handling of stochasticities of these systems still pose a serious challenge to users.

Approximate numerical procedures which are applied in practice do not offer tangible solution to this problem. Other than the stochasticity issues, numerical methods employed to solve the 7-parameter datum transformation problem require some initial starting values, linearization and iterations as already mentioned. In Photogrammetry, where the rotation angles are very small, the initial starting values are often set to zero. This, unfortunately, may not be the case for other applications in geosciences. In Chaps. 8 and 15 we present powerful analytical and algebraic techniques developed from the fields of multidimensional scaling and abstract algebra to solve the problem. In particular, the Procrustes algorithm, which enjoys wide use in fields such as medicine and sociology is straightforward and easy to program. The advantages of these techniques are; the non-requirements of the conditions that underpin approximate numerical solutions, and their capability to take into consideration weights of the systems involved.

Solution of unknown parameters from nonlinear observational equations are only meaningful if the observations themselves are pure and uncontaminated by gross errors (outliers). This raises the issue of outlier detection and management. Traditionally, statistical procedures have been put forward for detecting outliers in observational data sample. Once detected, observations that are contaminated with outliers are isolated and the remaining pure observations used to estimate unknown parameters. Huber [191] and Hampel et al [172] however point out the dangers that exist in such approach. These are; false rejection of the otherwise good observations, and false retention of contaminated observations. To circumvent these dangers, robust estimation procedures were proposed in 1964 by the father of robust statistics, P. J. Huber [189] to manage outliers without necessarily rejecting outlying observations. Since then, as we shall see in Chap. 14, several contributions to outlier management using robust techniques have been put forward. Chapter 14 deviates from the statistical approaches to present the non deterministic outlier diagnosis based on algebraic approaches which enjoy the advantages already discussed.

On the instrumentation front, there has been tremendous improvement in computer technology. Today, laptops are made with large storage capacity with high memory thus enabling faster computations. Problems can now be solved using algebraic methods that would have been impossible to solve by hand. The advances in computer technology has also propelled development of precise and accurate measuring devices. With the improvement in computer technology and the manufacture of precise and accurate measuring devices, modern challenges facing those in fields of geosciences and engineering include:

- Handling in an efficient and manageable way the nonlinear systems of equations that relate observations to unknowns. In GPS meteorology for example, more than 1000 satellite occultations are obtained on daily basis, from which bending angles of the signals are to be computed. In practice, the nonlinear system of equations for bending angles is often solved using Newton's method iteratively. An explicit formula could be derived from the nonlinear system of equations as presented in Chap. 13.
- Obtaining a unified closed form solution (e.g., Awange et al. [36]) for different problems. For a particular problem, several procedures are often put forward in an attempt to offer exact solution. The GPS pseudo-range problem for example, has attracted several exact solution procedures as evidenced in the works of [47, 150, 198, 199, 221, 296]. It is desirable in such a case to have a unified solution approach which can easily be applied to all problems in general.
- Controlling approximate numerical algorithms that are widely used.
- Obtaining computational procedures that are time saving.
- Having computational procedures that do not peg their operations on approximate starting values, linearization or iterations.
- Take advantage of the large storage capacity and fast speed of modern computers to solve problems which have hitherto evaded solution.
- Prove the validity of theorems and formulae that are in use which were derived based on trial and error basis.
- Perform rigorous analysis of the nonlinearity effects on most models that are in operation but assume or ignore nonlinearity.

These challenges and many others had existed before, and earlier researchers had acknowledged the fact and realized the need for addressing them through developing explicit solutions. Merritt [249] had, for example, listed the advantages of explicit solutions as;

1. provision of satisfaction to the users (photogrammetrists and mathematicians) of the methods,
2. provision of data tools for checking the iterative methods,
3. desired by geodesists whose task of control network densification does not favour iterative procedures,

4. provision of solace and,

5. the requirement of explicit solutions rather than iterative by some applications.

Even though such advantages had been noted, their actual realization was out of reach as the equations involved were large and required more than a paper and a pen to solve. Besides, another drawback was that these exact solutions were like rare jewel. The reason for this was partly because the methods required extensive computations and partly because the resulting symbolic expressions were too large and required computers with large storage capacity. Until recently, computers that were available could hardly handle large computations due to lack of faster Central Processing Unit (CPU), shortage of Random Access Memory (RAM) and limited hard disk storage capacity. The other setback was that some of the methods, especially those from algebraic fields, were formulated based on theoretical concepts that were hard to realize or comprehend without the help of computers. For a long time therefore, these setbacks hampered progress of explicit procedures. The painstaking efforts to obtain exact solution discouraged practitioners to the extent that the use of numerical approaches were the order of the day. Most of these numerical approaches had no independent methods for validation, while other problems evaded numerical solutions and required closed form solutions.

The answer to these modern challenges, lies in the application of algebraic computational techniques. Algebra has been widely applied in fields such as robotics for kinematic modelling of robots, engineering for offset surface construction in solid modelling, computer science for automated theorem proving, Computer Aided Design (CAD) etc. The well known application of algebra in geodesy could perhaps be the use of Legendre polynomials in spherical harmonic expansion studies. More recent application of algebra in geodesy is evidenced in the works of Biagi and Sanso [63], Awange [11] and Lannes and Durand [214]. The latter proposes a new approach to differential GPS based on algebraic graph theory. The present book examines algebraic computational problems inherent in geodesy and geoinformatics which require algebraic solutions. Powerful tools for solving such problems are presented with numerous examples given on their applicability in practice. We focus on nonlinear systems of equations whose exact (algebraic) solutions have been a thorn in the flesh to users, and provide several examples that are encountered in practice.

2

Basics of Ring Theory

2.1 Some Applications to Geodesy and Geoinformatics

This chapter presents the concepts of *ring theory* from a geodetic and geoinformatics perspective. The presentation is such that the mathematical formulations are augmented with examples from the two fields. Ring theory forms the basis upon which polynomial rings operate. As we shall see later, exact solution of nonlinear systems of equations are pinned to the operations on polynomial rings. In Chap. 3, polynomials will be discussed in detail. In-order to understand the concept of polynomial rings, one needs first to be familiar with the basics of ring theory. This chapter is therefore a preparation for the understanding of the polynomial rings presented in Chap. 3. Ring of numbers which is presented in Sect. 2.2 plays a significant role in daily operations. They permit operations addition, subtraction, multiplication and division of numbers. For those engaged in data collection, ring of numbers play the following role;

- they specify the number of sets of observations to be collected,
- they specify the number of observations or measurements per set,
- they enable manipulation of these measurements to determine the unknown parameters.

We start by presenting *ring of numbers*. Elementary introduction of the sets of *natural numbers, integers, rational numbers, real numbers, complex numbers* and *quaternions* are first given before defining the ring. We strive to be as simple as possible so as to make the concepts clear to readers with less or no knowledge of rings.

2.2 Numbers from Operational Perspective

When undertaking operations such as measurements of angles, distances, gravity, photo coordinates, digitizing of points etc., numbers are often used. Measured values are normally assigned numbers. A measured distance for example can be assigned a value of 100 m to indicate the length. Numbers, e.g., $1, 2, ...$, also find use as;

- counters to indicate the frequency of taking measurements,
- counters indicating the number of points to be observed or,
- passwords to;
  - the processing hardware (e.g. computers),
  - softwares (such as those of Geographical Information Systems (GIS) packages) and,
  - accessing pin numbers in the bank!

In all these cases, one operates on a set of *natural numbers*

$$\mathbb{N} = \{0, 1, 2, ...\}, \tag{2.1}$$

with 0 added. The number 0 was invented by the Babylonians in the third century B.C.E, re-invented by the Mayans in the fourth century B.C.E and in India in the fifth century [193, p. 69]. The set $\mathbb{N}$ in (2.1) is closed under;

- addition, in which case the sum of two numbers is also a natural number (e.g., $3 + 6 = 9$) and,
- multiplication, in which case the product of two numbers is a natural number (e.g., $3 \times 6 = 18$).

Subtraction, i.e., the difference of two natural numbers is however not necessarily a natural number (e.g. $3 - 6 = -3$). To circumvent the failure of the natural numbers to be closed under subtraction, negative numbers were introduced and added in front of natural numbers. For a natural number n for example, $-n$ is written. This expanded set

$$\mathbb{Z} = \{-2, -1, 0, 1, 2, ...\}, \tag{2.2}$$

is the set of *integers*. The letter $\mathbb{Z}$ is adopted from the first letter of the German word for integers "**Z**ahl". The set $\mathbb{Z}$ is said to have:

- an "additive identity" number 0 which when added to any integer n preserves the "identity" of n, e.g., $0 + 13 = 13$,

- "additive inverse" $-n$ which when added to an integer n results in an identity 0, e.g., $-13 + 13 = 0$. The number -13 is an additive inverse of 13.

The set $\mathbb{Z}$ with the properties "addition" and "additive inverse" enables one to manipulate numbers by being able to add and subtract. This is particularly helpful when handling measured values. It allows for instance the solution of equations of type $y + m = 0$, where m is an integer. In-order to allow them to divide numbers as is the case with distance ratio observations, "multiplicative identity" and "inverse" have to be specified as:

- "multiplicative identity" is the integer 1 which when multiplied with any integer n preserves the "identity" of n, e.g., $1 \times 13 = 13$,
- "multiplicative inverse" is an integer m such that its multiplication with an integer n results in an identity 1, e.g., $m \times n = 1$.

For a non-zero integer n, therefore, a multiplicative inverse $\frac{1}{n}$ has to be specified. The multiplicative inverse of 5 for example is $\frac{1}{5}$. This leads to an expanded set comprising of both integers and their multiplicative inverses as

$$\mathbb{Q} = \{-2, -\frac{1}{2}, -1, 0, 1, 2, \frac{1}{2}, ...\}, \tag{2.3}$$

where a new number has been created for each number except $-1, 0, 1$. Except for 0, which is a special case, the set $\mathbb{Q}$ is closed under "additive" and "multiplicative inverses" but not "addition" and "multiplication". This is circumvented by incorporating all products of integers and multiplicative inverses $m \times \frac{1}{n} = \frac{m}{n}$, which are ratios of integers resulting into a set of *rational numbers* $\mathbb{Q}$. $\mathbb{Q}$ is the first letter of **Q**uotient and is closed since:

- For every rational number, there exist an additive inverse which is also a rational number, e.g., $-\frac{1}{13} + \frac{1}{13} = 0$.
- Every rational number except 0 has a multiplicative inverse which is also a rational number, e.g., $13 \times \frac{1}{13} = 1$.
- The set of rational numbers is closed under addition and multiplication, e.g., $\frac{1}{3} + \frac{1}{3} = \frac{2}{3}$ and $\frac{1}{3} \times \frac{1}{3} = \frac{1}{9}$.

The set $\mathbb{Q}$ is suitable as it permits addition, subtraction, multiplication and division. It therefore enables the solution of equations of the form

$ny - m = 0$, $\{m, n\}$ being arbitrary integers, with $n \neq 0$. This set is however not large enough as it leaves out the square root of numbers and thus cannot measure the Pythagorean length. In geodesy, as well as geoinformatics, the computation of distances from station coordinates by Pythagoras demands the use of square root of numbers. The set of quotient $\mathbb{Q}$ is thus enlarged to the set of real numbers $\mathbb{R}$, where the positive real numbers are the ones required to measure distances as shall be seen in Chaps. 9, 11 and 12. Negative real numbers are included to provide additive inverses. The set $\mathbb{R}$ also possesses multiplicative inverses. This set enables the solution of equations of the form $y^2 - 3 = 0 \Rightarrow y = \pm\sqrt{3}$, which is neither integer nor rational. The set $\mathbb{R}$ is however not large enough to provide a solution to an equation of the form $y^2 + 1 = 0$. It therefore gives way to the set $\mathbb{C}$ of complex numbers, where $i^2 = -1$.

The set $\mathbb{C}$ can be expanded further into a set $\mathbb{H}$ of *quaternions* which was discovered by W. R. Hamilton on the 16th of October 1843, having worked on the problem for 13 years (see Note 2.1 on p. 11). Even as he discovered the quaternions, it occurred to him that indeed Euler had known of the existence of the four square identity in 1748 and that quaternion multiplication had been used by Rodrigues in 1840 to compute the product of rotations in $\mathbb{R}^3$ [301]. Indeed as we shall see in Chap. 6, Gauss knew of quaternions even before Hamilton, but unfortunately, he never published his work. In geodesy and geoinformatics, quaternions have been used to solve the three-dimensional resection problem by [156]. They have also found use in the solution of the similarity transformation problem discussed in Chap. 15 as evidenced in the works of [295, 311, 312, 352]. Quaternion is defined as the matrix

$$\begin{bmatrix} (a + di) & (b + ci) \\ (-b + ci) & (a - di) \end{bmatrix} | \{a, b, c, d\} \in \mathbb{R}, \tag{2.4}$$

which is expressed in terms of unit matrices $\mathbf{1}, \mathbf{i}, \mathbf{j}, \mathbf{k}$ as

$$\begin{bmatrix} \begin{bmatrix} a + di & b + ci \\ -b + ci & a - di \end{bmatrix} = a \begin{bmatrix} 1 & 0 \\ 0 & 1 \end{bmatrix} + b \begin{bmatrix} 0 & 1 \\ -1 & 0 \end{bmatrix} + c \begin{bmatrix} 0 & i \\ i & 0 \end{bmatrix} + d \begin{bmatrix} i & 0 \\ 0 & -i \end{bmatrix} \\ a\mathbf{1} + b\mathbf{i} + c\mathbf{j} + d\mathbf{k}, \end{bmatrix}$$

$$\tag{2.5}$$

where $\mathbf{1}, \mathbf{i}, \mathbf{j}, \mathbf{k}$ are quaternions of norm 1 that satisfy

$$\begin{bmatrix} \mathbf{i}^2 = \mathbf{j}^2 = \mathbf{k}^2 = -\mathbf{1} \\ \mathbf{ij} = \mathbf{k} = -\mathbf{ji} \\ \mathbf{jk} = \mathbf{i} = -\mathbf{kj} \\ \mathbf{ki} = \mathbf{j} = -\mathbf{ik}. \end{bmatrix} \qquad (2.6)$$

The norm of the quaternions is the determinant of the matrix (2.4) and gives

$$det \begin{bmatrix} (a+di) & (b+ci) \\ (-b+ci) & (a-di) \end{bmatrix} = a^2 + b^2 + c^2 + d^2, \qquad (2.7)$$

which is a four square identity. This matrix definition is due to Cayley, while Hamilton wrote the rule $\mathbf{i}^2 = \mathbf{j}^2 = \mathbf{k}^2 = \mathbf{ijk} = -\mathbf{1}$ that define quaternion multiplication from which he derived the four square identity [301, p. 156].

We complete this section by defining algebraic integers as

Definition 2.1 (Algebraic). *A number $n \in \mathbb{C}$ is algebraic if*

$$a_n \alpha^n + a_{n-1} \alpha^{n-1} + ... + a_1 \alpha + a_0 = 0, \qquad (2.8)$$

and it takes on the degree n if it satisfies no such equation of lower degree and $a_0, a_1, ..., a_n \in \mathbb{Z}$.

We shall see in Chap. 3 that Definition (2.1) satisfies the definition of a univariate polynomial.

Note 2.1 (Hamilton's Letter). How can one dream about such a "quaternion algebra" $\mathbb{H}$? W. R. Hamilton (16th October 1843) invented *quaternion numbers as outlined in a letter (1865) to his son A. H. Hamilton for the following reason:*

> "If I may be allowed to speak of *myself* in connection with the subject, I might do so in a way which would bring *you* in, by referring to an *antequaternionic* time, when you were a mere *child*, but had caught from me the conception of a vector, as represented by a *triplet*; and indeed I happen to be able to put the finger of memory upon the year and month –October, 1843– when having recently returned from visits to Cork and Parsonstown, connected with a meeting of the British Association, the desire to discover the laws of the multiplication referred to regained with me a certain strength and earnestness, which had for years been dormant, but was then on the point of being gratified, and was occasionally talked of with you. Every morning in the early part of the above cited month, on my coming down to breakfast, your (then) little brother William Edwin, and yourself, used to ask me, "well, Papa, can you *multiply* triplets"? Whereto I was always obliged to reply, with a sad shake of the head: "No, I can only *add* and subtract them.""

But on the 16th day of the same month – which happened to be a Monday, and a Council day of the Royal Irish Academy – I was walking in to attend and preside, and your mother was walking with me, along the Royal Canal, to which she had perhaps driven; and although she talked with me now and then, yet an *under-current* of thought was going on in my mind, which gave at last a *result*, whereof it is not too much to say that I felt *at once* the importance. An *electric* circuit seemed to *close*; and a spark flashed forth. The herald (as I *foresaw, immediately*) of many long years to come of definitely directed thought and work, by *myself* if spared, and at all events on the part of *others*, if should even be allowed to live long enough distinctly to communicate the discovery. Nor could I resist the impulse – unphilosophical as it may have been – to cut with a knife on a stone of Brougham Bridge, as we passed it, the fundamental formula with the symbols, i, j, k; namely

$$\mathbf{i}^2 = \mathbf{j}^2 = \mathbf{k}^2 = \mathbf{ijk} = -\mathbf{1},$$

which contains the *solution of the problem*, but of course, as an inscription, has long since mouldered away. A more durable notice remains, however, on the Council Books of the Academy for that day (October 16th, 1843), which records the fact, that I then asked for and obtained base to read a paper on *quaternion*, at the *First General Meeting* of the Session: which reading took place accordingly, on Monday the 13th of the November following."

2.3 Number Rings

In everyday lives of geodesists and geoinformatists, rings are used albeit without being noticed: A silent tool without which perhaps they might find the going tough. In the preceding section, the sets of integers $\mathbb{Z}$, rational numbers $\mathbb{Q}$, real numbers $\mathbb{R}$ and complex numbers $\mathbb{C}$ were introduced as being closed under addition and multiplication. Loosely speaking, a system of numbers that is closed under addition and multiplication is a ring. A more precise definition of a ring based on linear algebra will be given later.

It suffices at this point to think of the sets $\mathbb{Z}$, $\mathbb{Q}$, $\mathbb{R}$ and $\mathbb{C}$, upon which we manipulate numbers, as being a collection of numbers that can be added, multiplied, have additive identity 0 and multiplicative identity 1. In addition, every number in these sets has an additive inverse thus forming a ring. Measurements of distances, angles, directions, photo coordinates, gravity etc., comprise the set $\mathbb{R}$ of real numbers. This set as we saw earlier is closed under addition and multiplication. Its elements were seen to possess additive and multiplicative identities, and also additive inverses, thus qualifying to be a ring.

In algebra books, one often encounters the term *field* which seems somewhat confusing with the term *ring*. In the brief outline of the number ring above, whereas the sets $\mathbb{Z}$, $\mathbb{Q}$, $\mathbb{R}$ and $\mathbb{C}$ qualified as rings, the set $\mathbb{N}$ of natural numbers failed as it lacked additive inverse. The sets $\mathbb{Q}$, $\mathbb{R}$ and $\mathbb{C}$ also have an additional property that every number $n \neq 0$ in the ring has a multiplicative inverse. A ring in which every $n \neq 0$ has a multiplicative inverse is called a field. The set $\mathbb{Z}$ therefore is not a field as it does not have multiplicative inverse. In this book, the terms ring and field will be used interchangeably to refer to the sets $\mathbb{Q}$, $\mathbb{R}$ and $\mathbb{C}$ which qualify both as rings and as fields.

A curious reader will note that the term number ring was selected as the heading for this section and used in the discussion. This is because we have several other types of rings that do not use numbers as objects. In our examples, we used numbers to clarify closeness under addition and multiplication. We will see later in Chap. 3 that polynomials, which are objects and not numbers, also qualify as rings. For daily measurements and manipulation of observations, number rings and polynomial rings suffices. Other forms of rings such as fruit rings, modular arithmetic rings and congruence rings are elaborately presented in algebra books such as [193] and [246]. In-order to give a precise definition of a ring, we begin by considering the definition of *linear algebra*. Detailed treatment of linear algebra is presented in [55, 56, 251, 314].

Definition 2.2 (Linear algebra). *Algebra can be defined as a set S of elements and a finite set M of operations. In linear algebra the elements of the set S are vectors over the field $\mathbb{R}$ of real numbers, while the set M is basically made up of two elements of internal relation namely "additive" and "multiplicative". An additional definition of the external relation expounds on the term linear algebra as follows: A linear algebra over the field of real numbers $\mathbb{R}$ consists of a set R of objects, two internal relation elements (either "additive" or "multiplicative") and one external relation as follows:*

$$(opera)_1 =: \alpha : R \times R \to R$$
$$(opera)_2 =: \beta : \mathbb{R} \times R \to R \, or \, R \times \mathbb{R} \to R$$
$$(opera)_3 =: \gamma : R \times R \to R.$$

The three cases are outlined as follows:
* With respect to the internal relation α ("join"), R as a linear space in a vector space over $\mathbb{R}$, an Abelian group written "additively" or "multiplicatively":

$$\mathbf{a}, \mathbf{b}, \mathbf{c} \in R$$

Axiom	"Additively" written Abelian group	"Multiplicatively" written Abelian group
	$\alpha(\mathbf{a}, \mathbf{b}) =: \mathbf{a} + \mathbf{b}$	$\alpha(\mathbf{a}, \mathbf{b}) =: \mathbf{a} \circ \mathbf{b}$
1 Associativity	$G1+ : (\mathbf{a} + \mathbf{b}) + \mathbf{c} =$ $= \mathbf{a} + (\mathbf{b} + \mathbf{c})$ (additive assoc.)	$G1\circ : (\mathbf{a} \circ \mathbf{b}) \circ \mathbf{c} =$ $= \mathbf{a} \circ (\mathbf{b} \circ \mathbf{c})$ (multiplicative assoc.)
2 Identity	$G2+ : \mathbf{a} + \mathbf{0} = \mathbf{a}$ (additive identity, neutral element)	$G2\circ : \mathbf{a} \circ \mathbf{1} = \mathbf{a}$ (multiplicative identity neutral element)
3 Inverse	$G3+ : \mathbf{a} + (-\mathbf{a}) = \mathbf{0}$ (additive inverse)	$G3\circ : \mathbf{a} \circ \mathbf{a}^{-1} = \mathbf{1}$ (multiplicative inverse)
4 Commutativity	$G4+ : \mathbf{a} + \mathbf{b} = \mathbf{b} + \mathbf{a}$ (additive commutativity, Abelian axiom)	$G4\circ : \mathbf{a} \circ \mathbf{b} = \mathbf{b} \circ \mathbf{a}$ (multiplicative comm., Abelian axiom)

with the triplet of axioms $\{G1+, G2+, G3+\}$ or $\{G1\circ, G2\circ, G3\circ\}$ constituting the set of *group axioms* and $\{G4+, G4\circ\}$ the *Abelian axioms*. Examples of groups include:

1. The group of integer $\mathbb{Z}$ under addition.
2. The group of non-zero rational number $\mathbb{Q}$ under multiplication.
3. The set of rotation about the origin in the Euclidean plane under the operation of composite function.

* With respect to the external relation β the following compatibility conditions are satisfied

$$\mathbf{a}, \mathbf{b} \in R, \, t, u \in \mathbb{R}$$
$$\beta(t, \mathbf{a}) =: t \times \mathbf{a}$$

1 distr. $D1+ : t \times (\mathbf{a} + \mathbf{b}) = (\mathbf{a} + \mathbf{b}) \times t =$ $= t \times \mathbf{a} + t \times \mathbf{b} = \mathbf{a} \times t + \mathbf{b} \times t$ 1st additive distributivity	$D1\circ : t \times (\mathbf{a} \circ \mathbf{b}) = (\mathbf{a} \circ \mathbf{b}) \times t$ $= (t \times \mathbf{a}) \circ \mathbf{b} = \mathbf{a} \circ (\mathbf{b} \times t)$ 1st multiplicative distributivity
2 distr. $D2+ : (t + u) \times \mathbf{a} = \mathbf{a} \times (t + u) =$ $= t \times \mathbf{a} + u \times \mathbf{a} = \mathbf{a} \times t + \mathbf{a} \times u$ 2nd additive distributivity	$D2\circ : (t \circ u) \times \mathbf{a} = \mathbf{a} \times (t \circ u)$ $= t \circ (u \times \mathbf{a}) = (\mathbf{a} \times t) \circ u$ 2nd multiplicative distributivity

$$D3 : 1 \times \mathbf{a} = \mathbf{a} \times 1 = \mathbf{a} \, (left \ and \ right \ identity)$$

* With respect to the *internal relation* γ (*"meet"*) the following conditions are satisfied

$$\mathbf{a}, \mathbf{b}, \mathbf{c} \in R,\ t \in \mathbb{R}$$
$$\gamma(\mathbf{a}, \mathbf{b}) =: \mathbf{a} * \mathbf{b}$$

Axiom		Comments
1 Ass.	$G1* : (\mathbf{a} * \mathbf{b}) * \mathbf{c} = \mathbf{a} * (\mathbf{b} * \mathbf{c})$	Associativity w.r.t internal multiplication
1 dist.	$D1 * +;\ \mathbf{a} * (\mathbf{b} + \mathbf{c}) = \mathbf{a} * \mathbf{b} + \mathbf{a} * \mathbf{c}$ $(\mathbf{a} + \mathbf{b}) * \mathbf{c} = \mathbf{a} * \mathbf{c} + \mathbf{b} * \mathbf{c}$	Left and Right additive dist. w.r.t internal multiplication
1 dist.	$D1 * \circ;\ \mathbf{a} * (\mathbf{b} \circ \mathbf{c}) = (\mathbf{a} * \mathbf{b}) \circ \mathbf{c}$ $(\mathbf{a} \circ \mathbf{b}) * \mathbf{c} = \mathbf{a} \circ (\mathbf{b} * \mathbf{c})$	left and right multiplicative dist. w.r.t internal multiplication
2 dist.	$D2 * \times;\ t \times (\mathbf{a} * \mathbf{b}) = (t \times \mathbf{a}) * \mathbf{b}$ $(\mathbf{a} * \mathbf{b}) \times t = \mathbf{a} * (\mathbf{b} \times t)$	left and right dist. of internal and external multiplication

Definition 2.3 (Ring). *A sub-algebra is called a ring with identity if the following two conditions encompassing (seven conditions) hold:*
(a) The set R is an Abelian group with respect to addition, i.e. four conditions $\{G1+, G2+, G3+, G4+\}$ *of Abelian group hold.*
(b) The set R is a semi-group with respect to multiplication; that is, $\{G1*, G2*\}$ *holds. In other words, the set R comprises a monoid (i.e. a set with two operations, associativity and identity with respect to multiplication* (*)). *The last condition is the left and right additive distributivity with respect to internal multiplication* $\{D1 * +\}$ *which connects the Abelian group and the* monoid. *In total the four conditions forming the Abelian group (a) and the three forming the semi-group in (b) add up to form seven conditions enclosed in a ring in Fig. 2.1.*

Condition $G2*$ makes R a *"ring with identity"*, while the inclusion of $G3*$ makes the ring be known as the *"division ring"* if every non-zero element of the ring has a multiplicative inverse. The ring becomes a *"commutative ring"* If it has the commutative multiplicative $G4*$. Examples of *rings* include:

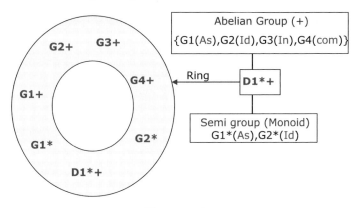

Fig. 2.1. Ring

- Field k of real numbers $\mathbb{R}$, complex numbers $\mathbb{C}$ and rational numbers
  $\mathbb{Q}$. In particular, a *ring* becomes a field if every non zero element of
  the *ring* has a multiplicative inverse as already discussed.
- Integers $\mathbb{Z}$.
- The set $\mathbb{H}$ of quaternions that are non commutative.
- Polynomial function P in n variables over a ring R expressed as
  $P = R[x_1, \ldots, x_n]$.

For the solution of algebraic computational problems in geodesy and
geoinformatics, it suffices to consider a ring as being commutative and
to include identity element.

2.4 Concluding Remarks

The concept of numbers and ring of numbers have been presented from
a geodetic and geoinformatics perspective. In the next chapter, the
number ring will provide the framework for discussing polynomial rings,
the main algebraic tool that permits the solution of nonlinear systems of
equations. The basics of ring algebra discussed provides fundamentals
required to understand the materials that will be presented in latter
chapters. For more detailed coverage of rings, we refer to [213].

3

Basics of Polynomial Theory

3.1 Polynomial Equations

In geodesy and geoinformatics, most observations are related to unknowns parameters through equations of algebraic (polynomial) type. In cases where the observations are not of polynomial type, as exemplified by the GPS meteorology problem of Chap. 13, they are converted via Theorem 3.1 on p. 20 into polynomials. The unknown parameters are then be obtained by solving the resulting polynomial equations. Such solutions are only possible through application of operations addition and multiplication on polynomials which form elements of polynomial rings. This chapter discusses polynomials and the properties that characterize them. Starting from the definitions of *monomials*, basic polynomial aspects that are relevant for daily operations are presented. A monomial is defined as

Definition 3.1 (Monomial). *A monomial is a multivariate product of the form* $x_1^{\alpha_1} x_2^{\alpha_2} \ldots x_n^{\alpha_n}$, $(\alpha_1, \ldots, \alpha_n) \in \mathbb{Z}_+^n$ *in the variables* $x_1, \ldots, x_n$.

In Definition 3.1 above, the set $\mathbb{Z}_+^n$ comprises positive elements of the set of integers (2.2) that we saw in Chap. 2, p. 8.

Example 3.1 (Monomial). Consider the system of equations for solving distances in the three-dimensional resection problem given as (see e.g., (11.44) on p. 180)

$$\begin{bmatrix} x_1^2 + 2a_{12}x_1x_2 + x_2^2 + a_o = 0 \\ x_2^2 + 2b_{23}x_2x_3 + x_3^2 + b_o = 0 \\ x_3^2 + 2c_{31}x_3x_1 + x_1^2 + c_o = 0 \\ where \ \ x_1 \in \mathbb{R}^+, x_2 \in \mathbb{R}^+, x_3 \in \mathbb{R}^+. \end{bmatrix}$$

The variables $\{x_1, x_2, x_3\}$ are unknowns while the other terms are known constants. The products of variables $\{x_1^2, x_1 x_2, x_2^2, x_2 x_3, x_3^2, x_3 x_1\}$ are monomials in $\{x_1, x_2, x_3\}$.

Summation of monomials form polynomials defined as

Definition 3.2 (Polynomial). *A polynomial $f \in k[x_1, \ldots, x_n]$ in variables $x_1, \ldots, x_n$ with coefficients in the field k is a finite linear combination of monomials with pairwise different terms expressed as*

$$f = \sum_\alpha a_\alpha x^\alpha, \quad a_\alpha \in k, \quad x^\alpha = (x^{\alpha_1}, \ldots, x^{\alpha_n}), \ \alpha = (\alpha_1, \ldots, \alpha_n), \quad (3.1)$$

where a_α are coefficients in the field k, e.g., $\mathbb{R}$ or $\mathbb{C}$ and x^α the monomials.

Example 3.2 (Polynomials). Equations

$$\begin{bmatrix} x_1^2 + 2a_{12}x_1 x_2 + x_2^2 + a_o = 0 \\ x_2^2 + 2b_{23}x_2 x_3 + x_3^2 + b_o = 0 \\ x_3^2 + 2c_{31}x_3 x_1 + x_1^2 + c_o = 0, \end{bmatrix}$$

in Example 3.1 are *multivariate polynomials*. The first expression is a multivariate polynomial in two variables $\{x_1, x_2\}$ and a linear combination of monomials $\{x_1^2, x_1 x_2, x_2^2\}$. The second expression is a *multivariate polynomial* in two variables $\{x_2, x_3\}$ and a linear combination of the monomials $\{x_2^2, x_2 x_3, x_3^2\}$, while the third expression is a *multivariate polynomial* in two variables $\{x_3, x_1\}$ and a linear combination of the monomials $\{x_3^2, x_3 x_1, x_1^2\}$.

In Example 3.2, the coefficients of the polynomials are elements of the set $\mathbb{Z}$. In general, the coefficients can take on any sets $\mathbb{Q}, \mathbb{R}, \mathbb{C}$ of number rings or other rings such as modular arithmetic rings. These coefficients can be added, subtracted, multiplied or divided, and as such play a key role in determining the solutions of polynomial equations. The definition of the set to which the coefficients belong determines whether a polynomial equation is solvable or not. Consider the following example:

Example 3.3. Given an equation $9w^2 - 1 = 0$ with the coefficients in the integral domain, obtain the integer solutions. Since the coefficient $9 \in \mathbb{Z}$, the equation does not have a solution. If instead the coefficient $9 \in \mathbb{Q}$, then the solution $w = \pm\dfrac{1}{3}$ exist.

From Definition 2.1 of algebraic, polynomials become algebraic once (3.1) is equated to 0. The fundamental problem of algebra can thus be stated as the solution of equations of form (3.1) equated to 0.

3.2 Polynomial Rings

In Sect. 2.3 of Chap. 2, the theory of rings was introduced with respect to numbers. Apart from the number rings, polynomials are objects that also satisfy ring axioms leading to *"polynomial rings"* upon which operations *"addition"* and *"multiplication"* are implemented.

3.2.1 Polynomial Objects as Rings

Polynomial rings are defined as

Definition 3.3 (Polynomial ring). *Consider a ring R say of real numbers $\mathbb{R}$. Given a variable $x \notin R$, a univariate polynomial $f(x)$ is formed (see Definition 3.2 on p. 18) by assigning coefficients $a_i \in R$ to the variable and obtaining summation over finite number of distinct integers. Thus*

$$f(x) = \sum_{\alpha} c_\alpha x^\alpha, c_\alpha \in R, \alpha \geq 0$$

is said to be a univariate polynomial over R. If two polynomials are given such that $f_1(x) = \sum_i c_i x^i$ and $f_2(x) = \sum_j d_j x^j$, then two binary operations "addition" and "multiplication" can be defined on these polynomials such that:

(a) Addition: $f_1(x) + f_2(x) = \sum_k e_k x^k, e_k = c_k + d_k, e_k \in R$

(b) Multiplication: $f_1(x).f_2(x) = \sum_k g_k x^k, g_k = \sum_{i+j=k} c_i d_j, g_k \in R$.

A collection of polynomials with these "additive" and "multiplicative" rules form a commutative ring with zero element and identity 1. A univariate polynomial $f(x)$ obtained by assigning elements c_i belonging to the ring R to the variable x is called a polynomial ring and is expressed as $f(x) = R[x]$. In general the entire collection of all polynomials in $x_1, \ldots, x_n$, with coefficients in the field k that satisfy the definition of a ring above are called a polynomial rings.

Designated P, polynomial rings are represented by n unknown variables x_i over k expressed as $P := k[x_1, \ldots, x_n]$. Its elements are polynomials known as *univariate* when $n = 1$ and *multivariate* otherwise. The distinction between a polynomial ring and a polynomial is that the latter is the sum of a finite set of monomials (see e.g., Definition 3.1 on p. 17) and is an element of the former.

Example 3.4. Equations

$$\begin{bmatrix} x_1^2 + 2a_{12}x_1x_2 + x_2^2 + a_o = 0 \\ x_2^2 + 2b_{23}x_2x_3 + x_3^2 + b_o = 0 \\ x_3^2 + 2c_{31}x_3x_1 + x_1^2 + c_o = 0 \end{bmatrix}$$

of Example 3.1 are said to be polynomials in three variables $[x_1, x_2, x_3]$ forming elements of the polynomial ring P over the field of real numbers $\mathbb{R}$ expressed as $P := \mathbb{R}[x_1, x_2, x_3]$.

Polynomials that we use in solving unknown parameters in various problems, as we shall see later, form elements of polynomial rings. Polynomial rings provide means and tools upon which to manipulate the polynomial equations. They can either be added, subtracted, multiplied or divided. These operations on polynomial rings form the basis of solving systems of equations algebraically as will be made clear in the chapters ahead. Next, we state the theorem that enables the solution of nonlinear systems of equations in geodesy and geoinformatics.

Theorem 3.1. *Given n algebraic (polynomial) observational equations, where n is the dimension of the observation space $\mathbb{Y}$ of order l in m unknown variables , and m is the dimension of the parameter space $\mathbb{X}$, the application of least squares solution (LESS) to the algebraic observation equations gives $(2l - 1)$ as the order of the set of nonlinear algebraic normal equations. There exists m normal equations of the polynomial order $(2l - 1)$ to be solved.*

Proof. Given nonlinear algebraic equations $f_i \in k\{\xi_1, \ldots, \xi_m\}$ expressed as

$$\begin{bmatrix} f_1 \in k\{\xi_1, \ldots, \xi_m\} \\ f_2 \in k\{\xi_1, \ldots, \xi_m\} \\ \cdot \\ \cdot \\ \cdot \\ f_n \in k\{\xi_1, \ldots, \xi_m\}, \end{bmatrix} \tag{3.2}$$

with the order considered as l, we write the objective function to be minimized as

$$\|f\|^2 = f_1^2 + \ldots + f_n^2 \mid \forall f_i \in k\{\xi_1, \ldots, \xi_m\}, \tag{3.3}$$

and obtain the partial derivatives (first derivatives of 3.3) with respect to the unknown variables $\{\xi_1, \ldots, \xi_m\}$. The order of (3.3) which is l^2 then reduces to $(2l - 1)$ upon differentiating the objective function with respect to the variables $\xi_1, \ldots, \xi_m$. Thus resulting in m normal equations of the polynomial order $(2l - 1)$.

□

Example 3.5 (Pseudo-ranging problem). For pseudo-ranging or distance equations, the order of the polynomials in the algebraic observational equations is $l = 2$. If we take the "pseudo-ranges squared" or "distances squared", a necessary procedure in-order to make the observation equations "algebraic" or "polynomial", and implement least squares solution (LESS), the objective function which is of order $l = 4$ reduces by one to order $l = 3$ upon differentiating once. The normal equations are of order $l = 3$ as expected.

The significance of Theorem 3.1 is that all observational equations of interest are successfully converted to *"algebraic"* or *"polynomial"* equations. This implies that problems requiring exact algebraic solutions must first have their equations converted into algebraic. This will be made clear in Chap. 13 where trigonometric nonlinear system on equations are first converted into algebraic.

3.2.2 Operations "Addition" and "Multiplication"

Definition 3.3 implies that a polynomial ring qualifies as a ring based on the applications of operations "addition" and "multiplication" on its coefficients. In this case, the axioms that follow the Abelian group with respect to "addition" and the semi group with respect to "multiplication" readily follow. Of importance in manipulating polynomial rings using operations "addition" and "multiplication" is the concept of division of polynomials defined as

Definition 3.4 (Polynomial division). *Consider the polynomial ring $k[x]$ whose elements are polynomials $f(x)$ and $g(x)$. There exists unique polynomials $p(x)$ and $r(x)$ also elements of polynomial ring $k[x]$ such that*

$$f(x) = g(x)p(x) + r(x),$$

with either $r(x) = 0$ or degree of $r(x)$ is less than the degree of $g(x)$.

For univariate polynomials, as in Definition 3.4, the Euclidean algorithm employs operations "addition" and "multiplication" to factor polynomials in-order to reduce them to satisfy the definition of division algorithm.

3.3 Factoring Polynomials

In-order to understand the factorization of polynomials, it is essential to revisit some of the properties of prime numbers of integers. This is due to the fact that polynomials behave much like integers. Whereas for integers, any integer $n > 1$ is either prime (i.e., can only be factored by 1 and n itself) or a product of prime numbers, a polynomial $f(x) \in k[x]$ is either irreducible in $k[x]$ or factors as a product of irreducible polynomials in the field $k[x]$. The polynomial $f(x)$ has to be of positive degree. Factorization of polynomials play an important role as it enables solution of polynomial roots as will be seen in the next section. Indeed, the Groebner basis algorithm presented in Chap. 4 makes use of the factorization of polynomials. In general, computer algebra systems discussed in Chap. 16 offers possibilities of factoring polynomials.

3.4 Polynomial Roots

More often than not, the most encountered interaction with polynomials is perhaps the solution of its roots. Finding the roots of polynomials is essential for most computations that we undertake in practice. As an example, consider a simple planar ranging case where distances have been measured from two known stations to an unknown station (see e.g, Fig. 4.1 on p. 30). In such a case, the measured distances are normally related to the coordinates of the unknown station by multivariate polynomial equations. If for instance a station P_1, whose coordinates are $\{x_1, y_1\}$ is occupied, the distance s_1 can be measured to an unknown station P_0. The coordinates $\{x_0, y_0\}$ of this unknown station are desired and have to be determined from distance measurements. The relationship between the measured distance and the coordinates is given by

$$s_1 = \sqrt{(x_1 - x_0)^2 + (y_1 - y_0)^2}. \tag{3.4}$$

Applying Theorem 3.1, a necessary step to convert (3.4) into polynomial, (3.4) is squared to give a multivariate quadratic polynomial

$$s_1^2 = (x_1 - x_0)^2 + (y_1 - y_0)^2. \tag{3.5}$$

Equation (3.5) has two unknowns thus necessitating a second distance measurement to be taken. Measuring this second distance s_2 from station P_2, whose coordinates $\{x_2, y_2\}$ are known, to the unknown station P_0 leads to a second multivariate quadratic polynomial equation

$$s_2^2 = (x_2 - x_0)^2 + (y_2 - y_0)^2. \tag{3.6}$$

The intersection of the two equations (3.5) and (3.6) results in two quadratic equations $ax_0^2 + bx_0 + c = 0$ and $dy_0^2 + ey_0 + f = 0$ whose roots give the desired coordinates x_0, y_0 of the unknown station P_0. In Sect. 4.1, we will expound further on the derivation of these multivariate quadratic polynomial equations.

In Sect. 3.6, we will discuss the types of polynomials with real coefficients. Suffice to mention at this point that polynomials, as defined in Definition 3.2 with the coefficients in the field k, has a solution ξ such that on replacing the variable x^α, one obtains

$$a_n\xi^n + a_{n-1}\xi^{n-1} + \dots + a_1\xi + a_0 = 0. \tag{3.7}$$

From high school algebra, we learnt that if ξ is a solution of a polynomial $f(x)$, also called the root of $f(x)$, then $(x - \xi)$ divides the polynomial $f(x)$. This fact enables the solution of the remaining roots of the polynomial as we already know. The division of $f(x)$ by $(x - \xi)$ obeys the division rule discussed in Sect. 3.2.2. In a case where $f(x) = 0$ has many solutions (i.e., multiple roots $\xi_1, \xi_2, ..., \xi_m$), then $(x - \xi_1), (x - \xi_2), ..., (x - \xi_m)$ all divide $f(x)$ in the field k.

In general, a polynomial of degree n will have n roots that are either real or complex. If one is operating in the real domain, i.e., the polynomial coefficients are real, the complex roots normally results in a pair of conjugate roots. Polynomial coefficients play a significant role in the determination of the roots. A slight change in the coefficients would significantly alter the solutions. For ill-conditioned polynomials, such a change in the coefficients can lead to disastrous results. Methods of determining polynomial roots have been elaborately presented by [269]. We should point out that for polynomials of degree n in the field of real numbers $\mathbb{R}$ however, the solutions exist only for polynomials up to degree 4. Above this, Niels Henrick Abel (1802-1829) proved through his *impossibility theorem* that the roots are insolvable, while Evariste Galois (1811-1832) gave a more concrete proof that for every integer n greater than 4, there can not be a formula for the roots of a general $n^t h$ degree polynomial in terms of coefficients.

3.5 Minimal Polynomials

In Sect. 2.3, we presented the number rings concept and extended the sets from that of natural numbers $\mathbb{N}$ to the complex number $\mathbb{C}$ in-order to cater for expanded operations. For polynomials, roots may fail to exist in one set say $\mathbb{Q}$ but exist in another set $\mathbb{R}$ as we saw in Sect. 2.2. The polynomial $y^2 - 12 = 0$, for example, has no roots in $\mathbb{Q}[y]$ but the roots ± 12 exist in $\mathbb{R}$. The expansion of the set from $\mathbb{Q}$ to $\mathbb{R}$ is also called *field extension* of k. It may occur however that in the polynomial ring $k[x]$, the solution ξ satisfy not only the polynomial $p(x)$ but also another polynomial $h(x)$, where $p(x)$ and $h(x)$ are both elements of $k[x]$. In case several polynomials in $k[x]$ have ξ as a root, and the polynomials are multiples of a polynomial of least degree that also contains ξ as root, this polynomial of least degree is termed the *minimal polynomial*.

In dealing with Groebner basis in Chap. 4 for example, it will be seen that several polynomials in the field $k[x]$ contain the same root ξ. This property will be used to reduce several multivariate polynomials to univariate polynomials whose solutions fulfill the multivariate polynomials.

3.6 Polynomials with Real Coefficients

In this section we revisit the various types of univariate polynomials with the coefficients in the field $\mathbb{R}$ of reals, which we often use to manipulate measurements. We recapture the basic high school mathematics of inferring the roots of polynomials from the coefficients.

3.6.1 Quadratic Polynomials

In Sect. 3.4 we introduced the quadratic equations and demonstrated their association with distance measurements. In general, the simplest polynomial is the linear polynomial $cx + d = 0$ which is solved for x by simply multiplying both sides of the equation by the inverse of c, provided that $c \neq 0$ holds. The solution thus becomes $x = -c^{-1}d$. Linear polynomials, i.e., polynomials of degree 1 find use in manipulating levelling and gravimetric observations. In these cases, they are manipulated in vector space through the solution of linear algebraic equation $\mathbf{Ax} + \mathbf{y} = \mathbf{0}$ to give the solution $\mathbf{x} = (\mathbf{A}'\mathbf{A}^{-1})\mathbf{A}'\mathbf{y}$, provided that $\mathbf{A}'\mathbf{A}$ is regular.

Polynomials of degree 2 are known as quadratic polynomials. For univariate cases, they take the form $ax^2 + bx + c = 0$. For simple cases such as $x^2 + 2x = 0$, the solution can be obtained by factorization, e.g., $x(x + 2)$ leading to $x = 0$ or $x = -2$. The general solution of quadratic equations of the form $ax^2 + bx + c = 0$ with real coefficients $\{a, b, c\}$ is given by the quadratic formulae

$$x = \frac{-b \pm \sqrt{b^2 - 4ac}}{2a}, \tag{3.8}$$

or

$$x = \frac{2c}{-b \pm \sqrt{b^2 - 4ac}}. \tag{3.9}$$

Press [269] discourages the use of (3.8) or (3.9) in the determination of the two roots for cases where a or c (or both) are small since this leads to inaccurate solutions. The main reason cited is that when either the coefficient a or c (or both) is small, one of the roots involves the subtraction b from a very nearly equal value. They instead propose the formular

$$q = -\frac{1}{2}[b + sgn(b)\sqrt{b^2 - 4ac}], \tag{3.10}$$

where the two roots are then given by

$$x_1 = \frac{q}{a}, \quad x_2 = \frac{c}{q}. \tag{3.11}$$

In computer algebra software of Matlab and Mathematica discussed in detail in Chap. 16, the roots of a quadratic polynomial are obtained via

- Matlab: $x = roots([a\ b\ c])$, where $[a\ b\ c]$ is a vector containing the coefficients in the field $\mathbb{R}$ of reals. The quadratic equation can also be solved using the solve command, e.g., $solve('ax^2 + bx + c = 0',' x')$, where x indicates the variable to be solved.
- Mathematica: $x = Root[f, k]$, where f is the quadratic equation and k the kth root. The quadratic equation can also be solved using the solve command, e.g., $Solve[ax^2 + bx + c == 0, x]$.

In general, every quadratic polynomial has exactly two real or two complex roots. From the coefficients, if $b^2 - 4ac > 0$, the roots are real but if $b^2 - 4ac < 0$ the roots are a pair of non real complex numbers. The case where $b^2 - 4ac = 0$ gives real and identical roots and is also known as the bifurcation point upon which the roots change sign.

3.6.2 Cubic Polynomials

These are polynomials of degree 3 and take the form $ax^3 + bx^2 + cx + d = 0$. Like quadratic polynomials, simple cases can also be solved via factorization e.g., $x^3 - 2x = 0$ is factored as $x(x^2 - 2)$ to give the solutions $x = 0, x = -\sqrt{2}$ or $x = +\sqrt{2}$. Another approach would be to reduce the cubic polynomial such that the polynomials of degree 2 are eliminated to give a simplified version of the form $y^3 + ey + f = 0$ known as a reduced cubic polynomial. The simplified version can then be solved for the roots via Cardano's formula as

$$y = \sqrt[3]{-\frac{f}{2} + \sqrt{T}} + \sqrt[3]{-\frac{f}{2} - \sqrt{T}}, \tag{3.12}$$

where $T = (\frac{e}{3})^3 + (\frac{f}{2})^2$. Once one real root say ξ_1 has been obtained, the polynomial $y^3 + ey + f = 0$ is divided by $(y - \xi_1)$ and the resulting quadratic polynomial solved for the remaining roots. An alternative approach is presented by [269] who proceed as follows: Let $\{a, b, c\}$ be the real coefficients of a cubic polynomial. Compute

$$\left[\begin{array}{l} K \equiv \dfrac{a^2 - 3b}{g} \\[3mm] L \equiv \dfrac{2a^3 - gab + 27c}{54}. \end{array} \right. \tag{3.13}$$

If K and L are real, and $L < K$, then the cubic polynomial has three real roots computed by

$$\left[\begin{array}{l} x_1 = -2\sqrt{K}\cos(\dfrac{\Theta}{3}) - \dfrac{a}{3} \\[3mm] x_2 = -2\sqrt{K}\cos(\dfrac{\Theta + 2\pi}{3}) - \dfrac{a}{3} \\[3mm] x_3 = -2\sqrt{K}\cos(\dfrac{\Theta - 2\pi}{3}) - \dfrac{a}{3}, \end{array} \right. \tag{3.14}$$

where

$$\Theta = \cos^{-1}(\frac{L}{\sqrt{K^3}})^1.$$

---

[1]The origin of the equation is traced by the authors to chapter VI of François Viète's treatise "De emendatione" Published in 1615

Using computer algebra software of Matlab and Mathematica the roots of a cubic polynomial are obtained via

- Matlab: $x = roots([a\ b\ c\ d])$, where $[a\ b\ c\ d]$ is a vector containing the coefficients in the field $\mathbb{R}$ of reals. The quadratic equation can also be solved using the solve command, e.g., $solve('ax^3 + bx^2 + cx + d = 0', 'x')$.
- Mathematica: $x = Root[f, k]$, where f is the cubic equation, and k, the kth root. The quadratic equation can also be solved using the solve command, e.g., $Solve[ax^3 + bx^2 + cx + d == 0, x]$.

In general, if ξ_1, ξ_2, ξ_3 are the roots of a cubic polynomial, the discriminant D can de defined as

$$D = (\xi_1 - \xi_2)^2(\xi_1 - \xi_3)^2(\xi_2 - \xi_3)^2, \tag{3.15}$$

and computed from the coefficients a, b, c, d to infer on the nature of the roots. Considering $a = 1$, [193, p. 156, Exercise 10.17] gives the formula of the discriminant D from the coefficients b, c, d as

$$D = 18bcd - 4b^3d + b^2c^2 - 4c^3 - 27d^2. \tag{3.16}$$

If $D > 0$ then the roots of the cubic polynomial are real and distinct. If $D < 0$, then one of the roots is real and the remaining two are non real complex conjugate. In a case where $D = 0$, multiple roots all which are real are given. In case the coefficients b, c, d are all positive, then all the three roots will be negative, while if b, d are negative and c positive, all the roots will be positive.

3.6.3 Quartic Polynomials

Quartic polynomials are those of degree 4. In a case where one root ξ_1 exist for a polynomial $p(x) = 0$, the division algorithm can be applied to obtain the factor $(x - \xi_1)f(x)$. Here, $f(x)$ is a cubic polynomial that can be solved as discussed in Sect. 3.6.2 to give at least one real root. The quartic polynomial $ax^4 + bx^3 + cx^2 + dx + e = 0$ therefore has at least two real roots. The following conditions may apply for a quartic polynomial:

- $p(x)$ has four real roots.
- $p(x)$ has two real roots and two complex conjugate roots.
- $p(x)$ has no real roots.

The solution of a quartic polynomial proceeds via substitution approach in-order to reduce it. Considering a case where $a = 1$, the quartic polynomial $x^4 + bx^3 + cx^2 + dx + e = 0$ is reduced by substituting $x = z + a$, with $a \in \mathbb{R}$, to $Z^4 + CZ^2 + EZ + F = 0$ which is solved for $g(Z) = 0$. The solutions of $g(Z) = 0$ satisfies those of $p(x) = 0$ (see Sect. 3.5). $Z^4 + CZ^2 + EZ + F = 0$ is called the reduced quartic polynomial which can be solved as discussed by [193, pp. 159-166].

Solution of the roots of quartic polynomials using Computer algebra software of Matlab and Mathematica is as follows:

- Matlab: $x = roots\,([a\ b\ c\ d\ e])$, where $[a\ b\ c\ d\ e]$ is a vector containing the coefficients in the field $\mathbb{R}$ of reals. The quadratic equation can also be solved using the solve command, e.g., $solve('ax^4 + bx^3 + cx^2 + dx + e = 0',' x')$
- Mathematica: $x = Root[f, k]$, where f is the quartic equation and k the kth root. The quadratic equation can also be solved using the solve command, e.g., $Solve[ax^4 + bx^3 + cx^2 + dx + e == 0, x]$.

In general, if $\xi_1, \xi_2, \xi_3, \xi_4$ are the roots of a quartic polynomial, the discriminant D can de defined as

$$D = (\xi_1 - \xi_2)^2(\xi_1 - \xi_3)^2(\xi_1 - \xi_4)^2(\xi_2 - \xi_3)^2(\xi_2 - \xi_4)^2(\xi_3 - \xi_4)^2, \quad (3.17)$$

and computed from the coefficients b, c, d, e to infer on the nature of the roots. Considering $a = 1$, [193, p. 171] gives the formula of the discriminant D from the coefficients b, c, d, e as

$$D = \begin{bmatrix} 18bcd^3 + 18b^3cde - 80bc^2de - 6b^2d^2e + 144cd^2e \\ \\ +144b^2ce^2 - 128c^2e^2 - 192bde^2 + b^2c^2d^2 - 4b^3d^3 - 4c^3d^2 \\ \\ -4b^2c^3e + 16c^4e - 27d^4 - 27b^4e^2 + 256e^3. \end{bmatrix}$$
$$(3.18)$$

If $D > 0$ then all the roots of the quartic polynomial are real and distinct or all the four roots are pairs of non real complex conjugates. If $D < 0$, then two roots are real and distinct while the other two are complex conjugates. For a case where $D = 0$, at least two of the roots coincide.

3.7 Concluding Remarks

What we have presented is just a nutshell of the topic "polynomials". Several books, e.g., [49, 246, 270, 353] are dedicated specifically to it.

4

Groebner Basis

"There are no good, general methods for solving systems of more than one nonlinear equation. Furthermore, it is not hard to see why (very likely) there never will be any good, general methods:..." W. H. Press et al.

4.1 The Origin

This chapter presents you the reader with one of the most powerful computer algebra tools, besides the polynomial resultants (discussed in the next chapter), for solving nonlinear systems of equations which you may encounter. The basic tools that you will require to develop your own algorithms for solving problems requiring closed form (exact) solutions are presented. This powerful tool is the "Gröbner basis" written in English as Groebner basis. It was first suggested by *W. Groebner* in 1949 and developed by his student *B. Buchberger* in 1965. In 1964, H. Hironaka (1931-) had independently used the same tool in connection with his work on resolution of singularities in algebraic geometry and named it *standard basis* [217, p. 187]. B. Buchberger decided to honour his thesis supervisor W. Groebner (1899-1980) by naming the standard basis for *Ideals* in polynomial rings $k[x_1, \ldots, x_n]$ as *Groebner basis* [78]. In this book, as in modern books, we will adopt the term Groebner basis and present the subject in the simplest form that can easily be understood from geodetic as well as geoinformatics perspective.

As a recipe, consider that most problems in nature, here in geodesy, geoinformatics, machine vision, robotics, surveying etc., can be modelled by *nonlinear systems of equations*. Let us consider a simple case of planar distance measurements in Fig. 4.1. Equations relating these measured distances to the coordinates of an unknown station were already presented in Sect. 3.4.

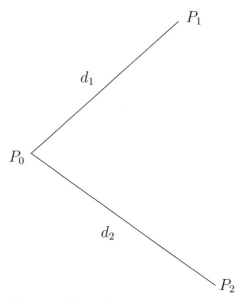

Fig. 4.1. Planar distance observations

In that section, we did relate the measured distances $\{s_i,\ i = 1, 2\}$ to the coordinates of the unknown station by (3.5) and (3.6). We stated that the intersection of these two equations lead to univariate polynomials whose solution give the desired position of an unknown station. We did not however give any explanation on how the univariate polynomials are derived from the set of multivariate quadratic polynomials (3.5) and (3.6). The derivation of the univariate polynomials from systems of nonlinear equations form one of the major tasks of Groebner basis. Let us denote the distance $\{s_i,\ i = 1, 2\}$ by $\{d_i,\ i = 1, 2\}$ and re-write (3.5) and (3.6) respectively as

$$d_1^2 = (x_1 - x_0)^2 + (y_1 - y_0)^2 \tag{4.1}$$

and

$$d_2^2 = (x_2 - x_0)^2 + (y_2 - y_0)^2. \tag{4.2}$$

The task confronting us now is to obtain from these two nonlinear equations the coordinates $\{x_0, y_0\}$ of the unknown station P_0. In case (4.1) and (4.2) were linear, the solution for $\{x_0, y_0\}$ would have been much easier. One could simply solve them using either matrix inversion, graphically, Gauss-Jordan or Gauss elimination techniques. Unfortunately they are nonlinear and can not be solved using the procedures above. Groebner basis and polynomial resultant approaches are algebraic techniques that are proposed to offer solutions to nonlinear systems of equations such as (4.1) and (4.2).

4.2 Basics of Groebner Basis

Groebner basis is the greatest common divisors of a multivariate system of equations. Its direct application is the elimination of variables in nonlinear systems of equations. Let us start by the problem of finding the greatest common divisors in Example 4.1:

Example 4.1 (Greatest common divisors (gcd)). Given the numbers 12, 20, and 18, find their greatest common divisor. We proceed by writing the factors as

$$\left.\begin{array}{l} 12 = 2^2.3^1.5^0 \\ 20 = 2^2.3^0.5^1 \\ 20 = 2^1.3^2.5^0 \end{array}\right] \rightarrow 2^1.3^0.5^0 = 2, \qquad (4.3)$$

leading to 2 as the greatest common divisor of 12, 20 and 18. Next, let us consider the case of univariate polynomials $f_1, f_2 \in k[x]$ in (4.4).

$$\left.\begin{array}{l} f_1 = 3x^4 - 3x^3 + 8x^2 + 2x - 5 \\ f_2 = 5x^4 - 4x^2 - 9x + 21 \end{array}\right] \rightarrow Euclidean\,algorithm = f \in k[x].$$
$$(4.4)$$

Equation (4.4) employs the Euclidean algorithm which obtains one univariate polynomial as the gcd of the two univariate polynomials f_1 and f_2. If on the other hand expressions in (4.4) were not univariate but multivariate, e.g., $g_1, g_2 \in k[x, y]$ as in equation (4.5), then one applies the Buchberger algorithm which is discussed in Sect. 4.3.

$$\left.\begin{array}{l} g_1 = xy + x - y - 1 \\ g_2 = xy - x - y + 1 \end{array}\right] \rightarrow Buchberger\;algorithm = Groebner\;basis.$$
$$(4.5)$$

Groebner basis therefore, is the greatest common divisors of a multivariate system of polynomial equations $\{g_1, g_2\}$.

Groebner basis as stated earlier is useful for eliminating variables in nonlinear systems of equations. Gauss elimination technique on the other hand is applicable for linear cases as shown in Example 4.2.

Example 4.2 (Gauss elimination technique). Solve the linear system of equations

$$\left[\begin{array}{l} -x + y + 2z = 2 \\ 3x - y + z = 6 \\ -x + 3y + 4z = 4. \end{array}\right. \tag{4.6}$$

The first step is to eliminate x in the second and third expressions of (4.6). This is achieved by multiplying the first expression by 3 and adding to the second expression to give the second expression of (4.7). The third expression of (4.7) is obtained by subtracting the first expression from the third expression in (4.6).

$$\left[\begin{array}{l} -x + y + 2z = 2 \\ 2y + 7z = 12 \\ 2y + 2z = 2. \end{array}\right. \tag{4.7}$$

The second step is to eliminate y in the second and third expressions of (4.7). This is achieved by subtracting the second expression from the third expression in (4.7) to give (4.8).

$$\left[\begin{array}{l} -x + y + 2z = 2 \\ 2y + 7z = 12 \\ -5z = -10. \end{array}\right. \tag{4.8}$$

The solution of $z = 2$ in (4.8) can now be substituted back into the second equation $2y + 7z = 12$ to give the value of $y = -1$, which together with the value of $z = 2$ are substituted into the first equation to give the value of $x = 1$ to complete the Gauss elimination technique.

In many applications however, equations relating unknown variables to the measured (observed) quantities are normally nonlinear and often consist of many variables (multivariate). In such cases, the Gauss elimination technique for the univariate polynomial equations employed in Example 4.2 gives way to Groebner basis as illustrated in Examples 4.3 and 4.4. In general, the Groebner basis algorithm reduces a system of multivariate polynomial equations. This is done by employing operations "*addition*" and "*multiplication*" on a polynomial ring (see Sect. 3.2.2) to give more simplified expressions. Given a system of polynomial equations which are to be solved explicitly for unknowns, e.g., (4.1) and (4.2), Groebner basis algorithm is applied to reduce the set of polynomials into another set (e.g., from a system $F(x, y, z)$ to another system $G(x, y, Z)$) of polynomials with suitable properties that allow solution. If $F(x, y, z)$ is a set of nonlinear system of polynomial equations, Groebner basis eliminates variables in a manner similar to Gauss elimination technique for linear cases to reduce it to $G(x, y, z)$. With Lexicographic ordering of the monomials (see Definition A.2 in Appendix A-1 on p. 303), one expression in $G(x, y, z)$ always turns out to be a univariate polynomial. Its roots are easily obtained using algebraic software of Matlab, Mathematica or Maple, and can be substituted in the other elements of the set $G(x, y, z)$ to obtain a complete solution which also satisfy the original set $F(x, y, z)$. Examples 4.3 and 4.4 elaborate on the application of Groebner basis.

Example 4.3 (Groebner basis computation). Let us consider a simple example from [81]. Consider a set $F(x, y) = \{f_1, f_2\}$ to have as its elements

$$\left[\begin{array}{l} f_1 = xy - 2y \\ f_2 = 2y^2 - x^2, \end{array} \right. \tag{4.9}$$

where $\{f_1, f_2\} \in I$ are the generators of the Ideal I (see definition of Ideal on p. 34). We now seek a simplified set of generators of this Ideal using Buchberger algorithm. By employing operations "addition" and "multiplication", the Groebner basis algorithm (also called Buchberger algorithm) reduces the system of nonlinear equations (4.9) into another set G of F as

$$G := \{-2x^2 + x^3, -2y + xy, -x^2 + 2y^2\}. \tag{4.10}$$

In Mathematica software, using the lexicographic order $x > y$, i.e., x comes before y, the Groebner basis could simply be computed by entering the command

$$GroebnerBasis[F, \{x, y\}]. \tag{4.11}$$

The set G in (4.10) contains one univariate polynomial $-2x^2 + x^3$, which can easily be solved using roots command in Matlab for solutions $\{x = 0, x = 0, x = 2\}$ and substituted in any of the remaining elements of the set G to solve for y. The solutions of G, i.e., the roots $\{x = 0, x = 0, x = 2\}$) and those of y satisfy polynomials in F. This can be easily tested by substituting these solutions into (4.9) to give 0.

Let us consider as a second example an optimization problem.

Example 4.4 (Minimum and maximization problem). Find the minimum and maximum of $f(x, y, z) = x^3 + 2xyz - z^2$, such that $g(x, y, z) =$

$x^2 + y^2 + z^2 - 1$. First, we obtain the partial derivatives of $f - Lg = 0$ with respect to $\{x, y, z, L\}$, where L is the lagrangean multiplier as

$$\frac{\partial f}{\partial \{x, y, z, L\}} := F = \begin{bmatrix} 3x^2 + 2yz - 2xL = 0 \\ 2xz - 2yL = 0 \\ 2xy - 2z - 2zL = 0 \\ x^2 + y^2 + z^2 - 1 = 0. \end{bmatrix} \quad (4.12)$$

Groebner basis is invoked in Mathematica by

$$GroebnerBasis[\{F\}, \{x, y, L, z\}],$$

which leads to

$$G = \begin{bmatrix} L - 1.5x - 1.5yz - 43.7z^6 - 62.2z^4 - 17.5z^2 = 0 \\ x^2 + y^2 + z^2 - 1 = 0 \\ y^2z - 1.8z^5 + 2.8z^3 - z = 0 \\ z^7 - 1.5z^5 + 0.6z^3 - 0.04z = 0. \end{bmatrix} \quad (4.13)$$

The solution of z in (4.13) can then be substituted into the third equation $y^2z - 1.8z^5 + 2.8z^3 - z = 0$ to give the value of y. The obtained values of z and y are then substituted into the second equation to give the value of x, and thus complete the Groebner basis solution. Later in the chapter, we will introduce the *reduced Groebner basis* which can be used to obtain directly the last expression of (4.13), i.e., the univariate polynomial in z.

The theory behind the operation of Groebner basis is however not so simple. In the remainder of this chapter, we will try to present in a simplified form the algorithm behind the computation of Groebner bases. In Chap. 3, we learnt that polynomials are elements of a ring and that they satisfy the ring axioms of addition and subtraction. The computation of Groebner basis is achieved by the capability to manipulate the polynomials to generate *Ideals* defined as

Definition 4.1 (Ideal). *An Ideal is generated by a family of generators as consisting of the set of linear combinations of these generators with polynomial coefficients. Let f_1, ..., f_s and c_1, ..., c_s be polynomials in $k[x_1, ..., x_n]$, then*

$$< f_1, ..., f_s > = \sum_{i=1}^{s} c_i f_i. \quad (4.14)$$

In (4.14), $< f_1, ..., f_s >$ is an Ideal and if a subset $I \subset k[x_1, ..., x_n]$ is an Ideal, it must satisfy the following conditions [94, p. 29];

- $0 \in I$,
- If $f, g \in I$, then $f + g \in I$ (i.e., I is an additive subgroup of the additive group of the field k),
- If $f \in I$ and $c \in k[x_1, \ldots, x_n]$, then $cf \in I$ (i.e., I is closed under multiplication ring element).

Example 4.5 (Ideal). Equations (4.1) and (4.2) are expressed algebraically as

$$\left[\begin{array}{l} f_1 := (x_1 - x_0)^2 + (y_1 - y_0)^2 - d_1^2 \\ f_2 := (x_2 - x_0)^2 + (y_2 - y_0)^2 - d_2^2, \end{array} \right. \tag{4.15}$$

where polynomials $\{f_1, f_2\}$ belong to the polynomial ring $\mathbb{R}[x_0, y_0]$. If the polynomials

$$\left[\begin{array}{l} c_1 := 4x_0 + 6 \\ c_2 := x_0 + y_0 \end{array} \right. \tag{4.16}$$

also belong to the same polynomial ring $\mathbb{R}[x_0, y_0]$, an Ideal is generated by a linear combination

$$I := \left[\begin{array}{l} < f_1, f_2 >= c_1 f_1 + c_2 f_2 \\ \\ = (4x_0 + 6)f_1 + (x_0 + y_0)f_2. \end{array} \right. \tag{4.17}$$

In this case, $\{f_1, f_2\}$ are said to be generators of the *Ideal I*.

Definition (4.1) of an *Ideal* can be presented in terms of polynomial equations $f_1, \ldots, f_s \in k[x_1, \ldots, x_n]$. This is done by expressing the system of polynomial equations as

$$\left[\begin{array}{l} f_1 = 0 \\ f_2 = 0 \\ \quad . \\ \quad . \\ f_s = 0, \end{array} \right. \tag{4.18}$$

and using them to derive others by multiplying each individual equation f_i by another polynomial $c_i \in k[x_1, \ldots, x_n]$ and summing to get $c_1 f_1 + c_2 f_2 + \ldots + c_s f_s = 0$ (cf., 4.14). The *Ideal* $< f_1, \ldots, f_s >$ thus consists of a system of equations $f_1 = f_2 = \ldots = f_s = 0$, thus indicating that if $f_1, \ldots, f_s \in k[x_1, \ldots, x_n]$, then $< f_1, \ldots, f_s >$ is an *Ideal* generated by $f_1, \ldots, f_s$, i.e., being the *basis* of the *Ideal I*.

In this case, a collection of these *nonlinear algebraic equations* forming *Ideals* are referred to as the set of polynomials generating the *Ideal* and forms the elements of this *Ideal*. Perhaps a curious reader may begin to wonder why the term *Ideal* is used. To quench this curiosity we refer to [256, p. 220] and quote from [55, p. 59] who wrote:

> "On the origin of the term *Ideal*, the concept is attributed to *Dedekind* who introduced it as a set theoretical version of *Kummer's* "Ideal number" to circumvent the failure of unique factorization in certain natural extension of the domain $\mathbb{Z}$. The relevance of *Ideal* in the theory of *polynomial rings* was highlighted by *Hilbert Basis Theorem*. The systematic development of *Ideal* theory; in more general rings is largely due to *E. Noether*. In the older literature, the term "module" is sometimes used for "Ideal" (cf., [230]). The term "ring" seems to be due to *D. Hilbert*; *Kronecker* used the term "order" for ring".

Example 4.6 (Ideal). Consider example (4.3) with polynomials in $\mathbb{R}[x, y]$. The *Ideal* $I = <xy - 2y, 2y^2 - x^2>$.

The generators of an Ideal can be computed using the *division algorithm* defined as (cf., Definition 3.4 of polynomial division on p. 21)

Definition 4.2 (Division algorithm). *Fix a monomial order of polynomials say $x > y$ for polynomials $F = (h_1, ..., h_s)$. Then every $f \in k[x, y]$ can be written in the form $f = a_1 h_1 + a_2 h_2 + ... + a_s h_s + r$, where $a_i, r \in k[x, y]$ and either $r = 0$ or a linear combination with coefficients in k of monomials, none of which is divisible by any of $LT(f_1), ..., LT(f_s)$ (see Definition A.5 on p. 305 for leading term LT).*

Example 4.7 (Division algorithm in a univariate case). Divide the polynomial $f = x^3 + 2x^2 + x + 5$ by $h = x^2 - 2$. We proceed as follows:

$$
\begin{array}{r}
x + 2 \\
x^2 - 2 \,|\, x^3 + 2x^2 + x + 5 \\
x^3 - 2x \\
\hline
2x^2 + 3x + 5 \\
2x^2 - 4 \\
\hline
3x + 1,
\end{array}
\tag{4.19}
$$

implying
$x^3 + 2x^2 + x + 5 = (x + 2)(x^2 - 2) + (3x + 1)$, with $a = (x + 2)$ and $r = (3x + 1)$.

The *division algorithm* given in definition (4.2) fits well to the case of *univariate polynomials* as the remainder r can uniquely be determined. For *multivariate polynomials*, the remainder may not be uniquely determined as this depends on the order of the divisors. The division of the polynomial F by $\{f_1, f_2\}$ where f_1 comes before f_2 may not necessarily give the same remainder as the division of F by $\{f_2, f_1\}$ in whose case the order has been changed. This problem is overcome if we pass over to Groebner basis where the existence of every *Ideal* is assured by the Hilbert Basis Theorem [94, pp. 47–61]. The Hilbert Basis Theorem assures that every Ideal $I \subset k[x_1, \ldots, x_n]$ has a finite generating set, that is $I = < g_1, \ldots, g_s >$ for some $\{g_1, \ldots, g_s\} \in I$. The finite generating set G in *Hilbert Basis Theorem* is what is known as a *basis*. Suppose every non-zero polynomial is written in decreasing order of its monomials:

$$\sum_{i=1}^{n} d_i x_i, \ d_i \neq 0, \ x_i > x_{i+1}, \tag{4.20}$$

if we let the system of generators of the *Ideal* be in a set G, a polynomial f is reduced with respect to G if no leading monomial of an element of G (LM (G)) divides the leading monomial of f (LM(f)). The polynomial f is said to be *completely reduced* with respect to G if no monomials of f is divisible by the leading monomial of an element of G [102, pp. 96–97].

The *basis* G, which completely reduces the polynomial f and uniquely determines the remainder r is also known as the Groebner basis and is defined as follows:

Definition 4.3 (Groebner basis). *A system of generators G of an Ideal I is called a Groebner basis (with respect to the order $<$) if every reduction of $f \in I$ to a reduced polynomial (with respect to G) always gives zero as a remainder. This definition is a special case of a more general definition given as: Fix a monomial order and let $G = \{g_1, \ldots, g_t\} \subset k[x_1, \ldots, x_n]$. Given $f \in k[x_1, \ldots, x_n]$, then f reduces to zero Modulo G, written as*

$$f \rightarrow_G 0, \tag{4.21}$$

if f can be written in the form (cf., 4.18 on p. 35)

$$f = a_1 g_1 + \ldots + a_t g_t \tag{4.22}$$

such that whenever $a_i g_i \neq 0$, we have multideg(f)$\geq$ multideg($a_i g_i$) (see Definition A.5 on p. 305 for leading term LT, LM and Multideg).

Following the Definition 4.3, the reader can revisit Examples (4.3) and (4.4) which present the Groebner basis G of the original system F of equations.

Groebner basis has become a household name in algebraic manipulations and finds application in fields such as mathematics and engineering for solving partial differential equations e.g., [222, p. 432]. It has found use as a tool for discovering and proving theorems to solving systems of polynomial equations as elaborated in publications by [82]. Groebner basis also give a solution to the *Ideal* membership problem. By reducing a given polynomial f with respect to the Groebner basis G, f is said to be a member of the *Ideal* if zero remainder is obtained. This implies that if $G = \{g_1, \ldots, g_s\}$ is a Groebner basis of an *Ideal* $I \subset k[x_1, \ldots, x_n]$ and $f \in k[x_1, \ldots, x_n]$ a polynomial, $f \in I$ if and only if the remainder on division of f by G is zero. Groebner bases can also be used to show the equivalence of polynomial equations. Two sets of polynomial equations will generate the same *Ideal* if and only if their Groebner bases are equal with respect to any term ordering, e.g., the solutions of (4.10) satisfy those of (4.9). This property is important in that the solutions of the Groebner basis will satisfy the original system formed by the generating set of nonlinear equations. It implies that a system of polynomial equations $f_1(x_1, \ldots, x_n) = 0, \ldots, f_s(x_1, \ldots, x_n) = 0$ will have the same solutions with a system arising from any Groebner basis of $f_1, \ldots, f_s$ with respect to any term ordering. This is the main property of Groebner basis that is used to solve systems of polynomial equations as will be explained in the next section.

4.3 Buchberger Algorithm

The *B. Buchberger algorithm* is the algorithm that computes Groebner bases from given systems of polynomial equations by cancelling the *leading terms* of these polynomials. With the *lexicographic* ordering chosen, one of the elements of the resulting Groebner basis is often a univariate polynomial whose roots can be obtained using Matlab's *"roots"* command. Given polynomials $g_1, \ldots, g_s \in I$, the algorithm seeks to derive the Groebner basis of this *Ideal*. Systems of equations $g_1 = 0, \ldots, g_s = 0$ to be solved in practice are normally formed by these same polynomials which here generate the *Ideal*. The algorithm computes the Groebner basis by making use of pairs of polynomials from the original polynomials $g_1, \ldots, g_s \in I$ and computes the sub-traction polynomial known as the $S-polynomial$ defined [94, p. 81] as:

Definition 4.4 (S–polynomial[1]). *Let $f, g \in k[x_1, \ldots x_n]$ be two non-zero polynomials. If multideg $(f) = \alpha$ and multideg $(g) = \beta$, then let $\gamma = \gamma_1, \ldots, \gamma_n$, where $\gamma_i = \max\{\alpha_i, \beta_i\}$ for each i. x^γ is called the Least Common Multiple (LCM) of $LM(f)$ and $LM(g)$ expressed as $x^\gamma = LCM\{LM(f), LM(g)\}$. The $S-polynomial$ of f and g is given as*

$$S(f, g) = \frac{x^\gamma}{LT(f)} f - \frac{x^\gamma}{LT(g)} g. \tag{4.23}$$

Expression (4.23) gives S as a linear combination of the monomials

$$\frac{x^\gamma}{LT(f)}, \frac{x^\gamma}{LT(g)},$$

with polynomial coefficients f and g and thus belongs to the *Ideal* generated by f and g (e.g., Definition (4.1) for Ideal on p. 34).

Example 4.8 (S–Polynomial). Consider two polynomials in variables $\{x, y, z\}$ as

$$\begin{bmatrix} g_1 = x^2 + 2a_{12}xy + y^2 + a_{oo} \\ g_2 = y^2 + 2b_{23}yz - +z^2 + b_{oo}. \end{bmatrix} \tag{4.24}$$

with the lexicographic ordering defined as $x > y > z$, the $S-$ polynomial $S(g_1, g_2)$ is computed as

$$\begin{bmatrix}
LM(g_1) = x^2, \; LM(g_2) = y^2, LT(g_1) = x^2, \; LT(g_2) = y^2 \\
LCM(LM(g_1), LM(g_2)) = x^2 y^2 \\
\\
S = \frac{x^2 y^2}{x^2}(x^2 + 2a_{12}xy + y^2 + a_{oo}) - \frac{x^2 y^2}{y^2}(y^2 + 2b_{23}yz + x_3^2 + b_{oo}) \\
\\
= y^2 x^2 + 2a_{12}xy^3 + y^4 + a_{oo}y^2 - x^2 y^2 - 2b_{23}x^2 yx_3 - x^2 x_3^2 - b_{oo}x^2) \\
\\
= -b_{oo}x^2 - 2b_{23}x^2 yx_3 - x^2 x_3^2 + 2a_{12}xy^3 + y^4 + a_{oo}y^2
\end{bmatrix}$$

$$\tag{4.25}$$

---

[1]For the terms appearing in this definition, refer to Appendix A-1, Definition A.5 on p. 305

Definition 4.5 (Groebner basis in terms of $S-$ polynomial). *A basis G is Groebner basis if and only if for every pair of polynomials f and g of G, $S(f,g)$ reduces to zero with respect to G. More generally a basis $G = \{g_1, \ldots, g_s\}$ for an Ideal I is a Groebner basis if and only if $S(f,g) \rightarrow_G 0, \quad i \neq j$.*

The implication of Definition (4.5) is the following: Given two polynomials $f, g \in G$ such that $LCM\{LM(f), LM(g)\} = LM(f).LM(g)$, the leading monomials of f and g are relatively prime leading to $S(f,g) \rightarrow_G 0$. The concept of prime integer is documented in [192, pp. 1–17].

Example 4.9 (Computation of Groebner basis from the $S-$polynomials). By completing the example given by [102, pp. 101–102], we illustrate how the Buchberger algorithm works. Let us consider the Ideal generated by the polynomial equations

$$\begin{bmatrix} g_1 = x^3yz - xz^2 \\ g_2 = xy^2z - xyz \\ g_3 = x^2y^2 - z, \end{bmatrix} \tag{4.26}$$

with the lexicographic ordering $x > y > z$ adopted. The $S-$polynomials to be formed are $S(g_1, g_2)$, $S(g_2, g_3)$ and $S(g_1, g_3)$. We consider first $S(g_2, g_3)$ and show that the result is used to suppress g_1. Consequently any pair $S(g_1, g_i)$ (e.g., $S(g_1, g_2)$ and $S(g_1, g_3)$) containing g_1 will not be considered. With $LT(g_2) = xy^2z$ and $LT(g_3) = x^2y^2$ the $LCM(g_2, g_3) = x^2y^2z$. The $S-$polynomials is then computed as

$$\begin{bmatrix} S(g_2, g_3) = \dfrac{x^2y^2z}{xy^2z}g_2 - \dfrac{x^2y^2z}{x^2y^2}g_3 \\[2mm] = (x^2y^2z - x^2yz) - (x^2y^2z - z^2) \\[2mm] = -x^2yz + z^2. \end{bmatrix} \tag{4.27}$$

One immediately notes that the leading term of the resulting polynomial $LT(S(g_2, g_3))$ is not divisible by any of the leading terms of the elements of G. The remainder upon the division of $S(g_2, g_3)$ by the polynomials in G is not zero (i.e., when reduced with respect to G). The set G therefore is *not* a Groebner basis. The resulting polynomial is denoted g_4, and its negative (to make calculations more reliable) added to the initial set of G leading to

$$\begin{bmatrix} g_1 = x^3yz - xz^2 \\ g_2 = xy^2z - xyz \\ g_3 = x^2y^2 - z \\ g_4 = x^2yz - z^2. \end{bmatrix} \tag{4.28}$$

The S–polynomials to be formed are now $S(g_1, g_2)$, $S(g_1, g_3)$, $S(g_1, g_4)$, $S(g_2, g_4)$ and $S(g_3, g_4)$. In the set of G, one can write $g_1 = xg_4$ leading, without any change, to the suppression of g_1 leaving only $S(g_2, g_4)$ and $S(g_3, g_4)$ to be considered. Then

$$\begin{bmatrix} S(g_2, g_4) = xg_2 - yg_4 \\ = -x^2yz + yz^2, \end{bmatrix} \tag{4.29}$$

is reduced by adding g_4 to give $g_5 = yz^2 - z^2$, a non zero value. The set G, which is still *not* a Groebner basis now becomes

$$\begin{bmatrix} g_2 = xy^2z - xyz, \\ g_3 = x^2y^2 - z, \\ g_4 = x^2yz - z^2, \\ g_5 = yz^2 - z^2. \end{bmatrix} \tag{4.30}$$

The S–polynomials to be considered are now $S(g_3, g_4)$, $S(g_2, g_5)$, $S(g_3, g_5)$ and $S(g_4, g_5)$. We have

$$\begin{bmatrix} S(g_3, g_4) = zg_3 - yg_4 \\ = yz^2 - z^2, \end{bmatrix} \tag{4.31}$$

which upon subtraction from g_5 reduces to zero. Further,

$$\begin{bmatrix} S(g_2, g_5) = zg_2 - xyg_5 \\ = -xyz^2 + xyz^2 \\ = 0 \end{bmatrix} \tag{4.32}$$

and

$$\begin{bmatrix} S(g_4, g_5) = zg_4 - x^2yg_5 \\ = x^2z^2 - z^3, \end{bmatrix} \tag{4.33}$$

which is added to G as g_6 giving

$$\begin{bmatrix} g_2 = xy^2z - xyz, \\ g_3 = x^2y^2 - z, \\ g_4 = x^2yz - z^2, \\ g_5 = yz^2 - z^2, \\ g_6 = x^2y^2 - z^3. \end{bmatrix} \tag{4.34}$$

The S polynomials to be formed next are $S(g_3, g_5)$, $S(g_2, g_6)$, $S(g_3, g_6)$, $S(g_4, g_6)$ and $S(g_5, g_6)$. We now complete the example by illustrating that all these $S - polynomials$ reduce to zero as follows:

$$
\begin{bmatrix}
S(g_3, g_5) = z^2 g_3 - x^2 y g_5 = x^2 y z^2 - z^3 - z g_4 = 0 \\
S(g_2, g_6) = x z g_2 - y^2 g_6 = -x^2 y^2 z^2 + y^2 z^3 + y^2 g_4 = 0 \\
S(g_3, g_6) = z^2 g_3 - y^2 g_6 = y^2 z^3 - z^3 - (yz - z) g_5 = 0 \\
S(g_4, g_6) = z g_4 - y g_6 = y z^3 - z^3 - z g_5 = 0 \\
S(g_5, g_6) = x^2 g_5 - y g_6 = -x^2 z^2 + y z^3 + g_6 - z g_5 = 0,
\end{bmatrix}
\tag{4.35}
$$

comprising the Groebner basis of the original set in (4.26).

The importance of S–polynomials is that they lead to the cancellation of the leading terms of the polynomial pairs involved. In so doing, polynomial variables are systematically eliminated according to the ordering chosen. For example if the lexicographic ordering $x > y > z$ is chosen, x will be eliminated first, followed by y and the final expression may consist only of the variable z. Cox et al [95, p. 15] has indicated the advantage of lexicographic ordering as being the ability to produce Groebner basis with systematic elimination of variables. *Graded lexicographic ordering* (see Definition A.3 of Appendix A-1 on p. 304), on the other hand has the advantage of minimizing the amount of computational space needed to produce the Groebner basis.

Buchberger algorithm is therefore a *generalization* of the *Gauss elimination procedure* for linear systems of equations as shown in Examples 4.2, 4.3 and 4.4. If we now put our system of polynomial equations to be solved in a set G, S–pair combinations can be formed from the set of G as illustrated in Examples 4.5 and 4.9. The *theorem*, known as the *Buchberger's S*–pair *polynomial criterion*, gives the criterion for deciding whether a given basis is a Groebner basis or not. It suffices to compute all the S–polynomials and check whether they reduce to zero. Should one of the polynomials not reduce to zero, then the basis fails to be a Groebner basis. Since the reduction is a linear combination of the elements of G, it can be added to the set G without changing the *Ideal* generated. Buchberger [80] gives an *optimization criterion* that reduces the number of the S–polynomials already considered in the algorithm. The criterion states that if there is an element h of G such that the leading monomial of h, i.e., LM(h, divides the LCM($f, g \in G$), and if $S(f, h)$, $S(h, g)$ have already been considered, then there is no need of considering $S(f, g)$ as this reduces to zero.

The essential observation in using Groebner bases to solve systems of polynomial equations is that the variety (simultaneous solution of systems of polynomial equations) does not depend on the original system of the polynomials $F := \{f_1, \ldots, f_s\}$, but instead on the *Ideal I* generated by F. This therefore means that the variety $V = V(I)$. One makes use of the special generating set (Groebner basis) instead of the actual system F. Since the Ideal is generated by F, the solutions obtained by solving the affine variety of this Ideal satisfies the original system F of equations as already stated. Buchberger [79] proved that $V(I)$ is void, and thus giving a test as to whether a system of polynomial F can be solved. The solution can be obtained if and only if the computed Groebner basis of Ideal I has 1 as its element. Buchberger [79] further gives the criterion for deciding if $V(I)$ is finite. If the system has been proved to be solvable and finite then [337, theorem 8.4.4, p. 192] gives a theorem for deciding whether the system has finitely or infinitely many solutions. The Theorem states that if G is a Groebner basis, then a solvable system of polynomial equations has finitely many solutions if and only if for every x_i, $1 \leq i \leq n$, there is a polynomial $g_i \in G$ such that $LM(g_i)$ is a pure power of x_i. The process of addition of the remainder after the reduction by the S–polynomials, and thus expanding the generating set is shown by [79], [95, p. 88] and [102, p. 101] to terminate.

The Buchberger algorithm thus makes use of the subtraction polynomials known as the S–polynomials in Definition (4.4) to eliminate the leading terms of a pair of polynomials. In so doing, and if lexicographic ordering is chosen, the process ends up with one of the computed S–polynomials being a univariate polynomial which can be solved and substituted back in the other S–polynomials using the *extension theorem* [95, pp. 25–26] to obtain the other variables.

4.3.1 Mathematica Computation of Groebner Basis

Groebner basis can be computed using algebraic softwares of Mathematica Versions 2 onwards. The Groebner basis command is executed by writing

$$In[1] := GroebnerBasis[\{polynomials\}, \{variables\}], \qquad (4.36)$$

where $In[1]:=$ is the Mathematica prompt which computes the Groebner basis for the Ideal generated by the polynomials with respect to the *monomial order* specified by *monomial order options.*

Example 4.10 (Mathematica computation of Groebner basis). In Example (4.3) on p. 33, the systems of polynomial equations were given as

$$\left[\begin{array}{l} f_1 = xy - 2y \\ f_2 = 2y^2 - x^2. \end{array}\right.$$

Groebner basis is computed by

$$In[1] := GroebnerBasis[\{f_1, f_2\}, \{x, y\}], \tag{4.37}$$

leading to the values in (4.10).

With this approach, one gets too many elements of Groebner basis which may not be relevant. In a case where the solution of a specific variable is desired, one can avoid computing the undesired variables, and alleviate the need for back-substitution by simply computing the *reduced Groebner basis*. In this case (4.36) modifies to

$$In[1] := GroebnerBasis[\{polynomials\}, \{variables\}, \{options\}], \tag{4.38}$$

with the *variables* to be eliminated specified in the options part.

Example 4.11 (Mathematica computation of reduced Groebner basis). In Example 4.10, one would compute the reduced Groebner basis using (4.38) as

$$In[1] := GroebnerBasis[\{f_1, f_2\}, \{x, y\}, \{y\}], \tag{4.39}$$

which will return only $-2x^2 + x^3$. This univariate polynomial is solved for x using the roots command in Matlab (see e.g., [174, p. 146]) by

$$roots([1 \ -2 \ 0 \ 0]). \tag{4.40}$$

The values of the row vector in (4.40) are the coefficients of the cubic polynomial $x^3 - 2x^2 + 0x + 0 = 0$ obtained from (4.39) (see Sect. 3.6.2 for solutions of cubic polynomials).

The values of y from Example 4.3 can equally be computed from (4.39) by replacing y in the option part with x and thus removing the need for back substitution. We leave it for the reader to compute the values of y from Example 4.3 and also those of z in Example 4.4 using reduced Groebner basis (4.38) as an exercise. The reader should confirm that the solution of y leads to $y^3 - 2y$ with the roots $y = 0$ or $y = \pm 1.4142$. From experience, we recommend the use of reduced Groebner basis for applications in geodesy and geoinformatics. This will; fasten the computations, save on computer space, and alleviates the need for back-substitution.

4.3.2 Maple Computation of Groebner Basis

In Maple Version 5 the command is accessed by typing > *with (grob-ner);* (where > is the Maple prompt and the semicolon ends the Maple command). Once the Groebner basis package has been loaded, the execution command then becomes > *gbasis (polynomials, variables, ter-morder)* which computes the Groebner basis for the *ideal* generated by the *polynomials* with respect to the *monomial ordering* specified by *termorder* and *variables* in the executable command.

4.4 Concluding Remarks

Using the Groebner basis, most systems of nonlinear equations that are encountered in geodesy and geoinformatics can be solved. All that is required of the user is to write algorithms that can easily be run in Mathematica or Maple using the steps discussed. In latter chapters, we will demonstrate how algorithms using Groebner basis can be written for various tasks. Application of the technique in geodesy can be found in the works of [11, 12, 14, 21, 24, 28, 30]. Several publications exist on the subject, e.g., [55, 56, 82, 92, 94, 95, 102, 217, 266, 304, 320, 337]. For readers who may be interested in exploring the subject further, these literature and similar others are worth reading. The Groebner bases approach presented in this chapter adds to the treasures of methods that are useful for solving nonlinear algebraic systems of equations in geodesy, geoinformatics, machine vision, robotics and surveying.

Finally, we begun the chapter by a quote from [269]. We think that indeed, systems of more than one nonlinear equations are solvable, and the answer lies in commutative algebra!

5

Polynomial Resultants

5.1 Resultants: An Alternative to Groebner Basis

Besides Groebner basis approach discussed in Chap. 4, the other powerful algebraic tools for solving nonlinear systems of equations are the polynomial resultants approaches. While Groebner basis may require large storage capacity during its computations, polynomial resultants approaches presented herein offers remedy to users who may not be lucky to have computers with large storage capacities. This chapter presents polynomial resultants approaches starting from the resultants of two polynomials, known as the *"Sylvester resultants"*, to the resultants of more than two polynomials in several variables known as *"multipolynomial resultants"*. In normal matrix operations in linear algebra, one is often faced with the task of computing determinants. Their applications to least squares approach are well known.

For polynomial resultants approaches discussed herein, the ability to compute determinants of matrices is the essential requirement. We will look at how they are formed and applied to solve nonlinear systems of equations. Indeed [291] had already used the resultant technique to the $R^2 \rightarrow R^2$ mapping of gravitation lens. Such mapping describes the light rays which run from a deflector plane (lens) to an observer. For simple lenses such as point masses in galactic fields, [291] observed the global mapping to be an algebraic expression whose inversion led to the problem of solving a polynomial in two variables. Further use of polynomial resultants in geodesy is exemplified in the works of [226, pp. 72–76] and [19, 25, 31].

5.2 Sylvester Resultants

Sylvester resultants approach is useful for solving explicitly nonlinear systems of equations with two polynomials in two variables. Problems in this category could be those of two dimensional nature such as planar ranging, planar resection etc., as shall be seen in subsequent chapters. Polynomial resultants approach is based on homogeneous polynomials defined as

Definition 5.1 (Homogeneous polynomial). *If monomials of a polynomial p with non zero coefficients have the same total degree, the polynomial p is said to be homogeneous.*

Example 5.1 (Homogeneous polynomial equation). A homogeneous polynomial equation of total degree 2 is $s = x^2 + y^2 + z^2 + xy + xz + yz$, since the monomials $\{x^2, y^2, z^2, xy, xz, yz\}$ all have the sum of their powers (total degree) being 2.

To set the ball rolling, let us examine next the resultant of two univariate polynomials $s, t \in k[x]$ of *positive degree* as

$$\begin{cases} s = k_0 x^i + \ldots + k_i, \ k_0 \neq 0, \ i > 0 \\ t = l_0 x^j + \ldots + l_j, \ l_0 \neq 0, \ j > 0. \end{cases} \tag{5.1}$$

The resultant of s and t, denoted $\mathrm{Res}(s, t)$, is the $(i + j) \times (i + j)$ determinant

$$\mathrm{Res}\,(s, t) = \det \begin{bmatrix} k_0 & k_1 & k_2 & . & . & . & k_i & 0 & 0 & 0 & 0 & 0 \\ 0 & k_0 & k_1 & k_2 & . & . & . & k_i & 0 & 0 & 0 & 0 \\ 0 & 0 & k_0 & k_1 & k_2 & . & . & . & k_i & 0 & 0 & 0 \\ 0 & 0 & 0 & k_0 & k_1 & k_2 & . & . & . & k_i & 0 & 0 \\ 0 & 0 & 0 & 0 & k_0 & k_1 & k_2 & . & . & . & k_i & 0 \\ 0 & 0 & 0 & 0 & 0 & k_0 & k_1 & k_2 & . & . & . & k_i \\ l_0 & l_1 & l_2 & . & . & . & l_j & 0 & 0 & 0 & 0 & 0 \\ 0 & l_0 & l_1 & l_2 & . & . & . & l_j & 0 & 0 & 0 & 0 \\ 0 & 0 & l_0 & l_1 & l_2 & . & . & . & l_j & 0 & 0 & 0 \\ 0 & 0 & 0 & l_0 & l_1 & l_2 & . & . & . & l_j & 0 & 0 \\ 0 & 0 & 0 & 0 & l_0 & l_1 & l_2 & . & . & . & l_j & 0 \\ 0 & 0 & 0 & 0 & 0 & l_0 & l_1 & l_2 & . & . & . & l_j \end{bmatrix}, \tag{5.2}$$

where the coefficients of the first polynomial s in (5.1) occupy j rows, while those of the second polynomial t occupy i rows. The empty spaces are occupied by zeros as shown above such that a square matrix is obtained. This resultant is known as the *Sylvester resultant* and has the following properties [95, §3.5] and [305];

1. $Res(s, t)$ is a polynomial in $k_0, \ldots, k_i, l_0, \ldots, l_j$ with integer coefficients.
2. $Res(s, t) = 0$ if and only if $s(x)$ and $t(x)$ have a common factor in $k[x]$.
3. There exist a polynomial $q, r \in k[x]$ such that $qs + rt = Res(s, t)$.

Sylvester resultants can be used to solve systems of polynomial equations in two variables as shown in Example (5.2).

Example 5.2 (Sylvester resultants solution of systems of nonlinear equations). Consider the system of equations given in [305, p. 72] as

$$\begin{bmatrix} p := xy - 1 = 0 \\ q := x^2 + y^2 - 4 = 0. \end{bmatrix} \tag{5.3}$$

In-order to eliminate one variable e.g., x, the variable y is *hidden*, i.e., the variable say y is considered as a constant (polynomial of degree zero). We then have the *Sylvester resultant* from (5.2) as

$$Res\,(s, t, x) = det \begin{bmatrix} y & -1 & 0 \\ 0 & y & -1 \\ 1 & 0 & y^2 - 4 \end{bmatrix} = y^4 - 4y^2 + 1, \tag{5.4}$$

which can be readily solved for the variable y and substituted back in any of the equations in (5.3) to obtain the values of the variable x. Alternatively, the procedure can be applied to derive x directly. Hiding x, one obtains with (5.2)

$$Res\,(s, t, y) = det \begin{bmatrix} x & -1 & 0 \\ 0 & x & -1 \\ 1 & 0 & x^2 - 4 \end{bmatrix} = x^4 - 4x^2 + 1. \tag{5.5}$$

The roots of the univariate polynomials (5.4) and (5.5) are then obtained using the Matlab's root command as

$$\{x, y\} = roots([1\ 0\ -4\ 0\ 1]\ ) = \pm 1.9319\,or\, \pm 0.5176. \tag{5.6}$$

In (5.6), the row vector $[1\ 0\ -4\ 0\ 1]$ are the coefficients of the quartic polynomials in either (5.4) or (5.5). Zeros are the coefficients of the variables $\{x^3, y^3\}$ and $\{x, y\}$. The solutions in (5.6) satisfy the polynomials in (5.4) and (5.5). They also satisfy the original nonlinear system of equations (5.3). In (5.4) and (5.5), the determinant can readily be obtained from MATLAB software by typing $det(\mathbf{A})$, where $\mathbf{A}$ is the matrix whose determinant is desired.

For two polynomials in two variables, the construction of resultants is relatively simpler and algorithms for the execution are incorporated in computer algebra systems. Resultants of more than 2 polynomials of multiple variables are however complicated. For their construction, we turn to the *multipolynomial resultants*.

5.3 Multipolynomial Resultants

Whereas the resultant of two polynomials in two variables is well known and algorithms for computing it well incorporated into computer algebra packages such as Maple, multipolynomial resultants, i.e., the resultant of more than two polynomials still remain an active area of research. This section therefore extends on the use of Sylvester resultants to resultants of more than two polynomials of multiple variables, known as multipolynomial resultants.

The need for multipolynomial resultants method in geodesy and geoinformatics is due to the fact that many problems encountered require the solution of more than two polynomials of multiple variables. This is true since we are living in a three-dimensional world. We shall therefore understand the term multipolynomial resultants to mean resultants of more than two polynomials. We treat it as a tool besides Groebner bases, and perhaps more powerful to eliminate variables in systems of polynomial equations. In defining it, [237] writes:

> "Elimination theory, a branch of classical algebraic geometry, deals with conditions for common solutions of a system of polynomial equations. Its main result is the construction of a single resultant polynomial of n homogeneous polynomial equations in n unknowns, such that the vanishing of the resultant is a *necessary* and *sufficient* condition for the given system to have a non-trivial solution. We refer to this resultant as the multipolynomial resultant and use it in the algorithm presented in the paper".

In the formation of the design matrix whose determinants are needed, several approaches can be used as discussed in [233, 234, 235, 236, 237] who applies the eigenvalue-eigenvector approach, [83] who uses characteristic polynomial approach, and [303, 305] who proposes a more compact approach for solving the resultants of a ternary quadric using the Jacobian determinant approach. In this book, two approaches are presented; first the approach based on F. Macaulay [229] formulation (the pioneer of resultants approach) and then a more modern approach based on B. Sturmfels' [305] formulation.

5.3.1 F. Macaulay Formulation:

With n polynomials, the construction of the matrix whose entries are the coefficients of the polynomials $f_1, \ldots, f_n$ can be done in five steps as follows:

Step 1: The given polynomials $f_1 = 0, \ldots, f_n = 0$ are considered to be homogeneous equations in the variables $x_1, \ldots, x_n$ and if not, they are homogenized. Let the degree of the polynomial f_i be d_i. The first step involves the determination of the *critical degree* given by [43] as

$$d = 1 + \sum (d_i - 1). \tag{5.7}$$

Step 2: Once the critical degree has been established, the given monomials of the polynomial equations are multiplied with each other to generate a set X. The elements of this set consists of monomials whose total degree equals the critical degree. Thus if we are given polynomial equations $f_1 = 0, \ldots, f_n = 0$, each monomial of f_1 is multiplied by those of $f_2, \ldots, f_n$, those of f_2 are multiplied by those of $f_3, \ldots, f_n$ until those of f_{n-1} are multiplied by those of f_n. The set X of monomials generated in this form is

$$X^d = \{x^d \mid d = \alpha_1 + \alpha_2 + \ldots + \alpha_n\}, \tag{5.8}$$

with the variable $x^d = x_1^{\alpha_1} \ldots x_n^{\alpha_n}$.

Step 3: The set X containing monomials each of total degree d is now partitioned according to the following criteria [83, p. 54]

$$\begin{bmatrix} X_1^d = & \{x^\alpha \in X^d \mid \alpha_1 \geq d_1\} \\ X_2^d = & \{x^\alpha \in X^d \mid \alpha_2 \geq d_2 \text{ and } \alpha_1 < d_1\} \\ \cdot & \cdot \\ \cdot & \cdot \\ \cdot & \cdot \\ X_n^d = \{x^\alpha \in X^d \mid & \alpha_n \geq d_n \text{ and } \alpha_i < d_i, \text{ for } i = 1, \ldots, n-1\}. \end{bmatrix}$$

(5.9)

The resulting sets of X_i^d are disjoint and every element of X^d is contained in exactly one of them.

Step 4: From the resulting subsets $X_i^d \subset X^d$, a set of polynomials F_i which are homogeneous in n variables are defined as

$$F_i = \frac{X_i^d}{x_i^{d_i}} f_i. \tag{5.10}$$

From (5.10), a *square matrix* $\mathbf{A}$ is now formed with the row elements being the *coefficients* of the monomials of the polynomials $F_i \mid_{i=1,\ldots,n}$ and the columns corresponding to the N monomials of the set X^d. The formed square matrix $\mathbf{A}$ is of the order

$$\binom{d+n-1}{d} \times \binom{d+n-1}{d},$$

and is such that for a given polynomial F_i in (5.10), the row of the square matrix $\mathbf{A}$ is made up of the symbolic coefficients of each polynomial. The square matrix $\mathbf{A}$ has a special property that the non trivial solution of the homogeneous equations F_i which also form the solution of the original equations f_i are in its null space. This implies that the matrix must be singular or its determinant, $det(\mathbf{A})$, must be zero. For the determinant to vanish, therefore, the original equations f_i and their homogenized counterparts F_i must have the same non trivial solutions.

Step 5: After computing the determinant of the square matrix $\mathbf{A}$ above, [229] suggests the computation of *extraneous factor* in-order to obtain the resultant. Cox et al. [95, Proposition 4.6, p. 99] explains the *extraneous factors* to be integer polynomials in the coefficients of $\bar{F}_0, \ldots, \bar{F}_{n-1}$, where $\bar{F}_i = F_i(x_0, \ldots, x_{n-1}, 0)$. It is related to the determinant via

$$determinant = Res(F_1, \ldots, F_n).Ext, \tag{5.11}$$

with the determinant computed as in step 4, $Res(F_1, \ldots, F_n)$ being the multipolynomial resultant and Ext the extraneous factor. This expression was established as early as 1902 by F. Macaulay [229] and this procedure of *resultant* formulation thus named after him. Macaulay [229] determines the extraneous factor from the sub-matrix of the $N \times N$ square matrix $\mathbf{A}$ and calls it a factor of minor obtained by deleting rows and columns of the $N \times N$ matrix $\mathbf{A}$. A monomial x^α of total degree d is said to be reduced if $x_i^{d_i}$ divides x^α for exactly one i. The extraneous factor is obtained by computing the determinant of the sub-matrix of the coefficient matrix $\mathbf{A}$, obtained by deleting rows and columns corresponding to reduced monomials x^α.

For our purpose, it suffices to solve for the unknown variable hidden in the coefficients of the polynomials f_i by obtaining the determinant of the $N \times N$ square matrix $\mathbf{A}$ and equating it to zero neglecting the extraneous factor. This is because the extraneous factor is an integer polynomial and as such not related to the variable in the determinant of $\mathbf{A}$. The existence of the non-trivial solutions provides the *necessary* and *sufficient* conditions for the vanishing of the determinant.

5.3.2 B. Sturmfels' Formulation

Given three homogeneous equations of degree 2 as

$$
\begin{aligned}
F_1 &:= a_{11}x^2 + a_{12}y^2 + a_{13}z^2 + a_{14}xy + a_{15}xz + a_{16}yz = 0 \\
F_2 &:= a_{21}x^2 + a_{22}y^2 + a_{23}z^2 + a_{24}xy + a_{25}xz + a_{26}yz = 0 \\
F_3 &:= a_{31}x^2 + a_{32}y^2 + a_{33}z^2 + a_{34}xy + a_{35}xz + a_{36}yz = 0,
\end{aligned}
\tag{5.12}
$$

the Jacobian determinant is computed by

$$
J = det
\begin{bmatrix}
\dfrac{\partial F_1}{\partial x} & \dfrac{\partial F_1}{\partial y} & \dfrac{\partial F_1}{\partial z} \\[2ex]
\dfrac{\partial F_2}{\partial x} & \dfrac{\partial F_2}{\partial y} & \dfrac{\partial F_2}{\partial z} \\[2ex]
\dfrac{\partial F_3}{\partial x} & \dfrac{\partial F_3}{\partial y} & \dfrac{\partial F_3}{\partial z}
\end{bmatrix},
\tag{5.13}
$$

resulting in a cubic polynomial in the coefficients $\{x, y, z\}$. Since the determinant polynomial J in (5.13) is a cubic polynomial, its partial derivatives will be quadratic polynomials in variables $\{x, y, z\}$ and are written in the form

$$\frac{\partial J}{\partial x} := b_{11}x^2 + b_{12}y^2 + b_{13}z^2 + b_{14}xy + b_{15}xz + b_{16}yz = 0$$

$$\frac{\partial J}{\partial y} := b_{21}x^2 + b_{22}y^2 + b_{23}z^2 + b_{24}xy + b_{25}xz + b_{26}yz = 0 \qquad (5.14)$$

$$\frac{\partial J}{\partial z} := b_{31}x^2 + b_{32}y^2 + b_{33}z^2 + b_{34}xy + b_{35}xz + b_{36}yz = 0.$$

The coefficients b_{ij} in (5.14) are cubic polynomials in a_{ij} of (5.12). The final step in computing the resultant of the initial system (5.12) involves the computation of the determinant of a 6×6 matrix given by

$$Res_{222}(F_1, F_2, F_3) = det \begin{bmatrix} a_{11} & a_{12} & a_{13} & a_{14} & a_{15} & a_{16} \\ a_{21} & a_{22} & a_{23} & a_{24} & a_{25} & a_{26} \\ a_{31} & a_{32} & a_{33} & a_{34} & a_{35} & a_{36} \\ b_{11} & b_{12} & b_{13} & b_{14} & b_{15} & b_{16} \\ b_{21} & b_{22} & b_{23} & b_{24} & b_{25} & b_{26} \\ b_{31} & b_{32} & b_{33} & b_{34} & b_{35} & b_{36} \end{bmatrix}. \qquad (5.15)$$

The resultant (5.15) vanishes if and only if (5.12) have a common solution $\{x, y, z\}$, where $\{x, y, z\}$ are complex numbers or real numbers not all equal zero. The subscripts on the left-hand-side of (5.15) indicate the degree of the polynomials in (5.12) whose determinants are sought.

5.4 Concluding Remarks

With modern computers, polynomial resultants approaches discussed can easily be used to develop algorithms for solving systems of nonlinear equations. Compared to Groebner basis, it has the advantage of not computing extra parameters thus requiring less of computer's space. Its shortcoming, however, lies in the formation of the design matrix which become more complicated and cumbersome as the number of polynomials and variables increases. Unless Groebner basis fails, we recommend it for solving geodetic and geoinformatics nonlinear systems of equations. On the other hand, polynomial resultants approach comes in handy when computer's space is limited or when the practitioner has time. With modern computer storage capacity, though, most problems requiring algebraic solutions in the fields above can easily be handled by Groebner basis without the fear of computer breakdown. Publications on the subject include: [19, 25, 31, 43, 83, 84, 85, 95, 104, 118, 119, 163, 209, 228, 229, 230, 231, 233, 234, 235, 236, 237, 285, 238, 239, 240, 241, 253, 303, 305, 326, 327].

Besides Groebner bases and polynomial resultants techniques, there exists another approach for eliminating variables developed by W. WU [340] using the ideas proposed by [280]. This approach is based on Ritt's characteristic set construction and was successfully applied to automated geometric theorem by Wu. This algorithm is referred as the *Ritt-Wu's algorithm* [241].

6

Gauss-Jacobi Combinatorial Algorithm

"Pauca des Matura" –a few but ripe – C. F. Gauss

6.1 Estimating Unknown Parameters

In geodesy and geoinformatics, field observations are normally collected with the aim of estimating parameters. In geodynamics for example, GPS and gravity measurements are undertaken with the aim of determining crustal deformation. With improvement in instrumentation, more observations are often collected than the unknowns. Let us consider a simple case of measuring structural deformation. For deformable surfaces, such as mining areas, or structures (e.g., bridges), several observable points are normally marked on the surface of the body. These points would then be observed from a network of points set up on a non-deformable stable surface. Measurements taken are distances, angles or directions which are normally more than the unknown positions of the points marked on the deformable surface leading to redundant observations.

Procedures that are often used to estimate the unknowns from the measured values will depend on the nature of the equations relating the observations to the unknowns. If these equations are linear, then the task is much simpler. In such cases, any procedure that can invert the normal equation matrix such as least squares, linear Gauss-Markov model etc., would suffice. Procedures for estimating parameters in linear models have been documented in [200]. Press et al. [269] present algorithms for solving linear systems of equations. If the equations relating the observations to the unknowns are nonlinear, they have first to be linearized and the unknown parameters estimated iteratively using numerical methods. The operations of these numerical methods require some approximate starting values. At each iteration step, the preceding estimated values of the unknowns are improved. The iteration steps are repeated until the difference between two consecutive estimates of the unknowns satisfies a specified threshold. Procedures for solving nonlinear problems such as the Steepest-descent, Newton's, Newton-Rapson and Gauss-Newton's have been discussed in [269, 308]. In particular, [308] recommends the Gauss-Newton's method as it exploits the structure of the objective function (sum of squares) that is to be minimized. In [310], the manifestation of the nonlinearity of a function during the various stages of adjustment is considered. While extending the work of [207] on *nonlinear adjustment* with respect to geometric interpretation, [147, 148] have presented the *necessary* and *sufficient* conditions for least squares adjustment of *nonlinear Gauss-Markov model*, and provided the geometrical interpretation of these conditions. Another geometrical approach include the work of [67], while non geometrically treatment of nonlinear problems have been presented by [53, 210, 245, 267, 283, 287].

6.2 Combinatorial Approach: The Origin

Presented in this chapter is an alternative approach to traditional iterative numerical procedures for solving overdetermined problems, i.e., where more observations than unknown exist. This approach, which we call the *Gauss-Jacobi combinatorial* has the following advantages:

1. From the start, the objective is known.
2. It does not require linearization.
3. The need for iteration does not exist.
4. The variance-covariance matrices of all parameters are considered.
5. It can be exploited for outlier diagnosis.

The combinatorial approach traces its roots to the work of *C. F. Gauss* which was published posthumously (see Appendix A-2). Whereas the procedures presented in Chaps. 4 and 5 solve nonlinear systems of equations where the number of observations n and unknowns m are equal, i.e., $n = m$, Gauss-Jacobi combinatorial solves the case where $n > m$. In Fig. 4.1 on p. 30 for example, two distance measurements from known stations P_1 and P_2 were used to determine the position of unknown station P_0. Let us assume that instead of the two known stations, a third distance was measured from point P_3 as depicted in Fig. 6.1. In such a case, there exist three possibilities (combi-

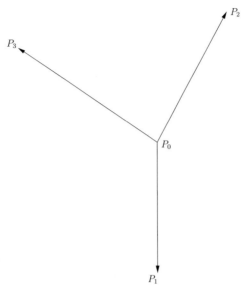

Fig. 6.1. Combinatorial possibilities for determining unknown station P_0

nations) for determining the position of the unknown station P_0. Recall that for Fig. 4.1 on p. 30, two nonlinear distance equations were written (e.g., 4.1 and 4.2). For Fig. 6.1, systems of distance equations could be written for combinations $\{P_1P_0P_2\}$, $\{P_1P_0P_3\}$ and $\{P_2P_0P_3\}$. For combination $\{P_1P_0P_2\}$ for example, one writes

$$d_1^2 = (x_1 - x_0)^2 + (y_1 - y_0)^2 \tag{6.1}$$

and

$$d_2^2 = (x_2 - x_0)^2 + (y_2 - y_0)^2. \tag{6.2}$$

Equations (6.1) and (6.2) lead to solutions $\{x_0, y_0\}_{1,2}$ as position of the unknown station P_0, where the subscripts indicate the combinations used. Combination $\{P_1P_0P_3\}$ gives

$$d_1^2 = (x_1 - x_0)^2 + (y_1 - y_0)^2 \tag{6.3}$$

and

$$d_3^2 = (x_3 - x_0)^2 + (y_3 - y_0)^2, \tag{6.4}$$

leading to solutions $\{x_0, y_0\}_{1,3}$ as the position of the unknown station P_0. The last combination $\{P_2 P_0 P_3\}$ has

$$d_2^2 = (x_2 - x_0)^2 + (y_2 - y_0)^2 \qquad (6.5)$$

and

$$d_3^2 = (x_3 - x_0)^2 + (y_3 - y_0)^2, \qquad (6.6)$$

as its system of equations leading to solutions $\{x_0, y_0\}_{2,3}$. The solutions $\{x_0, y_0\}_{1,2}$, $\{x_0, y_0\}_{1,3}$ and $\{x_0, y_0\}_{2,3}$ from these combinations are however not the same due to unavoidable effects of random errors. It is in attempting to harmonize these solutions to give the correct position of point P_0 that *C. F. Gauss* proposed the combinatorial approach. He believed that plotting these three combinatorial solutions resulted in an error figure with the shape of a triangle. He suggested the use of weighted arithmetic mean to obtain the final position of point P_0. In this regard the weights were obtained from the products of squared distances $\overline{P_0 P_1}$, $\overline{P_0 P_2}$ and $\overline{P_0 P_3}$ (from unknown station to known stations) and the square of the perpendicular distances from the sides of the error triangle to the unknown station. According to [256, pp. 272–273], the motto in Gauss seal read *"pauca des matura"* meaning *few but ripe*. This belief led him not to publish most of his important contributions. For instance, [256, pp. 272–273] writes

> "Although not all his results were recorded in the diary (many were set down only in letters to friends), several entries would have each given fame to their author if published. Gauss knew about the quaternions before Hamilton...".

Unfortunately, the combinatorial method, like many of his works, was later to be published after his death (see e.g., Appendix A-2). Several years later, the method was independently developed by *C. G. I. Jacobi* [194] who used the square of the determinants as the weights in determining the unknown parameters from the arithmetic mean. Werkmeister [332] later established the relationship between the area of the error figure formed from the combinatorial solutions and the standard error of the determined point. In this book, the term combinatorial is adopted since the algorithm uses combinations to get all the finite solutions from which the optimum value is obtained. The optimum value is obtained by minimizing the sum of square of errors of pseudo-observations formed from the combinatorial solutions. For combinatorial optimization techniques, we refer to [113]. We will refer to this *combinatorial approach* as the *Gauss-Jacobi combinatorial algorithm* in appreciation of the work done by both *C. F. Gauss* and *C. G. I. Jacobi.*

In the approaches of *C. F. Gauss* and later *C. G. I. Jacobi*, one difficulty however remained unsolved. This was the question of how the various nonlinear systems of equations, e.g., (6.1 and 6.2), (6.3 and 6.4) or (6.5 and 6.6) could be solved explicitly! The only option they had was to linearize these equations, which in essence was a negation of what they were trying to avoid in the first place. Had they been aware of algebraic techniques that we saw in Chaps. 4 and 5, they could have succeeded in providing a complete algebraic solution to the overdetermined problem. In this chapter, we will complete what was started by these two gentlemen and provide a complete algebraic algorithm which we name in their honour. This algorithm is designed to provide a solution to the *nonlinear Gauss-Markov model.* First we define both the linear and nonlinear Gauss-Markov model and then formulate the *Gauss-Jacobi combinatorial algorithm* in Sect. 6.4.

6.3 Linear and Nonlinear Gauss-Markov Models

Linear and *nonlinear Gauss-Markov models* are commonly used for parameter estimation. Koch [200] presents various models for estimating parameters in linear models, while [149] divide the models into nonstochastic, stochastic and mixed models. We limit ourselves in this book to the simple or special Gauss Markov model with full rank. For readers who want extensive coverage of parameter estimation models, we refer to the books of [149, 200]. The use of the Gauss-Jacobi combinatorial approach proposed as an alternative solution to the nonlinear Gauss-Markov model will require only the special linear Gauss-Markov model during optimization. We start by defining the *linear Gauss-Markov model* as follows:

Definition 6.1 (Special linear Gauss-Markov model). *Given a real $n \times 1$ random vector $\mathbf{y} \in \mathbb{R}^n$ of observations, a real $m \times 1$ vector $\boldsymbol{\xi} \in \mathbb{R}^m$ of unknown fixed parameters over a real $n \times m$ coefficient matrix $\mathbf{A} \in \mathbb{R}^{n \times m}$, a real $n \times n$ positive definite dispersion matrix $\boldsymbol{\Sigma}$, the functional model*

$$\mathbf{A}\boldsymbol{\xi} = E\{\mathbf{y}\}, E\{\mathbf{y}\} \in R(\mathbf{A}), rk\mathbf{A} = m, \boldsymbol{\Sigma} = D\{\mathbf{y}\}, rk\boldsymbol{\Sigma} = n \quad (6.7)$$

is called special linear Gauss-Markov model with full rank.

The unknown vector $\boldsymbol{\xi}$ of fixed parameters in the *special linear Gauss-Markov model* (6.7) is normally estimated by **B**est **L**inear **U**niformly **U**nbiased **E**stimation **BLUUE**, defined in [149, p. 93] as

Definition 6.2 (Best Linear Uniformly Unbiased Estimation BLUUE). *An $m \times 1$ vector $\hat{\boldsymbol{\xi}} = \mathbf{L}\mathbf{y} + \boldsymbol{\kappa}$ is $V - BLUUE$ for $\boldsymbol{\xi}$ (Best Linear Uniformly Unbiased Estimation) respectively the $(V - Norm)$ in (6.7) when on one hand it is uniformly unbiased in the sense of*

$$E\{\hat{\boldsymbol{\xi}}\} = E\{\mathbf{L}\mathbf{y} + \boldsymbol{\kappa}\} = \boldsymbol{\xi} \, for \, all \, \boldsymbol{\xi} \in \mathbb{R}^m, \tag{6.8}$$

and on the other hand in comparison to all other linear uniformly unbiased estimators give the minimum variance and therefore the minimum mean estimation error in the sense of

$$\begin{aligned} tr D\{\hat{\boldsymbol{\xi}}\} = E\{(\hat{\boldsymbol{\xi}} - \boldsymbol{\xi})'(\hat{\boldsymbol{\xi}} - \boldsymbol{\xi})\} = \\ = \sigma^2 \mathbf{L}\boldsymbol{\Sigma}\mathbf{L} = \|\mathbf{L}\|_V^2 = \min_{L}, \end{aligned} \tag{6.9}$$

where $\mathbf{L}$ is a real $m \times n$ matrix and $\boldsymbol{\kappa}$ an $m \times 1$ vector.

Using (6.9) to estimate the unknown fixed parameters' vector $\boldsymbol{\xi}$ in (6.7) leads to

$$\hat{\boldsymbol{\xi}} = (\mathbf{A}'\boldsymbol{\Sigma}^{-1}\mathbf{A})^{-1}\mathbf{A}'\boldsymbol{\Sigma}^{-1}\mathbf{y}, \tag{6.10}$$

with its regular dispersion matrix

$$D\{\hat{\boldsymbol{\xi}}\} = (\mathbf{A}'\boldsymbol{\Sigma}^{-1}\mathbf{A})^{-1}. \tag{6.11}$$

Equations (6.10) and (6.11) are the two main equations that are applied during the combinatorial optimization. The dispersion matrix (variance-covariance matrix) $\boldsymbol{\Sigma}$ is *unknown* and is obtained by means of estimators of type MINQUE, BIQUUE or BIQE as in [129, 271, 272, 273, 274, 275, 286]. In Definition 6.1, we used the term 'special'. This implies the case where the matrix $\mathbf{A}$ has full rank and $\mathbf{A}'\boldsymbol{\Sigma}^{-1}\mathbf{A}$ is invertible, i.e., regular. In the event that $\mathbf{A}'\boldsymbol{\Sigma}^{-1}\mathbf{A}$ is not regular (i.e., $\mathbf{A}$ has a rank deficiency), the rank deficiency can be overcome by procedures such as those presented by [149, pp. 107–165], [200, pp. 181–197] and [77, 145, 146, 247, 250, 264] among others.

Definition 6.3 (Nonlinear Gauss-Markov model). *The model*

$$E\{\mathbf{y}\} = \mathbf{y} - \mathbf{e} = \mathbf{A}(\boldsymbol{\xi}), D\{\mathbf{y}\} = \boldsymbol{\Sigma}, \qquad (6.12)$$

with a real $n \times 1$ random vector $\mathbf{y} \in \mathbb{R}^n$ of observations, a real $m \times 1$ vector $\boldsymbol{\xi} \in \mathbb{R}^m$ of unknown fixed parameters, $n \times 1$ vector $\mathbf{e}$ of random errors (with zero mean and dispersion matrix $\boldsymbol{\Sigma}$), $\mathbf{A}$ being an injective function from an open domain into $n-$dimensional space $\mathbb{R}^n (m < n)$ and E the "expectation" operator is said to be a nonlinear Gauss-Markov model.

While the solution of the linear Gauss-Markov model by **B**est **L**inear **U**niformly **U**nbiased **E**stimator (BLUUE) is straight forward, the solution of the nonlinear Gauss-Markov model is not straight forward owing to the *nonlinearity* of the *injective function* (or map function) $\mathbf{A}$ that maps $\mathbb{R}^m$ to $\mathbb{R}^n$. The difference between the linear and nonlinear Gauss-Markov models therefore lies on the injective function $\mathbf{A}$. For the linear Gauss-Markov model, the injective function $\mathbf{A}$ is linear and thus satisfies the algebraic axiom discussed in Chap. 2, i.e.,

$$\mathbf{A}(\alpha\xi_1 + \beta\xi_2) = \alpha\mathbf{A}(\xi_1) + \beta\mathbf{A}(\xi_2), \alpha, \beta \in \mathbb{R}, \xi_1, \xi_2 \in \mathbb{R}^m. \qquad (6.13)$$

The $m-$dimensional manifold traced by $\mathbf{A}(.)$ for varying values of $\boldsymbol{\xi}$ is flat. For the nonlinear Gauss-Markov model on the other hand, $\mathbf{A}(.)$ is a *nonlinear* vector function that maps $\mathbb{R}^m$ to $\mathbb{R}^n$ tracing an $m-$dimensional manifold that is curved. The immediate problem that presents itself is that of obtaining a global minimum. Procedures that are useful for determining global minimum and maximum can be found in [269, pp. 387–448].

In geodesy and geoinformatics, many nonlinear functions are normally assumed to be moderately nonlinear thus permitting linearization by Taylor series expansion and then applying the linear model (Definition 6.1, Eqs. 6.10 and 6.11) to estimate the unknown fixed parameters and their dispersions [200, pp. 155–156]. Whereas this may often hold, the effect of nonlinearity of these models may still be significant on the estimated parameters. In such cases, the Gauss-Jacobi combinatorial algorithm presented in Sect. 6.4 can be used as we will demonstrate in the chapters ahead.

6.4 Gauss-Jacobi Combinatorial Formulation

The *C. F. Gauss* and *C. G. I Jacobi* [194] *combinatorial Lemma* is stated as follows:

Lemma 6.1 (Gauss-Jacobi combinatorial). *Given n algebraic observation equations in m unknowns, i.e.,*

$$
\begin{aligned}
a_1 x + b_1 y - y_1 &= 0 \\
a_2 x + b_2 y - y_2 &= 0 \\
a_3 x + b_3 y - y_3 &= 0,
\end{aligned}
\tag{6.14}
$$

$$\dots\dots$$

for the determination of the unknowns x and y, there exist no set of solutions $\{x, y\}_{i,j}$ from any combinatorial pair in (6.14) that satisfy the entire system of equations. This is because the solutions obtained from each combinatorial pair of equations differ from the others due to the unavoidable random measuring errors. If the solutions from the pair of the combinatorial equations are designated $x_{1,2}, x_{2,3}, \dots$ and $y_{1,2}, y_{2,3}, \dots$ with the subscripts indicating the combinatorial pairs, then the combined solutions are the sum of the weighted arithmetic mean

$$
x = \frac{\pi_{1,2} x_{1,2} + \pi_{2,3} x_{2,3} + \dots}{\pi_{1,2} + \pi_{2,3} + \dots}, \; y = \frac{\pi_{1,2} y_{1,2} + \pi_{2,3} y_{2,3} + \dots}{\pi_{1,2} + \pi_{2,3} + \dots},
\tag{6.15}
$$

with $\{\pi_{1,2}, \pi_{2,3}, \dots\}$ being the weights of the combinatorial solutions given by the square of the determinants as

$$
\begin{aligned}
\pi_{1,2} &= (a_1 b_2 - a_2 b_1)^2 \\
\pi_{2,3} &= (a_2 b_3 - a_3 b_2)^2
\end{aligned}
\tag{6.16}
$$

$$\dots\dots$$

The results are identical to those of least squares solution.

The proof of Lemma 6.1 is given in [188] and [328, pp. 46–47]. For nonlinear cases however, the results of the combinatorial optimization may not coincide with those of least squares as will be seen in the coming chapters. This could be attributed to the remaining traces of nonlinearity following linearization of the nonlinear equations in the least squares approach or the generation of weight matrix by the combinatorial approach. We will later see that the combinatorial approach permits linearization only for generation of the weight matrix during optimization process.

Levelling is one of the fundamental tasks carried out in engineering, geodynamics, geodesy and geoinformatics for the purpose of determining heights of stations. In carrying out levelling, one starts from a point whose height is known and measures height differences along a levelling route to a closing point whose height is also known. In case where the starting point is also the closing point, one talks of loop levelling. The heights of the known stations are with respect to the mean sea level as a reference. In Example 6.1, we use loop levelling network to illustrate Lemma 6.1 of the Gauss-Jacobi combinatorial approach.

Example 6.1 (Levelling network). Consider a levelling network with four-points in Fig. 6.2 below.

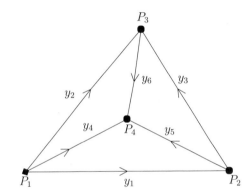

Fig. 6.2. Levelling Network

Let the known height of point P_1 be given as h_1. The heights h_2 and h_3 of points P_2 and P_3 respectively are unknown. The task at hand is to carry out loop levelling from point P_1 to determine these unknown heights. Given three stations with two of them being unknowns, there exist

$$\binom{3}{2} = \frac{3!}{2!(3-2)!} = 3$$

number of combinatorial routes that can be used to obtain the heights of points P_2 and P_3. If station P_4 is set out for convenience along the loop, the levelling routes are $\{P_1 - P_2 - P_4 - P_1\}$, $\{P_2 - P_3 - P_4 - P_2\}$, and $\{P_3 - P_1 - P_4 - P_3\}$. These combinatorials sum up to the outer loop $P_1 - P_2 - P_3 - P_1$. The observation equations formed by the height difference measurements are written as

$$
\begin{aligned}
x_2 - h_1 &= y_1 \\
x_3 - h_1 &= y_2 \\
x_3 - x_2 &= y_3 \\
x_4 - h_1 &= y_4 \\
x_4 - x_2 &= y_5 \\
x_4 - x_3 &= y_6,
\end{aligned}
\tag{6.17}
$$

which can be expressed in the form of the special linear Gauss-Markov model (6.7) on p. 61 as

$$
E\left\{\begin{bmatrix}
y_1 + h_1 \\
y_2 + h_1 \\
y_3 \\
y_4 + h_1 \\
y_5 \\
y_6
\end{bmatrix}\right\} = \begin{bmatrix}
1 & 0 & 0 \\
0 & 1 & 0 \\
-1 & 1 & 0 \\
0 & 0 & 1 \\
-1 & 0 & 1 \\
0 & -1 & 1
\end{bmatrix}\begin{bmatrix}
x_2 \\
x_3 \\
x_4
\end{bmatrix},
\tag{6.18}
$$

where $y_1, y_2, \ldots, y_6$ are the observed height differences and x_2, x_3, x_4 the unknown heights of points P_2, P_3, P_4 respectively. Let the dispersion matrix $D\{\mathbf{y}\} = \mathbf{\Sigma}$ be chosen such that the correlation matrix is unit (i.e., $\mathbf{\Sigma} = \mathbf{I}_3 = \mathbf{\Sigma}^{-1}$ positive definite, $rk\mathbf{\Sigma}^{-1} = 3 = n$), the decomposition matrix $\mathbf{Y}$ and the normal equation matrix $\mathbf{A}'\mathbf{\Sigma}^{-1}\mathbf{A}$ are given respectively by

$$
\mathbf{Y} = \begin{bmatrix}
y_1 + h_1 & 0 & 0 \\
0 & 0 & y_2 + h_1 \\
0 & y_3 & 0 \\
-(y_4 + h_1) & 0 & y_2 + h_1 \\
y_5 & -y_5 & 0 \\
0 & y_6 & -y_6
\end{bmatrix}, \quad \mathbf{A}'\mathbf{\Sigma}^{-1}\mathbf{A} = \begin{bmatrix}
3 & -1 & -1 \\
-1 & 3 & -1 \\
-1 & -1 & 3
\end{bmatrix}.
\tag{6.19}
$$

The columns of $\mathbf{Y}$ correspond to the vectors of observations $\mathbf{y}_1$, $\mathbf{y}_2$ and $\mathbf{y}_3$ formed from the combinatorial levelling routes. We compute the heights of points P_2 and P_3 using (6.10) for each combinatorial levelling routes as follows:

- Combinatorials route(1):=$P_1 - P_2 - P_4 - P_1$. Equations (6.19) and (6.10) leads to the partial solutions

$$\hat{\boldsymbol{\xi}}_{route(1)} = \frac{1}{2} \begin{bmatrix} y_1 + \dfrac{h_1}{2} - \dfrac{y_5}{2} - \dfrac{y_4}{2} \\[2ex] \dfrac{y_1}{2} - \dfrac{y_4}{2} \\[2ex] \dfrac{y_1}{2} - \dfrac{h_1}{2} + \dfrac{y_5}{2} - y_4 \end{bmatrix} \qquad (6.20)$$

- Combinatorials route(2):$=P_2 - P_3 - P_4 - P_2$ gives

$$\hat{\boldsymbol{\xi}}_{route(2)} = \frac{1}{2} \begin{bmatrix} \dfrac{y_5}{2} - \dfrac{y_3}{2} \\[2ex] \dfrac{y_3}{2} - \dfrac{y_6}{2} \\[2ex] \dfrac{y_6}{2} - \dfrac{y_5}{2} \end{bmatrix}. \qquad (6.21)$$

- Combinatorials route(3):$=P_3 - P_1 - P_4 - P_3$ gives

$$\hat{\boldsymbol{\xi}}_{route(3)} = \frac{1}{2} \begin{bmatrix} \dfrac{y_4}{2} - \dfrac{y_2}{2} \\[2ex] \dfrac{y_4}{2} + \dfrac{y_6}{2} - \dfrac{h_1}{2} - y_2 \\[2ex] \dfrac{h_1}{2} - \dfrac{y_2}{2} - \dfrac{y_6}{2} + y_4 \end{bmatrix}. \qquad (6.22)$$

The heights of the stations x_2, x_3, x_4 are then given by the summation of the combinatorial solutions

$$\begin{bmatrix} x_2 \\ x_3 \\ x_4 \end{bmatrix} = \hat{\boldsymbol{\xi}}_l = \hat{\boldsymbol{\xi}}_{route(1)} + \hat{\boldsymbol{\xi}}_{route(2)} + \hat{\boldsymbol{\xi}}_{route(3)} = \frac{1}{2} \begin{bmatrix} y_1 + \dfrac{h_1}{2} - \dfrac{y_3}{2} - \dfrac{y_2}{2} \\[2ex] \dfrac{y_1}{2} + \dfrac{y_3}{2} - \dfrac{h_1}{2} - y_2 \\[2ex] \dfrac{y_1}{2} - \dfrac{y_2}{2} \end{bmatrix}. \qquad (6.23)$$

If one avoids the combinatorial routes and carries out levelling along the outer route $P_1 - P_2 - P_3 - P_1$, the heights could be obtained directly using (6.10) as

$$
\begin{bmatrix} x_2 \\ x_3 \\ x_4 \end{bmatrix} = \hat{\boldsymbol{\xi}}_l = (\mathbf{A}'\boldsymbol{\Sigma}^{-1}\mathbf{A})^{-1}\mathbf{A}'\boldsymbol{\Sigma}^{-1}
\begin{bmatrix} y_1 + h_1 \\ -(y_2 + h_1) \\ y_3 \\ 0 \\ 0 \end{bmatrix}
= \frac{1}{2}
\begin{bmatrix} y_1 + \dfrac{h_1}{2} - \dfrac{y_3}{2} - \dfrac{y_2}{2} \\[2mm] \dfrac{y_1}{2} + \dfrac{y_3}{2} - \dfrac{h_1}{2} - y_2 \\[2mm] \dfrac{y_1}{2} - \dfrac{y_2}{2} \end{bmatrix},
$$

$$(6.24)$$

In which case the results are identical to (6.23). For linear cases therefore, the results of Gauss-Jacobi combinatorial algorithm gives solution (6.23) which is identical to that of least squares approach in (6.24), thus validating the postulations of Lemma 6.1.

□

6.5 Combinatorial Solution of Nonlinear Gauss-Markov Model

The Gauss-Jacobi combinatorial Lemma 6.1 on p. 64 and the levelling example were based on a linear case. In case of nonlinear systems of equations, such as (6.1 and 6.2), (6.3 and 6.4) or (6.5 and 6.6), the nonlinear Gauss-Markov model (6.12) is solved in two steps:

- Step 1: Combinatorial minimal subsets of observations are constructed and rigorously solved by means of either Groebner basis or polynomial resultants.
- Step 2: The combinatorial solution points obtained from step 1, which are now linear, are reduced to their final adjusted values by means of **B**est **L**inear **U**niformly **U**nbiased **E**stimator (BLUUE). The dispersion matrix of the real valued random vector of pseudo-observations from *Step 1* are generated via the *nonlinear error propagation law* also known as the *nonlinear variance-covariance propagation*.

Construction of Minimal Combinatorial Subsets

Since $n > m$ we construct minimal combinatorial subsets comprising m equations solvable in closed form using either Groebner basis or polynomial resultants. We begin by giving the following elementary definitions:

Definition 6.4 (Permutation). *Let us consider that a set S with elements $\{i, j, k\} \in S$ is given, the arrangement resulting from placing*

$\{i, j, k\} \in S$ in some sequence is known as *permutation*. If we choose any of the elements say i first, then each of the remaining elements j, k can be put in the second position, while the third position is occupied by the unused letter either j or k. For the set S, the following permutations can be made:

$$\begin{bmatrix} ijk \ ikj \ jik \\ jki \ kij \ kji. \end{bmatrix} \tag{6.25}$$

From (6.25) there exist three ways of filling the first position, two ways of filling the second position and one way of filling the third position. Thus the number of permutations is given by $3 \times 2 \times 1 = 6$. In general, for n different elements, the number of permutation is equal to $n \times \ldots \times 3 \times 2 \times 1 = n!$

Definition 6.5 (Combination). *If for n elements only m elements are used for permutation, then we have a combination of the mth order.* If we follow the definition above, then the first position can be filled in n ways, the second in $\{n-1\}$ ways and the mth in $\{n - (m-1)\}$ ways. In (6.25), the combinations are identical and contain the same elements in different sequences. If the arrangement is to be neglected, then we have for n elements, a combination of mth order being given by

$$C_k = \binom{n}{m} = \frac{n!}{m!(n-m)!} = \frac{n(n-1)\ldots(n-m+1)}{m \times \ldots \times 3 \times 2 \times 1}. \tag{6.26}$$

Given n nonlinear equations to be solved, we first form C_k minimal combinatorial subsets each consisting of m elements (where m is the number of the unknown elements). Each minimal combinatorial subset C_k is then solved using either of the algebraic procedures discussed in Chaps. 4 and 5.

Example 6.2 (Combinatorial). In Fig.6.1 for example, $n = 3$ and $m = 2$, which with (6.26) leads to three combinations given by (6.1 and 6.2), (6.3 and 6.4) and (6.5 and 6.6). Groebner basis or polynomial resultants approach is then applied to each combinatorial pair to give the combinatorial solutions $\{x_0, y_0\}_{1,2}$, $\{x_0, y_0\}_{1,3}$ and $\{x_0, y_0\}_{2,3}$.

Optimization of Combinatorial Solutions

Once the *combinatorial minimal subsets* have been solved using either Groebner basis or polynomial resultants, the resulting sets of solutions are considered as pseudo-observations. For each combinatorial, the obtained minimal subset solutions are used to generate the dispersion matrix via the nonlinear error propagation law/variance-covariance propagation e.g., [149, pp. 469–471] as follows:

From the nonlinear observation equations that have been converted into its algebraic (polynomial) via Theorem 3.1 on p. 20, the *combinatorial minimal subsets* consist of polynomials $f_1, \ldots, f_m \in k[x_1, \ldots, x_m]$, with $\{x_1, \ldots, x_m\}$ being the unknown variables (fixed parameters) to be determined. The variables $\{y_1, \ldots, y_n\}$ are the known values comprising the pseudo-observations obtained following closed form solutions of the minimum combinatorial subsets. We write the polynomials as

$$
\begin{bmatrix}
f_1 := g(x_1, \ldots, x_m, y_1, \ldots, y_n) = 0 \\
f_2 := g(x_1, \ldots, x_m, y_1, \ldots, y_n) = 0 \\
\cdot \\
\cdot \\
\cdot \\
f_m := g(x_1, \ldots, x_m, y_1, \ldots, y_n) = 0,
\end{bmatrix}
\tag{6.27}
$$

which are expressed in matrix form as

$$
\mathbf{f} := \mathbf{g}(\mathbf{x}, \mathbf{y}) = \mathbf{0}. \tag{6.28}
$$

In (6.28) the unknown variables $\{x_1, \ldots, x_m\}$ are placed in a vector $\mathbf{x}$ and the known variables $\{y_1, \ldots, y_n\}$ in $\mathbf{y}$. Error propagation is then performed from pseudo-observations $\{y_1, \ldots, y_n\}$ to parameters $\{x_1, \ldots, x_m\}$ which are to be explicitly determined. They are characterized by the *first moments*, the expectations $E\{\mathbf{x}\} = \boldsymbol{\mu}_x$ and $E\{\mathbf{y}\} = \boldsymbol{\mu}_y$, as well as the *second moments*, the variance-covariance matrices/dispersion matrices $D\{\mathbf{x}\} = \boldsymbol{\Sigma}_x$ and $D\{\mathbf{y}\} = \boldsymbol{\Sigma}_y$. From [149, pp. 470–471], we have up to nonlinear terms

$$
D\{\mathbf{x}\} = \mathbf{J}_x^{-1} \mathbf{J}_y \boldsymbol{\Sigma}_y \mathbf{J}_y' (\mathbf{J}_x^{-1})', \tag{6.29}
$$

with $\mathbf{J}_x, \mathbf{J}_y$ being the partial derivatives of (6.28) with respect to $\mathbf{x}, \mathbf{y}$ respectively at the Taylor points $(\boldsymbol{\mu}_x, \boldsymbol{\mu}_y)$. The approximate values of unknown parameters $\{x_1, \ldots, x_m\} \in \mathbf{x}$ appearing in the Jacobi matrices $\mathbf{J}_x, \mathbf{J}_y$ are obtained either from Groebner basis or polynomial resultants solution of the nonlinear system of equations (6.27).

Given $\mathbf{J}_i = \mathbf{J}_{x_i}^{-1} \mathbf{J}_{y_i}$ from the ith combination and $\mathbf{J}_j = \mathbf{J}_{x_j}^{-1} \mathbf{J}_{y_j}$ from the jth combination, the correlation between the ith and j^{th} combinations is given by

$$\boldsymbol{\Sigma}_{ij} = \mathbf{J}_j \boldsymbol{\Sigma}_{y_j y_i} \mathbf{J}'_i. \tag{6.30}$$

The sub-matrices variance-covariance matrix for the individual combinatorials $\boldsymbol{\Sigma}_1, \boldsymbol{\Sigma}_2, \boldsymbol{\Sigma}_3, \ldots, \boldsymbol{\Sigma}_k$ (where k is the number of combinations) obtained via (6.29) and the correlations between combinatorials obtained from (6.30) form the variance-covariance/dispersion matrix

$$\boldsymbol{\Sigma} = \begin{bmatrix} \boldsymbol{\Sigma}_1 & \boldsymbol{\Sigma}_{12} & . & . & . & \boldsymbol{\Sigma}_{1k} \\ \boldsymbol{\Sigma}_{21} & \boldsymbol{\Sigma}_2 & . & . & . & \boldsymbol{\Sigma}_{2k} \\ & & \boldsymbol{\Sigma}_3 & & & \\ . & & & & & \\ . & & & & . & \\ . & & & & & \\ \boldsymbol{\Sigma}_{k1} & . & & . & . & \boldsymbol{\Sigma}_k \end{bmatrix} \tag{6.31}$$

for the entire k combinations. This will be made clear by Example 6.4. The obtained dispersion matrix $\boldsymbol{\Sigma}$ is then used in the *linear Gauss-Markov model* (6.10) to obtain the estimates $\hat{\boldsymbol{\xi}}$ of the unknown parameters $\boldsymbol{\xi}$. The combinatorial solutions are considered as pseudo-observations and placed in the vector $\mathbf{y}$ of observations, while the design matrix $\mathbf{A}$ comprises of integer values 1 which are the coefficients of the unknowns as in (6.34). The procedure thus optimizes the combinatorial solutions by the use of BLUUE. Consider the following example.

Example 6.3. From Fig. 6.1 on p. 59, three possible combinations each containing two nonlinear equations necessary for solving the two unknowns are given and solved as discussed in Example 6.2 on p. 69. Let the combinatorial solutions $\{x_0, y_0\}_{1,2}$, $\{x_0, y_0\}_{1,3}$ and $\{x_0, y_0\}_{2,3}$ be given in the vectors $\mathbf{z}_I(y_1, y_2)$, $\mathbf{z}_{II}(y_1, y_3)$ and $\mathbf{z}_{III}(y_2, y_3)$ respectively. If the solutions are placed in a vector $\mathbf{z}_J = [\mathbf{z}_I \ \mathbf{z}_{II} \ \mathbf{z}_{III}]'$, the adjustment model is then defined as

$$E\{\mathbf{z}_J\} = \mathbf{I}_{6\times3}\boldsymbol{\xi}_{3\times1}, D\{\mathbf{z}_J\} \text{ from variance/covariance propagation.} \tag{6.32}$$

Let

$$\boldsymbol{\xi}^n = \mathbf{L}\mathbf{z}_J \text{ subject to } \mathbf{z}_J := \begin{bmatrix} \mathbf{z}_I \\ \mathbf{z}_{II} \\ \mathbf{z}_{III} \end{bmatrix} \in \mathbb{R}^{6\times1}, \tag{6.33}$$

such that the postulations $trD\{\boldsymbol{\xi}^n\} = min$, i.e., "*best*," and $E\{\boldsymbol{\xi}^n\} = \boldsymbol{\xi}$ for all $\boldsymbol{\xi}^n \in \mathbb{R}^m$ i.e., "*uniformly unbiased*" holds. We then have from (6.31), (6.32) and (6.33) the result

$$\hat{\boldsymbol{\xi}} = (\mathbf{I}'_{3\times6}\boldsymbol{\Sigma}_{\mathbf{z}_J}\mathbf{I}_{6\times3})\mathbf{I}'_{3\times6}\boldsymbol{\Sigma}_{\mathbf{z}_J}^{-1}\mathbf{z}_J \tag{6.34}$$

$$\hat{\mathbf{L}} = arg\{trD\{\boldsymbol{\xi}^n\} = tr\, \mathbf{L}\boldsymbol{\Sigma}_y\mathbf{L}' = min \mid UUE\}$$

The dispersion matrix $D\{\hat{\boldsymbol{\xi}}\}$ of the estimates $\hat{\boldsymbol{\xi}}$ is obtained via (6.11). The shift from arithmetic weighted mean to the use of *linear Gauss Markov model* is necessitated as we do not readily have the weights of the minimal combinatorial subsets but instead have their dispersion matrices obtained via *error propagation/variance-covariance propagation*. If the equivalence *Theorem* of [149, pp. 339–341] is applied, an adjustment using linear Gauss Markov model instead of weighted arithmetic mean in Lemma 6.1 is permissible.

Example 6.4 (Error propagation for planar ranging problem). For the unknown station $P_0(X_0, Y_0) \in \mathbb{E}^2$ of the planar ranging problem in Fig. 6.1 on p. 59, let distances S_1 and S_2 be measured to two known stations $P_1(X_1, Y_1) \in \mathbb{E}^2$ and $P_2(X_2, Y_2) \in \mathbb{E}^2$ respectively. The distance equations are expressed as

$$\begin{bmatrix} S_1^2 = (X_1 - X_0)^2 + (Y_1 - Y_0)^2 \\ S_2^2 = (X_2 - X_0)^2 + (Y_2 - Y_0)^2, \end{bmatrix} \tag{6.35}$$

which are written algebraically as

$$\begin{bmatrix} f_1 := (X_1 - X_0)^2 + (Y_1 - Y_0)^2 - S_1^2 = 0 \\ f_2 := (X_2 - X_0)^2 + (Y_2 - Y_0)^2 - S_2^2 = 0. \end{bmatrix} \tag{6.36}$$

On taking total differential of (6.36), we have

$$\begin{bmatrix} df_1 := 2(X_1 - X_0)dX_1 - 2(X_1 - X_0)dX + 2(Y_1 - Y_0)dY_1 - \\ \qquad -2(Y_1 - Y_0)dY - 2S_1dS_1 = 0 \\ \\ df_2 := 2(X_2 - X_0)dX_2 - 2(X_2 - X_0)dX + 2(Y_2 - Y_0)dY_2 - \\ \qquad -2(Y_2 - Y_0)dY - 2S_2dS_2 = 0. \end{bmatrix} \tag{6.37}$$

Arranging (6.37) with the unknown terms $\{X_0, Y_0\} = \{x_1, x_2\} \in \mathbf{x}$ on the left-hand-side and the known terms

$$\{X_1, Y_1, X_2, Y_2, S_1, S_2\} = \{y_1, y_2, y_3, y_4, y_5, y_6\} \in \mathbf{y},$$

on the right-hand-side leads to

$$\mathbf{J}_x \begin{bmatrix} dX_0 \\ dY_0 \end{bmatrix} = \mathbf{J}_y \begin{bmatrix} dS_1 \\ dX_1 \\ dY_1 \\ dS_2 \\ dX_2 \\ dY_2 \end{bmatrix}, \tag{6.38}$$

with

$$
\mathbf{J}_x =
\begin{bmatrix}
\dfrac{\partial f_1}{\partial X_0} & \dfrac{\partial f_1}{\partial Y_0} \\[2mm]
\dfrac{\partial f_2}{\partial X_0} & \dfrac{\partial f_2}{\partial Y_0}
\end{bmatrix}
=
\begin{bmatrix}
-2(X_1 - X_0) & -2(Y_1 - Y_0) \\
-2(X_2 - X_0) & -2(Y_2 - Y_0)
\end{bmatrix},
\tag{6.39}
$$

and

$$
\begin{bmatrix}
\mathbf{J}_y
\end{bmatrix}
=
\begin{bmatrix}
\dfrac{\partial f_1}{\partial S_1} & \dfrac{\partial f_1}{\partial X_1} & \dfrac{\partial f_1}{\partial Y_1} & 0 & 0 & 0 \\[2mm]
0 & 0 & & \dfrac{\partial f_2}{\partial S_2} & \dfrac{\partial f_2}{\partial X_2} & \dfrac{\partial f_2}{\partial Y_2}
\end{bmatrix}
=
$$

$$
=
\begin{bmatrix}
2S_1 & -2(X_1 - X_0) & -2(Y_1 - Y_0) & 0 & 0 & 0 \\
0 & 0 & 0 & 2S_2 & -2(X_2 - X_0) & -2(Y_2 - Y_0)
\end{bmatrix}.
\tag{6.40}
$$

If we consider that

$$
D\{\mathbf{x}\} = \boldsymbol{\Sigma}_x =
\begin{bmatrix}
\sigma^2_{X_0} & \sigma_{X_0 Y_0} \\
\sigma_{Y_0 X_0} & \sigma^2_{Y_0}
\end{bmatrix}
$$

$$
D\{\mathbf{y}\} = \boldsymbol{\Sigma}_y =
\begin{bmatrix}
\sigma^2_{S_1} & \sigma_{S_1 X_1} & \sigma_{S_1 Y_1} & \sigma_{S_1 X_2} & \sigma_{S_1 S_2} & \sigma_{S_1 Y_2} \\
\sigma_{X_1 S_1} & \sigma^2_{X_1} & \sigma_{X_1 Y_1} & \sigma_{X_1 S_2} & \sigma_{X_1 X_2} & \sigma_{X_1 Y_2} \\
\sigma_{Y_1 S_1} & \sigma_{Y_1 X_1} & \sigma^2_{Y_1} & \sigma_{Y_1 S_2} & \sigma_{Y_1 X_2} & \sigma_{Y_1 Y_2} \\
\sigma_{S_2 S_1} & \sigma_{S_2 X_1} & \sigma_{S_2 Y_1} & \sigma^2_{S_2} & \sigma_{S_2 X_2} & \sigma_{S_2 Y_2} \\
\sigma_{X_2 S_1} & \sigma_{X_2 X_1} & \sigma_{X_2 Y_1} & \sigma_{X_2 S_2} & \sigma^2_{X_2} & \sigma_{X_2 Y_2} \\
\sigma_{Y_2 S_1} & \sigma_{Y_2 X_1} & \sigma_{Y_2 Y_1} & \sigma_{Y_2 S_2} & \sigma_{Y_2 X_2} & \sigma^2_{Y_2}
\end{bmatrix},
\tag{6.41}
$$

we obtain with (6.38), (6.39) and (6.40) the dispersion (6.29) of the unknown variables $\{X_0, Y_0\} = \{x_1, x_2\} \in \mathbf{x}$.

The Gauss-Jacobi Combinatorial Program

The Gauss-Jacobi combinatorial program operates in *three* phases. In the *first* phase, one forms *minimal combinations* of the nonlinear equations using (6.26) on p. 69. Using either Groebner basis or polynomial resultants, the desired combinatorial solutions are obtained. The combinatorial results form pseudo-observations, which are within the solution space of the desired values. This first phase in essence *projects a nonlinear* case into a *linear* case.

Once the first phase is successfully carried out with the solutions of the various subsets forming pseudo-observations, the nonlinear variance-covariance/error propagation is carried out in the *second phase* to obtain the *weight matrix*. This requires that the stochasticity of the initial observational sample be known in-order to propagate them to the pseudo-observations.

The *final phase* entails the adjustment step, which is performed to obtain the *barycentric values*. Since the pseudo-observations are linearly independent, the special linear Gauss-Markov model (see Definition 6.1 on p. 61) is employed.

Stepwise, the *Gauss-Jacobi combinatorial algorithm* proceeds as follows:

- **Step 1**: Given an overdetermined system with n observations in m unknowns, using (6.26), form minimal combinations from the n observations that comprise m equations in m unknowns.
- **Step 2**: Solve each set of m equations from step 1 above for the m unknowns using either Groebner basis or polynomial resultant algebraic techniques.
- **Step 3**: Perform the nonlinear error/variance-covariance propagation to obtain the variance-covariance matrix of the combinatorial solutions obtained in Step 2.
- **Step 4**: Using the pseudo-observations of step 2, and the variance-covariance matrix from step 3, adjust the pseudo-observations via the special linear Gauss-Markov model to obtain the adjusted position of the unknown station.

Figure 6.3 summarizes the operations of the Gauss-Jacobi combinatorial algorithm which employs Groebner basis or polynomial resultants as computing engines to solve nonlinear systems of equations (e.g., Fig.6.4).

Gauss-Jacobi combinatorial algorithm

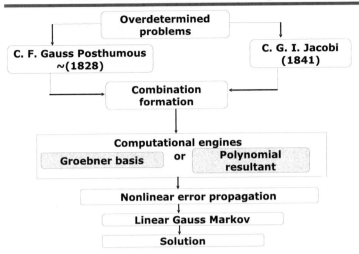

Fig. 6.3. Gauss-Jacobi combinatorial algorithm

Algebraic algorithms

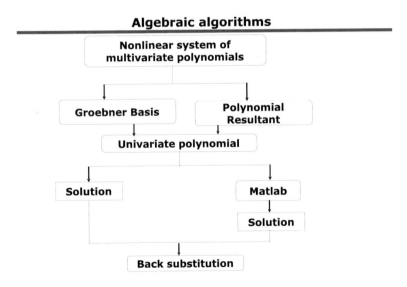

Fig. 6.4. Combinatorial computing engine

6.6 Concluding Remarks

In Chaps. 4 and 5, Groebner basis and polynomial resultants algorithms were presented for solving in exact form the nonlinear systems of equa-

tions. It was demonstrated in this chapter how they play a leading role
in overcoming the major difficulty that was faced by C. F. Gauss and C.
G. I. Jacobi. The key to success is to use these algebraic techniques as
the computing engine of the Gauss-Jacobi combinatorial algorithm. In
so doing, an alternative procedure to linearized and iterative numerical
procedures that peg their operations on approximate starting values
was presented. Such algebraic technique for solving overdetermined
problems requires neither approximate starting values nor linearization (except for the generation of the weight matrix). With modern
computing technology, the combinatorial formation and computational
time for geodetic or geoinformatics' algebraic computational problems
is immaterial. In the chapters ahead, the power of this technique will
be demonstrated. Fig. 6.5 gives a summary of the algebraic algorithms
and show when each procedure can be applied. Further materials on
the topic are presented in [167, 168].

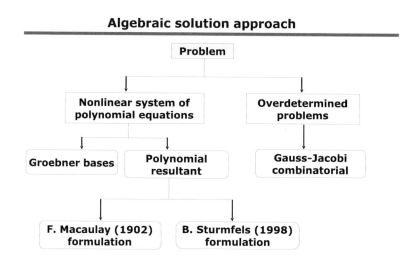

Fig. 6.5. Algebraic solution approach

Local versus Global Positioning Systems

7.1 Positioning Systems

In daily operations, geodesists and geoinformatists have at their disposal two operating systems namely:

- Global Positioning System; in this system, the practitioner operates at a global scale with positions referred to the global reference frame (e.g., World Geodetic System WGS-84). The tools employed comprise mainly satellites with global positioning capabilities. These satellites include; the US based Global Positioning System (GPS), Russian based Globalnaya Navigationnaya Sputnikovaya Sistema (or simply Global Navigation Satellite System) GLONASS and the proposed European Union's proposed Global Navigation Satellite System GALILEO which will be operational around 2008 [335]. Unlike GPS satellites which were designed for the US army, GALILEO satellites will be civilian owned. Information on GALILEO can be obtained from its web page.[1]

  Apart from positioning satellites, low flying satellites have been launched for various missions. These satellites, which work together with GPS satellites, include among others:

  - The twin satellites Gravity Recovery and Climate Experiment (GRACE) which were launched in March 2002. They are expected to make detailed measurements of the Earth's gravity field leading to discoveries about gravity and Earth's natural systems. GRACE satellites will also be used for atmospheric studies.

---

[1]http://europa.eu.int/comm/dgs/energy_transport/galileo/index_en.htm

- The German Low Earth Orbiting (LEO) satellite Challenging Minisatellite Payload (CHAMP) for geophysical research and applications. CHAMP satellite was launched on July 15, 2000 and similar to GRACE satellites, it is currently offering data for gravity and atmospheric study.
- The Gravity field and steady-state Ocean Circulation Explorer (GOCE) which is dedicated to measuring the Earth's gravity field and modelling the geoid with extremely high accuracy and spatial resolution.
- Local Positioning Systems (LPS); which are applicable at national levels. The main positioning tools include; total stations, theodolites, EDMs, photogrammetric cameras, laser scanners etc. Positions in these systems are referred to the local level reference frames. With these systems, for example, engineers have possibilities of setting horizontal and vertical networks for constructions. Those in geodynamics use them together with GPS for deformation monitoring.

The present chapter discusses these two systems in detail. In particular, for the LPS, the issue of local datum choice is addressed. The test network of "Stuttgart Central" which is applied to test the algorithms of Chaps. 4–6 is also presented.

7.2 Global Positioning System (GPS)

Global Positioning System (GPS) are satellites that were primarily designed for use of US military in the early 60's, with a secondary role of civilian navigation. The oscillators aboard the GPS satellites generate a fundamental frequency f_0 of 10.23 MHz. Two carrier signals in the L band denoted L_1 and L_2 are generated by integer multiplication of the fundamental frequency f_0. These carriers are modulated by codes to provide satellite clock readings measured by GPS receivers. Two types of codes; the coarse acquisition C/A and precise acquisition P/A are emitted. C/A code in the L_1 carrier is less precise and is often reserved for civilian use, while the P/A code is reserved for the use of US military and its allies. It is coded on both L_1 and L_2 [186]. The design comprises three segments namely; the space segment, user segment and the control segment. The space segment was designed such that the constellation consisted of 24 satellites (with a spare of four) orbiting at a height of about 20,200 km. The orbits are inclined at an angle of 55° from the equator with an orbiting period of about 12 hours. The user segment consists of a receiver that tracks signals from at least four satellites in order to position (see e.g., discussion on Chap. 9). The control segment consist 5 ground stations with the master station located at the Air force base in Colorado. The master station measures satellite signals which are incorporated in the orbital models for each satellite. The models compute ephemerids and satellite clock correction parameters which are transmitted to the satellites. The satellites then transmit the orbital data to the receivers.

The results of the *three-dimensional positioning* using GPS satellites are the three-dimensional geodetic coordinates $\{\lambda, \phi, h\}$ of a receiver station. These coordinates comprise the geodetic longitude λ, geodetic latitude ϕ and geodetic height h. When positioning with GPS, the outcome is the geocentric position for an individual receiver or the relative positions between co-observing receivers.

The *global reference frame* $\mathbb{F}^\bullet$ upon which the GPS observations are based is defined by the base vectors $\mathbb{F}_{1\bullet}, \mathbb{F}_{2\bullet}, \mathbb{F}_{3\bullet}$, with the origin being the center of mass. The fundamental vector is defined by the base vector $\mathbb{F}_{3\bullet}$ and coincides with the mean axis of rotation of the Earth and points to the direction of the Conventional International Origin (CIO). $\mathbb{F}_{1\bullet}$ is oriented such that the plane formed by $\mathbb{F}_{1\bullet}$ and $\mathbb{F}_{3\bullet}$ points to the direction of Greenwich in England. $\mathbb{F}_{2\bullet}$ completes the right handed system by being perpendicular to $\mathbb{F}_{1\bullet}$ and $\mathbb{F}_{3\bullet}$. The geocentric Cartesian coordinates of a positional vector $\mathbf{X}$ is given by

$$\mathbf{X} = \mathbb{F}_{1\bullet}X + \mathbb{F}_{2\bullet}Y + \mathbb{F}_{3\bullet}Z, \tag{7.1}$$

where $\{X, Y, Z\}$ are the components of the vector $\mathbf{X}$ in the system $\mathbb{F}_{1\bullet}, \mathbb{F}_{2\bullet}, \mathbb{F}_{3\bullet} \,|o$.

7.3 Local Positioning Systems (LPS)

Grafarend [133] defines a *local level system* as a three-dimensional reference frame *at the hand* of an experimenter in an engineering network. When one is positioning using a theodolite or a total Station, one first centers the instrument. When the instrument is properly centered and ready for operation, the vertical axis of the instrument at this moment coincides with the direction of the local gravity vector at that particular point, hence the term *direction of local gravity vector*. The vertical axis at the theodolite station however points in the direction opposite to that of the gravity vector (i.e., to the zenith). The instrument can now be used to measure observations of the type horizontal directions T_i, angles, vertical directions B_i or the spatial distances S_i. The triplet $\{S_i, T_i, B_i\}$ are measured in the *local level reference frame* and are used to form the spherical coordinates of a point. These systems as opposed to GPS are only used within the local networks and are referred to as the *Local Positioning Systems* (LPS). When one is operating in these systems, one is faced with two datum choices upon which to operate. The next section elaborates on these datum choices.

7.3.1 Local Datum Choice in an LPS 3-D Network

When measuring directions in the LPS system, one has two options, namely;

- orienting the theodolite to a station whose azimuth is known or,
- orienting the theodolite to an arbitrary station whose azimuth is unknown.

When the first option is adopted, one operates in the local level reference frame of type $\mathbb{E}^*$ discussed in (a) below. Should the second approach be chosen, then one operates in the local level reference frame of type $\mathbb{F}^*$ discussed in (b).

(a) Local level reference frame of type $\mathbb{E}^*$:
    The origin of the $\mathbb{E}^*$ system is a point P whose coordinates

$$\mathbf{X} = \begin{bmatrix} X \\ Y \\ Z \end{bmatrix}_P = \begin{bmatrix} 0 \\ 0 \\ 0 \end{bmatrix} \tag{7.2}$$

are defined by base vectors $\mathbb{E}_{1*}$, $\mathbb{E}_{2*}$, $\mathbb{E}_{3*}$ of type south, east, vertical. $\mathbb{E}_{3*}$ points to the direction opposite to that of the local gravity vector $\mathbf{\Gamma}$ at point P. $\mathbb{E}_{1*}$ points south, while $\mathbb{E}_{2*}$ completes the system by pointing east. The datum spherical coordinates of the direction point P_i in the local level reference frame $\mathbb{E}^*$ are

$$\left(\begin{array}{l} P \longrightarrow X^* = Y^* = Z^* = \qquad\qquad 0 \\[2mm] PP_i \longrightarrow \begin{bmatrix} X^* \\ Y^* \\ Z^* \end{bmatrix}_{\mathbb{E}^*} = S_i \begin{bmatrix} \cos A_i \, \cos B_i \\ \sin A_i \, \cos B_i \\ \sin B_i \end{bmatrix} \end{array} \right) , \tag{7.3}$$

with azimuths A_i, vertical directions B_i, and spatial distances S_i.

(b) Local level reference frame of type $\mathbb{F}^*$:

This system is defined by the base vectors $\mathbb{F}_{1*}$, $\mathbb{F}_{2*}$, $\mathbb{F}_{3*}$, with $\mathbb{F}_{1*}$ within the local horizontal plane spanned by the base vectors $\mathbb{E}_{1*}$ and $\mathbb{E}_{2*}$ directed from P to P_i in vacuo. The angle between the base vectors $\mathbb{E}_{1*}$ and $\mathbb{F}_{1*}$ is the *"unknown orientation parameter"* Σ in the horizontal plane. $\mathbb{E}_{1*}$, $\mathbb{E}_{2*}$, $\mathbb{E}_{3*}$ are related to $\mathbb{F}_{1*}$, $\mathbb{F}_{2*}$, $\mathbb{F}_{3*}$ by a *"Karussel-Transformation"* as follows

$$\begin{bmatrix} \mathbb{F}_{1*} = \mathbb{E}_{1*}\cos\Sigma + \mathbb{E}_{2*}\sin\Sigma \\ \mathbb{F}_{2*} = -\mathbb{E}_{1*}\sin\Sigma + \mathbb{E}_{2*}\cos\Sigma \\ \mathbb{F}_{3*} = \mathbb{E}_{3*}, \end{bmatrix} \qquad (7.4)$$

or

$$[\mathbb{F}_{1*}, \mathbb{F}_{2*}, \mathbb{F}_{3*}] = [\mathbb{E}_{1*}, \mathbb{E}_{2*}, \mathbb{E}_{3*}] \begin{bmatrix} \cos\Sigma & -\sin\Sigma & 0 \\ \sin\Sigma & \cos\Sigma & 0 \\ 0 & 0 & 1 \end{bmatrix}. \qquad (7.5)$$

From (7.5), one notes that the local level reference frame of type $\mathbb{F}^*$ is related to the local level reference frame of type $\mathbb{E}^*$ by

$$[\mathbb{E}_{1*}, \mathbb{E}_{2*}, \mathbb{E}_{3*}] = [\mathbb{F}_{1*}, \mathbb{F}_{2*}, \mathbb{F}_{3*}]\, \mathbf{R}_3^{\mathrm{T}}(\Sigma). \qquad (7.6)$$

The datum spherical coordinates of point P_i in the local level reference frame $\mathbb{F}^*$ are given as

$$\left(\begin{array}{l} P \to X^* = Y^* = Z^* = 0 \\ PP_i \to \begin{bmatrix} X^* \\ Y^* \\ Z^* \end{bmatrix}_{\mathbb{F}^*} = S_i \begin{bmatrix} \cos T_i \cos B_i \\ \sin T_i \cos B_i \\ \sin B_i \end{bmatrix}, \end{array} \right) \qquad (7.7)$$

where T_i and B_i are the horizontal and vertical directions respectively, while S_i are the spatial distances.

The local cartesian coordinates of a point whose positional vector is $\mathbf{x}$ in the $\mathbb{F}^*$ system is given by

$$\mathbf{x} = \mathbb{F}_{1*}x + \mathbb{F}_{2*}y + \mathbb{F}_{3*}z, \qquad (7.8)$$

where $\{x, y, z\}$ are the components of the vector $\mathbf{x}$ in the system $\{\mathbb{F}_{1*}, \mathbb{F}_{2*}, \mathbb{F}_{3*} | P\}$.

In the chapters ahead, the local level reference frame of type $\mathbb{F}^*$ will be adopted. This system arbitrarily defines the horizontal directions such that the orientation to the system $\mathbb{E}^*$, i.e., Σ, is treated as unknown besides the unknown positions. In case of the three-dimensional orientation problem, it is determined alongside the direction $\{\Lambda_\Gamma, \Phi_\Gamma\}$ of the local gravity vector $\boldsymbol{\Gamma}$. For position determination using three-dimensional resection method, it is determined alongside unknown coordinates $\{X, Y, Z\}$. This will become clear in Chaps. 11 and 15.

7.3.2 Relationship between Global and Local Level Reference Frames

In positioning within the LPS framework, one is interested not only in the geometrical position $\{X, Y, Z\}$, but also in the physical quantities $\{\Lambda_\Gamma, \Phi_\Gamma\}$ which define the direction of the local gravity vector $\boldsymbol{\Gamma}$ at the instrument station. This direction $\{\Lambda_\Gamma, \Phi_\Gamma\}$ of the local gravity vector $\boldsymbol{\Gamma}$ together with the unknown orientation Σ relate LPS and GPS systems. They are obtained by solving the *three-dimensional orientation problem*. This is achieved by transforming coordinates from the local level reference frame to the global reference frame (e.g., ITRF97). It is conventionally solved by a means of a 3×3 rotation matrix, which is represented by a triplet $\{\Lambda_\Gamma, \Phi_\Gamma, \Sigma_\Gamma\}$ of orientation parameters called the *astronomical longitude* Λ_Γ, *astronomical latitude* Φ_Γ, and the "orientation unknown" Σ_Γ in the horizontal plane. With respect to the local gravity vector $\boldsymbol{\Gamma}$, the triplets $\{\Lambda_\Gamma, \Phi_\Gamma, \Gamma = \|\boldsymbol{\Gamma}\|\}$ are its spherical coordinates, in particular $\{\Lambda_\Gamma, \Phi_\Gamma\}$ its direction parameters. The *three-dimensional orientation problem* therefore determines;

(i) the 3×3 rotation matrix and,
(ii) the triplet $\{\Lambda_\Gamma, \Phi_\Gamma, \Sigma_\Gamma\}$ of orientation parameters from GPS/LPS measurements.

After the *astronomical longitude* Λ_Γ and *astronomical latitude* Φ_Γ are determined via (i) and (ii) above- no astronomical observations are needed anymore - the vertical deflections with respect to a well-chosen reference frame, e.g., the ellipsoidal normal vector field can be obtained as discussed in Sect. 8.4 of Chap. 8. The three-dimensional orientation problem is formulated by relating the local level reference frame $\mathbb{F}^*$ to the global reference frame $\mathbb{F}^\bullet$ as follows:

$$[\mathbb{F}_{1^*}, \mathbb{F}_{2^*}, \mathbb{F}_{3^*}] = [\mathbb{F}_{1\bullet}, \mathbb{F}_{2\bullet}, \mathbb{F}_{3\bullet}] \, \mathbf{R}_E \left(\Lambda_\Gamma, \Phi_\Gamma, \Sigma_\Gamma \right), \qquad (7.9)$$

where the Euler rotation matrix $\mathbf{R}_E$ is parameterized by

$$\boxed{\mathbf{R}_E\left(\Lambda_\Gamma, \Phi_\Gamma, \Sigma_\Gamma\right) := \mathbf{R}_3\left(\Sigma_\Gamma\right) \mathbf{R}_2\left(\tfrac{\pi}{2} - \Phi_\Gamma\right) \mathbf{R}_3(\Lambda_\Gamma),} \qquad (7.10)$$

i.e., the three-dimensional orientation parameters; astronomical longitude Λ_Γ, astronomical latitude Φ_Γ, and the orientation unknown Σ in the horizontal plane. In terms of;

(a) Cartesian coordinates $\{x, y, z\}$ of the station point and $\{x_i, y_i, z_i\}$ target points in the local level reference frame $\mathbb{F},^*$ and,
(b) Cartesian coordinates $\{X, Y, Z\}$ of the station point and target points $\{X_i, Y_i, Z_i\}$ in the global reference frame $\mathbb{F}^\bullet$,

one writes

$$\begin{bmatrix} x_i - x \\ y_i - y \\ z_i - z \end{bmatrix}_{\mathbb{F}^*} = \mathbf{R}_E(\Lambda_\Gamma, \Phi_\Gamma, \Sigma_\Gamma) \begin{bmatrix} X_i - X \\ Y_i - Y \\ Z_i - Z \end{bmatrix}_{\mathbb{F}^\bullet}, \qquad (7.11)$$

with

$$\begin{bmatrix} x_i - x \\ y_i - y \\ z_i - z \end{bmatrix}_{\mathbb{F}^*} = S_i \begin{bmatrix} \cos T_i \cos B_i \\ \sin T_i \cos B_i \\ \sin B_i \end{bmatrix}, \forall i \in \{1, 2, \ldots, n\}. \qquad (7.12)$$

Equation (7.11) contains the orientation parameters $\mathbf{R}_E(\Lambda_\Gamma, \Phi_\Gamma, \Sigma_\Gamma)$ relating the local level reference frame $\mathbb{F}^*$ to the global reference frame $\mathbb{F}^\bullet$. These orientation parameters have been solved by:

1. Determining the direction $(\Lambda_\Gamma, \Phi_\Gamma)$ of the local gravity vector $\mathbf{\Gamma}$ at the origin of the network and the orientation unknown Σ in the horizontal plane from stellar astronomical observations.
2. Solving the three-dimensional resection problem as discussed in Chap. 11. In the approach proposed by [156], directional measurements are performed to the neighbouring 3 points in the global reference frame and used to derive distances by solving the *Grunert's equations*. From these derived distances, a closed form solution of the six unknowns $\{X, Y, Z, \Lambda_\Gamma, \Phi_\Gamma, \Sigma_\Gamma\}$ by means of the *Hamilton-quaternion* procedure is performed.
3. Using the *simple Procrustes algorithm* as discussed in Chap. 8 to determine the three-dimensional orientation parameters $\{\Lambda_\Gamma, \Phi_\Gamma, \Sigma_\Gamma\}$ and the deflection of the vertical for a point whose geometrical positional quantities $\{X, Y, Z\}$ are known.

4. By first determining the geometrical values $\{X, Y, Z\}$ of the unknown station using resection approach as discussed in Chap. 11. Once these geometrical values have been determined, they are substituted back in (7.11) to obtain the Euler rotation matrix $\mathbf{R}_E(\Lambda_\Gamma, \Phi_\Gamma, \Sigma_\Gamma)$. The Euler rotation angles can then be deduced via an inverse map presented in Lemma 15.1 on p. 264 of Chap. 15.

7.3.3 Observation Equations

Let us now have a look at the equations that we often encounter when positioning with a stationary theodolite. Elaborate exposition of three-dimensional observations is given by [128]. Stationed at the point $P_0 \in \mathbb{E}^3$, and with the theodolite properly centered, one sights the target points $P_i \in \mathbb{E}^3$, where $i = 1,2,3,\ldots,n$. There exist three types of measurements that will be taken from $P_0 \in \mathbb{E}^3$ to $P_i \in \mathbb{E}^3$ in the LPS system (i.e., local level reference frame $\mathbb{F}^*$). These are:

- Horizontal directions T_i whose equation is given by

$$T_i = \arctan\left(\frac{\Delta y_i}{\Delta x_i}\right)_{\mathbb{F}^*} - \Sigma_\Gamma(P_0), \qquad (7.13)$$

  where $\Sigma_\Gamma(P_0)$ is the unknown orientation in the horizontal plane after setting the zero reading of the theodolite in the direction $P \to P_i$.

- Vertical directions B_i given by

$$B_i = \arctan\left(\frac{\Delta z_i}{\sqrt{\Delta x_i^2 + \Delta y_i^2}}\right)_{\mathbb{F}^*}. \qquad (7.14)$$

- Spatial distances S_i, i.e.,

$$S_i = \sqrt{\Delta x_i^2 + \Delta y_i^2 + \Delta z_i^2}\,|_{\mathbb{F}}^*, \qquad (7.15)$$

and $\Delta x_i = (x_i - x)$, $\Delta y_i = (y_i - y)$, $\Delta z_i = (z_i - z)$ denote the coordinate difference in the local level reference frame $\mathbb{F}^*$.

The relationship between the local level reference frame $\mathbb{F}^*$ and the global reference frame $\mathbb{F}^\bullet$ is then given by (e.g., 7.11)

$$\begin{bmatrix} \Delta x_i \\ \Delta y_i \\ \Delta z_i \end{bmatrix}_{\mathbb{F}^*} = \mathbf{R}_E(\Lambda_\Gamma, \Phi_\Gamma, 0) \begin{bmatrix} \Delta X_i \\ \Delta Y_i \\ \Delta Z_i \end{bmatrix}_{\mathbb{F}^\bullet}, \qquad (7.16)$$

with

$$\mathbf{R}_E(\Lambda_\Gamma, \Phi_\Gamma, 0) = \begin{bmatrix} \sin\Phi_\Gamma\cos\Lambda_\Gamma & \sin\Phi_\Gamma\sin\Lambda_\Gamma & -\cos\Phi_\Gamma \\ -\sin\Lambda_\Gamma & \cos\Lambda_\Gamma & 0 \\ \cos\Phi_\Gamma\cos\Lambda_\Gamma & \cos\Phi_\Gamma\sin\Lambda_\Gamma & \sin\Phi_\Gamma \end{bmatrix}. \quad (7.17)$$

Observations in (7.13), (7.14) and (7.15) are now expressed in the global reference frame as

$$T_i = \arctan\left\{ \frac{-\sin\Lambda_\Gamma\Delta X_i + \cos\Lambda_\Gamma\Delta Y_i}{\sin\Phi_\Gamma\cos\Lambda_\Gamma\Delta X_i + \sin\Phi_\Gamma\sin\Lambda_\Gamma\Delta Y_i - \cos\Phi_\Gamma\Delta Z_i} \right\} - \Sigma_\Gamma(P),$$
$$(7.18)$$

and

$$B_i = \arctan\left\{ \frac{\cos\Phi_\Gamma\cos\Lambda_\Gamma\Delta X_i + \cos\Phi_\Gamma\sin\Lambda_\Gamma\Delta Y_i + \sin\Phi_\Gamma\Delta Z_i}{\sqrt{(\sin\Phi_\Gamma\cos\Lambda_\Gamma\Delta X_i + \sin\Phi_\Gamma\sin\Lambda_\Gamma\Delta Y_i - \cos\Phi_\Gamma\Delta Z_i)^2 + D_2}} \right\},$$
$$(7.19)$$

where $D_2 = (\cos\Lambda_\Gamma\Delta Y_i - \sin\Lambda_\Gamma\Delta X_i)^2$, $\Delta X_i = (X_i - X)$, $\Delta Y_i = (Y_i - Y)$, $\Delta Z_i = (Z_i - Z)$ in the global reference frame $\mathbb{F}^\bullet$ and $\{\Sigma_\Gamma(P_0), \Lambda_\Gamma(P_0), \Phi_\Gamma(P_0)\}$ are the three unknown orientation parameters at the unknown theodolite station P_0.

7.4 Test Network Stuttgart Central

Observations

The following experiment was performed at the center of Stuttgart on one of the pillars of Stuttgart University's building along Kepler Strasse 11 as depicted by Fig. 7.1. The test network *"Stuttgart Central"* consisted of 8 GPS points listed in Table 7.1. A theodolite was stationed at pillar $K1$ whose astronomical longitude Λ_Γ as well as astronomic latitude Φ_Γ were known from previous astrogeodetic observations made by the Department of Geodesy and Geoinformatics, Stuttgart University. Since theodolite observations of type horizontal directions T_i as well as vertical directions B_i from the pillar $K1$ to the target points i, $i = 1, 2, \ldots, 6, 7$, were only partially available, the horizontal and vertical directions were simulated from the given values of $\{\Lambda_\Gamma, \Phi_\Gamma\}$ as well as the Cartesian coordinates of the station point $\{X, Y, Z\}$

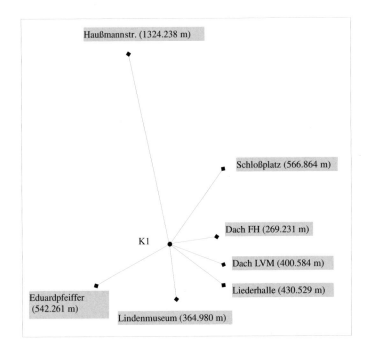

Fig. 7.1. Graph of the Test network *"Stuttgart Central"*

and target points $\{X_i, Y_i, Z_i\}$ using (7.18) and (7.19). The relationship between the observations of type horizontal directions T_i, vertical directions B_i, values of $\{\Lambda_\Gamma, \Phi_\Gamma\}$ and the Cartesian coordinates of the station point $\{X, Y, Z\}$ and target points $\{X_i, Y_i, Z_i\}$ enabled generation of the observation data sets in Table 7.3. Such a procedure had also an advantage in that we had full control of the algorithms that will be tested later in the book. In detail, the directional parameters $\{\Lambda_\Gamma, \Phi_\Gamma\}$ of the local gravity vector were adopted from the astrogeodetic observations $\phi_\Gamma = 48°46'54''.9$ and $\Lambda_\Gamma = 9°10'29''.8$ reported by [212, p. 46] with a root-mean-square error $\sigma_\Lambda = \sigma_\Phi = 10''$. Table 7.1 contains the $\{X, Y, Z\}$ coordinates obtained from a GPS survey of the test network Stuttgart Central, in particular with root-mean-square errors $(\sigma_X, \sigma_Y, \sigma_Z)$ neglecting the covariances $(\sigma_{XY}, \sigma_{YZ}, \sigma_{ZX})$. The spherical coordinates of the relative position vector, namely of the coordinate differences $\{x_i - x, y_i - y, z_i - z\}$, are called horizontal directions T_i, vertical directions B_i and spatial distances S_i and are given in Table 7.2. The standard deviations/root-mean-square errors were fixed to $\sigma_T = 6", \sigma_B = 6''$. Such root mean square errors can be obtained on the basis of a proper refraction model. Since the horizontal and vertical directions of Table 7.2 were simulated, with zero noise level, we used a random generator *randn* in Matlab e.g., [174, p. 84, p. 144] to produce additional observational data sets within the framework of the given root-mean-square errors. For each observable of type T_i and B_i, 30 randomly simulated data were obtained and the mean taken. Let us refer to the observational data sets $\{T_i, B_i\}, i = 1, 2, \ldots, 6, 7$, of Table 7.3 which were enriched by the root-mean-square errors of the individual randomly generated observations as well as by the differences $\Delta T_i := T_i - T_i(\text{generated})$, $\Delta B_i := B_i - B_i(\text{generated})$. Such differences $(\Delta T_i, \Delta B_i)$ indicate the difference between the ideal values of Table 7.2 and those randomly generated.

Observations are thus designed such that by observing the other seven GPS stations, the orientation of the local level reference frame $\mathbb{F}^*$ whose origin is station $K1$, to the global reference frame $\mathbb{F}^\bullet$ is obtained. The direction of Schlossplatz was chosen as the zero direction of the theodolite leading to the determination of the third component Σ_Γ of the three-dimensional orientation parameters. To each of the GPS target points i, the observations of type horizontal directions T_i and the vertical directions B_i are measured. The spatial distances $S_i^2(\mathbf{X}, \mathbf{X}_i) = \|\mathbf{X}_i - \mathbf{X}\|$ are readily obtained from the observation of type horizontal directions T_i and vertical directions B_i. The following symbols have been used: $\sigma_X, \sigma_Y, \sigma_Z$ are the standard errors of the GPS Cartesian coordinates. Covariances $\sigma_{XY}, \sigma_{YZ}, \sigma_{ZX}$ are neglected. σ_T, σ_B are the standard deviation of horizontal and vertical directions respectively after an adjustment, Δ_T, Δ_B are the magnitude of the noise on the horizontal and vertical directions, respectively.

7.5 Concluding Remarks

What is presented here is just a nutshell of GPS. For more exposition of its operations and techniques, we refer to related publications, e.g., [101, 161, 186, 218, 244, 294, 302, 341]. For LPS systems, more insight can be found in [139, 144] and [278, p. 28]. In particular, for cases where the theodolite moves from point to point, i.e., moving horizontal triad, [127, 130, 133] presents interesting materials.

Table 7.1. GPS Coordinates in the global reference frame $\mathbb{F}^{\bullet}(X, Y, Z)$, (X_i, Y_i, Z_i), $i = 1, 2, \ldots, 7$

Station[2]	$X(m)$	$Y(m)$	$Z(m)$	σ_X mm	σ_X mm	σ_X mm
Dach $K1$	4157066.1116	671429.6655	4774879.3704	1.07	1.06	1.09
1	4157246.5346	671877.0281	4774581.6314	0.76	0.76	0.76
2	4156749.5977	672711.4554	4774981.5459	1.77	1.59	1.61
3	4156748.6829	671171.9385	4775235.5483	1.93	1.84	1.87
4	4157066.8851	671064.9381	4774865.8238	1.38	1.29	1.38
5	4157266.6181	671099.1577	4774689.8536	1.29	1.28	1.34
6	4157307.5147	671171.7006	4774690.5691	0.20	0.10	0.30
7	4157244.9515	671338.5915	4774699.9070	2.80	1.50	3.10

Table 7.2. Ideal spherical coordinates of the relative position vector in the local level reference frame $\mathbb{F}^{*}$: Spatial distances, horizontal directions, vertical directions

Station Observed from $K1$	Distances (m)	Horizontal directions(gon)	Vertical directions(gon)
Schlossplatz (1)	566.8635	52.320062	-6.705164
Haussmanstr. (2)	1324.2380	107.160333	0.271038
Eduardpfeiffer (3)	542.2609	224.582723	4.036011
Lindenmuseum (4)	364.9797	293.965493	-8.398004
Liederhalle (5)	430.5286	336.851237	-6.941728
Dach LVM (6)	400.5837	347.702846	-1.921509
Dach FH (7)	269.2309	370.832476	-6.686951

Table 7.3. Randomly generated spherical coordinates of the relative position vector: horizontal directions T_i and vertical directions $B_i, i = 1, 2, \ldots, 6, 7$, root-mean-square errors of individual observations, differences $\Delta T_i := T_i - T_i(\text{generated})$, $\Delta B_i := B_i - B_i(\text{generated})$ with respect to (T_i, B_i) ideal data of Table 7.2

St.	H/dir.(gon)	V/dir.(gon)	$\sigma_T(gon)$	$\sigma_B(gon)$	$\Delta_T(gon)$	$\Delta_B(gon)$
1	0.000000	-6.705138	0.0025794	0.0024898	-0.000228	-0.000039
2	54.840342	0.271005	0.0028756	0.0027171	-0.000298	0.000033
3	172.262141	4.035491	0.0023303	0.0022050	0.000293	0.000520
4	241.644854	-8.398175	0.0025255	0.0024874	0.000350	0.000171
5	284.531189	-6.942558	0.0020781	0.0022399	-0.000024	0.000830
6	295.382909	-1.921008	0.0029555	0.0024234	0.000278	-0.000275
7	318.512158	-6.687226	0.0026747	0.0024193	-0.000352	0.000500

Partial Procrustes and the Orientation Problem

"It seems very strange that up to now Procrustes analysis has not been widely applied in geodetic literature. With this technique linearization problems of non linear equations system and iterative procedures of computation could be avoided, in general, with significant time saving and less analytical difficulties" F. Crosilla.

8.1 Motivation

This chapter presents the minimization approach known as *"Procrustes"* which falls within the multidimensional scaling techniques discussed in Sect. 8.2.2. Procrustes analysis is the technique of matching one configuration into another in-order to produce a measure of match. In adjustment terms, the partial Procrustes problem is formulated as the least squares problem of transforming a given matrix $\mathbf{A}$ into another matrix $\mathbf{B}$ by an orthogonal transformation matrix $\mathbf{T}$ such that the sum of squares of the residual matrix $\mathbf{E} = \mathbf{A} - \mathbf{BT}$ is minimum. This technique has been widely applied in shape and factor analysis. It has also been used for multidimensional rotation and also in scaling of different matrix configurations. In geodesy and geoinformatics, data analysis often require *scaling*, *rotation* and *translation* operations of different matrix configurations. Photogrammetrists, for example, have to determine the orientation of the camera during aerial photogrammetry and transform photo coordinates into ground coordinates. This is achieved by employing scaling, translation and rotation operations. These operations are also applicable to remote sensing and Geographical Information System (GIS) where map coordinates have to be transformed to those of the digitizing table. In case of robotics, the orientation of the robotic arm has to be determined, while for machine and computer visions, the orientation of the Charge-Coupled Device (CCD) cameras has to be established. In practice, positioning with satellites, particularly the Global Navigation Satellite Systems (GNSS) such us GPS and GLONASS has been on rise. The anticipated GALILEO satellites will further increase the use of satellites in positioning. This has necessitated the transformation of coordinates from the Global Positioning System (WGS 84) into local geodetic systems and vice versa.

A classical problem in geodesy and geoinformatics that would benefit from this technique is transformation, and in particular the 7–parameter datum transformation problem. The traditional approach of solving this problem, for instance, has been to linearize the nonlinear equations and then apply least squares method iteratively. With the proposed Procrustes approach, all that is required of the user is to insert the coordinates of one system (e.g., local coordinate system) in say, the matrix **A**, and those of the other system (e.g., GPS in WGS-84) into the matrix **B**. Using Procrustes analysis technique presented in this chapter, and later in Chap. 15, the desired scale, rotation and translation parameters can be obtained directly.

Although long applied in other fields such as; sociology, to map crime versus cities, and also in medicine as we will see in Sect. 8.2.3, Procrustes method is relatively new to the fields of geodesy and geoinformatics. Its first entry into geodesy can be traced back to the work of [97, 98] where the method was used in the creation of the criterion matrix used for deformation analysis. Further applications include the works of [13, 137, 138] who applies it to compute the three-dimension orientation parameters, deflection of the vertical, and 7-parameter datum transformation.

Recent application of the approach in geoinformatics can be found in the works of [100] who employs it to solve the photogrammetric block adjustment by independent models, [57] who applies it for size and shape three-dimensional object reconstructions, and [58] who uses the technique to update cadastral maps. At the beginning of the Chapter, we quoted F. Crosilla [99], the *father* of Procrustes in geodesy and geoinformatics. He wonders why such an amazing technique has not been widely applied in geodesy.

In this chapter, the partial, also called simple Procrustes algorithm which is sufficient for solving only the rotation elements is presented.

It will be demonstrated how the approach solves the three-dimensional orientation and the vertical deflection (direction of local gravity vector) problems. In Chap. 15, the *general Procrustes algorithm* will be presented and used to solve the 7-parameter similarity transformation problem which is often encountered in practice.

8.2 Procrustes: Origin and Applications

8.2.1 Procrustes and the Magic Bed

The origin of the name, and perhaps the concept is traced back to Greece. Somewhere in Attica in Greece lived a robber whose name was Procrustes. His house was so well positioned besides the road such that he was frequented by visitors who had to spend the night. In his house, Procrustes also known as Damastes kept a special bed: So special was the bed such that the visitors were required to fit in it. Unfortunately for Procrustes, neither were all his visitors of the same height nor length of the magic bed. All the same, the visitors were somehow forced in some "magic" way to fit into the magic bed. This was not done by adjusting the bed, but to the contrary its occupants! Procrustes devised ways to fit his quests onto his bed. Guests who were shorter for the bed were stretched by hammering or racking their bodies to fit the bed, while those who were longer had their legs chopped off! In both cases, the victims died. As fate would have it, Procrustes was himself adjusted to fit his own bed by Theseus, a young Attic hero whose mission was to eliminate robbers. The Encyclopedia of Greek Mythology writes[1]:

> "Procrustes (proh-KRUS-teez). A host who adjusted his guests to their bed. Procrustes, whose name means "he who stretches", was arguably the most interesting of Theseus's challenges on the way to becoming a hero. He kept a house by the side of the road where he offered hospitality to passing strangers, who were invited in for a pleasant meal and a night's rest in his very special bed (see Fig. 8.1[1]). Procrustes described it as having the unique property that its length exactly matched whomsoever lay down upon it. What Procrustes didn't volunteer was the method by which this "one-size-fits-all" was achieved, namely as soon as the guest lay down Procrustes went to work upon

---

[1]http://www.mythweb.com/encyc/gallery/procrustes_c.html ©Mythweb.com

him, stretching him on the rack if he was too short for the bed and chopping off his legs if he was too long. Theseus turned the tables on Procrustes, fatally adjusting him to fit his own bed."

Fig. 8.1. Procrustes and his "magical" bed ©Mythweb.com

This magic bed of Procrustes has become a saying for arbitrarily - and perhaps ruthlessly - *forcing someone or something to fit into an unnatural scheme or pattern.*

8.2.2 Multidimensional Scaling

Multidimensional scaling (MDS) is a method that represents measurements of similarity (or dissimilarity) among pairs of objects such as distances between points of low-dimensional multidimensional space. Let us consider for example that data consists of intelligence tests and that one desires to see the correlation between the tests. MDS can be used to represent these data in a plane such that the correlation can be studied. The more closer the points are (i.e., the shorter the distances between the points), the more correlated they are. MDS thus gives an advantage of graphical visualization of hidden adherent properties between objects. MDS has been described by [70] as:

- An approach for representing similarity and dissimilarity data as exemplified by distances of low dimensional space. This is done in-order to make this data accessible for visual inspection and exploration.

- An approach for testing if and how certain criteria by which one distinguishes among different objects of interest are mirrored in a corresponding empirical differences of this object (i.e., correlated).
- A data analytic approach that allows one to discover the three-dimensions that underlie judgements of dissimilarity and similarity.
- A psychological model that explains judgements of dissimilarity in terms of a rule that mimics a particular type of distance function.

Procrustes approach therefore is a procedure that is applied to realize the goals of MDS. In other words, it is a tool of MDS concerned with the fitting of one configuration into another as close as possible.

8.2.3 Applications of Procrustes in Medicine

As a motivational urge to embrace this long overdue powerful tool, this section presents briefly two areas where Procrustes procedure has found practical application. These are:

- Procrustes application software for gene recognition [120].
- Identification of malarial parasites [105].

The technique has also been applied in various fields ranging from biology, psychology, to structural analysis etc.

Gene Recognition

Gene recognition started as a statistical analysis and splicing sites. The statistical procedures however could not deal with other types of genes such as eukaryotic (i.e., a single-celled or multicellular organism whose cells contain a distinct membrane-bound nucleus). To solve this problem, researchers in the field developed PROCRUSTES software, which uses similarity-based approach to gene recognition [120]. This was achieved using spliced alignment technique. The software is reported by Human Genome News[2] to be able to identify with remarkable accuracy human version of genes that are in other forms of life. The human genes are broken into smaller segments known as exons. Searching for exons is analogous to following a magazine article that

---

[2] July-September 1996; 8:(1)

appears in, say, pp. 5, 23, 84, and 93, with almost identical advertisement and other articles appearing between. The software is applied to construct all these pages that contain the required article and automatically combine them into a best fitting set. The technique is said to work best when a "target protein" from the nonhuman sample guides the search, thus ensuring an accuracy that approaches 100%. In this technique, if a genomic sequence and a set of candidate exons are given, the algorithm explores all possible exon assemblies and finds a chain with the best fit to relate target protein. Instead of trying to identify the correct exons by statistical means (which predicts the correlation between the predicted and the actual gene to 70%, with just 40-50% exons predicted correctly), PROCRUSTES considers all possible chain with the maximum global similarity to the target protein. The procedure predicts a correlation of about 99% between the predicted and the actual gene [120]. The tool is useful in pinpointing elusive human version of cancer-causing gene!

Identification of Malaria Parasites

Dryden [105] applies Procrustes to identify proteins by comparing the electrophoretic gel images (Fig. 8.2[3]). The gels are obtained from strains of parasite which carry malaria. The procedure uses Procrustes matching and affine shape registration to match the gels. It applies some biological material to the left corner of the two images of gels A and B in Fig. 8.2. The material is then separated down the gel according to molecular weight (with the highest on top) and across the gel according to isoelectric point (with the highest on the right of the gel). Gel image is then used to identify strains of parasites using pattern of spots marked by (+). Dark spots appearing on the gels indicate the composition of protein and are marked by some expert in both gels A and B. Ten spots are marked in each gel and then classified as either invariant or variant spots.

The invariant spots are considered to be present for all parasites. The arrangement of the variant spots is of particular interest as it helps in the identification of malarial parasite. The field problem sighted by [105] however is that gels are prone to deformation such as *translation, scaling, rotation, affine transformation* and *smooth-linear bending*. Gel images therefore need to be registered by matching each image us-

---

[3]©Chapman and Hall Press

ing a set of transformation to alleviate the deformations above [105]. This is achieved through the use of Procrustes analysis.

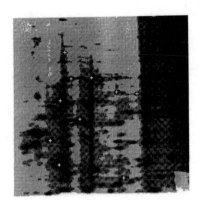 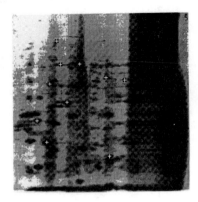

Fig. 8.2. The electrophoretic gels from gel A and gel B. Ten invariant spots have been marked by (+) in white above ©Chapman and Hall Press

8.3 Partial Procrustes Solution

8.3.1 Conventional Formulation

Procrustes being a technique of matching one configuration into another and producing a measure of match, seeks the *isotropic dilatation* and the rigid *translation*, *reflection* and *rotation* needed to best match one configuration to another [93, p. 92]. In this chapter, the term **partial** shall be used to mean ***optimal rotation*** in-order to avoid confusion since the term is used differently by different authors. For example, [165, 166] considers a case where the configuration matrix has several unknown elements in the minimization of the Frobenius norm as the partial Procrustes problem. Dryden [105] on the other hand uses the term partial Procrustes to refer to minimization of the Frobenius norm only over the translation and rotation. The general Procrustes solution is used as the minimization over the full set of similarity transformation as shall be seen in Chap. 15. In the solution of partial Procrustes problem, we refer to Table 8.1 for some matrix properties which will be of use.

The Procrustes problem is concerned with fitting a configuration **B** into **A** as close as possible. The simplest Procrustes case is one in which both configurations have the same dimensionality and the same number of points, which can be brought into a $1 - 1$ correspondence by substantive considerations [70, p. 339]. Let us consider the case where both **A** and **B** are of the same dimension. The partial Procrustes problem is then formulated as

$$\mathbf{A} = \mathbf{BT} \tag{8.1}$$

The rotation matrix **T** in (8.1) is then solved by measuring the distances between corresponding points in both configurations, square these values, and add them to obtain the sum of squares $\|\mathbf{A} - \mathbf{BT}\|^2$ which is then minimized. One proceeds via Frobenius norm as follows:

$$\min_{\mathbf{T}'\mathbf{T} = \mathbf{I}} \|\mathbf{X} - \mathbf{YT}\| := \sqrt{tr(\mathbf{X}' - \mathbf{T}'\mathbf{Y}')(\mathbf{X} - \mathbf{YT})} \tag{8.2}$$

In-order to obtain **T** in (8.2), the following properties of a matrix in Table (8.1) are essential.

Table 8.1. Matrix properties for procrustes analysis

(a)	$\operatorname{tr}\mathbf{A} = \sum_{i=1}^{n} a_{ii}$	Definition of trace function
(b)	$\operatorname{tr}\mathbf{A} = \operatorname{tr}\mathbf{A}'$	Invariant under transpose
(c)	$\operatorname{tr}\mathbf{ABC} = \operatorname{tr}\mathbf{C}\,\mathbf{AB} = \operatorname{tr}\mathbf{BCA}$	Invariant under 'cyclic' permutation
(d)	$\operatorname{tr}\mathbf{A}'\mathbf{B} = \operatorname{tr}(\mathbf{A}'\mathbf{B})' = \operatorname{tr}\mathbf{B}'\mathbf{A} = \operatorname{tr}\mathbf{AB}'$	Combining properties (b) and (c)
(e)	$\operatorname{tr}(\mathbf{A} + \mathbf{B}) = \operatorname{tr}\mathbf{A} + \operatorname{tr}\mathbf{B}$	Summation rule

Using (8.2) and the properties of Table 8.1, one writes

$$\left[\begin{array}{l} \|\mathbf{A} - \mathbf{BT}\|^2 := tr(\mathbf{A}' - \mathbf{T}'\mathbf{B}')(\mathbf{A} - \mathbf{BT}) \\ \mathbf{T}'\mathbf{T} = \mathbf{I} \\ = tr(\mathbf{A}'\mathbf{A} - 2\mathbf{A}'\mathbf{BT} + \mathbf{T}'\mathbf{B}'\mathbf{BT}) \\ = tr\mathbf{A}'\mathbf{A} - 2tr\mathbf{A}'\mathbf{BT} + tr\mathbf{B}'\mathbf{B} \\ tr\mathbf{T}'\mathbf{B}'\mathbf{BT} = tr\mathbf{TT}'\mathbf{B}'\mathbf{B} = tr\mathbf{B}'\mathbf{B}. \end{array} \right. \tag{8.3}$$

The simplification $tr\mathbf{T}'\mathbf{B}'\mathbf{BT} = tr\mathbf{B}'\mathbf{B}$ in (8.3) is obtained by using the property of invariance of the trace function under cyclic permutation (i.e., property (c) in Table 8.1). Since $tr(\mathbf{A}'\mathbf{A})$ and $tr(\mathbf{B}'\mathbf{B})$ are not dependent on **T**, we note from (8.3) that

$$\|\mathbf{A} - \mathbf{BT}\|^2 = \min \ \Leftrightarrow \ tr(\mathbf{A}'\mathbf{BT}) = \max$$
$$\mathbf{T}'\mathbf{T} = \mathbf{TT}' = \mathbf{I}_k. \tag{8.4}$$

If $\mathbf{U\Sigma V}'$ is the singular value decomposition of $\mathbf{A}'\mathbf{B}$ and $\mathbf{C} = \mathbf{A}'\mathbf{B}$, then we have

$$\left[\begin{array}{l} \mathbf{A}'\mathbf{B} = \mathbf{U\Sigma V}' \\ \text{if } \mathbf{C} = \mathbf{U\Sigma V}', \ \ \mathbf{U}, \mathbf{V}' \in SO(3) \\ \mathbf{\Sigma} = \mathrm{Diag}(\sigma_1, \ \ldots, \ \sigma_k) \text{then} \\ \mathrm{tr}(\mathbf{CT}) \le \sum_{i=1}^{k} \sigma_k, \\ \text{with} \\ k = 3. \end{array} \right. \tag{8.5}$$

The proof for (8.5) is given by [243, p. 34] as follows: Substituting for $\mathbf{C}$ from its singular value decomposition and with the property (c) in Table 8.1, one writes

$$\left[\begin{array}{l} \mathrm{tr}(\mathbf{CT}) = \mathrm{tr}(\mathbf{U\Sigma V}'\mathbf{T}) = \mathrm{tr}(\mathbf{\Sigma V}'\mathbf{TU}) \\ \text{taking} \\ \mathbf{R} = (ij)1 \le i, j \le k = \mathbf{V}'\mathbf{TU} \text{ orthogonal and } |r_{ii}| \le 1 \\ \text{then} \\ \mathrm{tr}(\mathbf{\Sigma V}'\mathbf{TU}) = \sum_{i=1}^{k} \sigma_i r_{ii} \le \sum_{i=1}^{k} \sigma_i. \end{array} \right. \tag{8.6}$$

From (8.5) and (8.6), one notes that

$$\mathrm{tr}(\mathbf{A}'\mathbf{BT}) = \max \ \Leftrightarrow \ \mathrm{tr}(\mathbf{A}'\mathbf{BT}) \le \sum_{i=1}^{k} \gamma_i, \tag{8.7}$$

subject to the singular value decomposition

$$\mathbf{A}'\mathbf{B} = \mathbf{U\Sigma V}', \ \ \mathbf{U}, \mathbf{V} \in SO(3) \text{ and orthogonal.} \tag{8.8}$$

Finally, the maximum value is obtained as

$$\max(\mathrm{tr}\mathbf{A}'\mathbf{BT}) = \sum_{i=1}^{k} \gamma_i \ \Leftrightarrow \ \mathbf{T} = \mathbf{VU}'. \tag{8.9}$$

Thus the solution of the rotation matrix by Procrustes method is

$$\mathbf{T} = \mathbf{VU}'. \tag{8.10}$$

8.3.2 Partial Derivative Formulation

This approach is attributed to P. H. Schonemann [288] as well as [289]. Proceeding from the Frobenius norm in (8.2) and using (8.1) leads to

$$d_1 = tr \, \mathbf{A}' \mathbf{A} - 2tr \, \mathbf{A}' \mathbf{BT}, + \mathbf{T}' \mathbf{B}' \mathbf{BT}, \qquad (8.11)$$

while the condition that $\mathbf{T}' \mathbf{T} = \mathbf{I}$ leads to

$$d_2 = \mathbf{\Lambda}(\mathbf{T}' \mathbf{T} - \mathbf{I}). \qquad (8.12)$$

where $\mathbf{\Lambda}$ is the m x m unknown matrix of Lagrange multipliers. Equations (8.11) and (8.12) are added to give

$$d = d_1 + d_2. \qquad (8.13)$$

The derivative of (8.13) are obtained with respect to $\mathbf{T}$ as

$$\begin{bmatrix} \dfrac{\partial d}{\partial \mathbf{T}} = \dfrac{\partial d_1}{\partial \mathbf{T}} + \dfrac{\partial d_2}{\partial \mathbf{T}} \\[2ex] = \dfrac{\partial \left(tr \, \mathbf{A}' \mathbf{A} - 2tr \, \mathbf{A}' \mathbf{BT} + tr \, \mathbf{T}' \mathbf{B}' \mathbf{BT} \right)}{\partial \mathbf{T}} + \dfrac{\partial \left(\mathbf{\Lambda} \mathbf{T}' \mathbf{T} - \mathbf{\Lambda} \mathbf{I} \right)}{\partial \mathbf{T}} \\[2ex] = -2 \, \mathbf{B}' \mathbf{A} + \mathbf{B}' \mathbf{BT} + \mathbf{B}' \mathbf{BT} + \mathbf{T} \mathbf{\Lambda} + \mathbf{T} \mathbf{\Lambda}' \\[1ex] = \left(\mathbf{B}' \mathbf{B} + \mathbf{B}' \mathbf{B} \right) \mathbf{T} - 2 \, \mathbf{B}' \mathbf{A} + \mathbf{T} \left(\mathbf{\Lambda} + \mathbf{\Lambda}' \right). \end{bmatrix}$$

$$(8.14)$$

From (8.14), let

$$\mathbf{B}' \mathbf{B} = \mathbf{B}^*, \quad \mathbf{B}' \mathbf{A} = \mathbf{C} \text{ and } \left(\mathbf{\Lambda} + \mathbf{\Lambda}' \right) = 2\mathbf{\Lambda}^*. \qquad (8.15)$$

For an extremum value of d, we set $\dfrac{\partial d}{\partial T} = 0$ such that

$$\begin{bmatrix} 2\mathbf{C} = 2\mathbf{B}^* \mathbf{T} + 2\mathbf{T} \mathbf{\Lambda}^* \\ \mathbf{C} = \mathbf{B}^* \mathbf{T} + \mathbf{T} \mathbf{\Lambda}^*, \end{bmatrix} \qquad (8.16)$$

leading to both $\mathbf{B}^*$ and $\mathbf{\Lambda}^*$ being symmetric. Hence

$$\mathbf{\Lambda}^* = \mathbf{T}' \mathbf{C} - \mathbf{T}' \mathbf{B}' \mathbf{T}. \qquad (8.17)$$

But $\mathbf{B}^* \rightarrow$ symmetric and thus $\mathbf{T}' \mathbf{B}^* \mathbf{T}$ is also symmetric. $\mathbf{T}' \mathbf{C}$ is therefore symmetric or

$$\left[\begin{array}{l} \mathbf{T}'\mathbf{C} \ = \ \mathbf{C}'\mathbf{T} \\ \text{from the side condition} \\ \mathbf{T}'\mathbf{T} = \mathbf{TT}' = \mathbf{I}_3 \\ \text{we have that} \\ \mathbf{C} \ = \ \mathbf{TC}'\mathbf{T}. \end{array}\right. \qquad (8.18)$$

From (8.8), we had $\mathbf{C} = \mathbf{A}'\mathbf{B} = \mathbf{U}\boldsymbol{\Sigma}\mathbf{V}'$ by SVD. In the present case we note that $\mathbf{C} = \mathbf{B}'\mathbf{A}$, thus $\mathbf{C} = \mathbf{B}'\mathbf{A} = \mathbf{V}\boldsymbol{\Sigma}\mathbf{U}'$. From (8.18) we have

$$\left[\begin{array}{l} \text{with } \mathbf{U}'\mathbf{U} \ = \ \mathbf{UU}' \ = \ \mathbf{V}'\mathbf{V} \ = \ \mathbf{V}\mathbf{V}' \ = \ \mathbf{I}_3 \\ \mathbf{C} \ = \mathbf{TC}'\mathbf{T} \\ \mathbf{V}\boldsymbol{\Sigma}\mathbf{U}' = \mathbf{TU}\boldsymbol{\Sigma}\mathbf{V}'\mathbf{T} \\ \mathbf{V} = \mathbf{TU} \\ \text{or } \ \mathbf{T} = \mathbf{VU}', \end{array}\right. \qquad (8.19)$$

which is identical to (8.10).

8.4 Practical Applications

8.4.1 Three-dimensional Orientation Problem

The transformation of coordinates from the local level reference frame to the global terrestrial reference frame (e.g., ITRF97) is a key, contemporary problem. In carrying out coordinate transformations, some of the sought parameters are those of orientation. Orientations are normally sought for; theodolites, cameras, and CCD sensors, etc. Procedures for solving explicitly the three-dimensional orientation problems in geoinformatics are presented in the works of [293, 311, 312, 352]. In geodesy, attempts to find closed form solution to the orientation problem have been carried out by [13, 137, 156] who proved that the three-dimensional orientation problem could be solved in a closed form through the integration of GPS and LPS systems.

The orientation problem is formulated by expressing (7.11) in Chap. 7 relating the two configurations, i.e., the local level reference frame and the global reference frame, with the left-hand-side in terms of spherical coordinates, as

$$\begin{bmatrix} s_i \begin{bmatrix} \cos T_i \cos B_i \\ \sin T_i \cos B_i \\ \sin B_i \end{bmatrix}_{\mathbb{F}^*} = \mathbf{R}(\Lambda_\Gamma, \Phi_\Gamma, \Sigma_\Gamma^i) \begin{bmatrix} X_i - X \\ Y_i - Y \\ Z_i - Z \end{bmatrix}_{\mathbb{F}^\bullet} \end{bmatrix} \quad (8.20)$$

$$with$$

$$s_i = \sqrt{(X_i - X)^{\bullet 2} + (Y_i - Y)^{\bullet 2} + (Z_i - Z)^{\bullet 2}}.$$

In (8.20), $X, Y, Z, X_i, Y_i, Z_i \; \forall_i \in N$ are GPS coordinates in the global reference frame $\mathbb{F}^\bullet$, while the spherical coordinates $T_i, B_i \; \forall_i \in N$ are used to derive the left-hand-side of (8.20) in the local level reference frame $\mathbb{F}^*$. The orientation problem (8.20) is conventionally solved by means of a 3×3 rotation matrix $\mathbf{R}$, which is represented by the triplet $\{\Lambda_\Gamma, \Phi_\Gamma, \Sigma_\Gamma\}$ of orientation parameters called the astronomical longitude Λ_Γ, astronomical latitude Φ_Γ, and the "orientation unknown" Σ_Γ in the horizontal plane. With respect to the local gravity vector $\mathbf{\Gamma}$, the triplet $\{\Lambda_\Gamma, \Phi_\Gamma, \Gamma = \|\mathbf{\Gamma}\|\}$ are its spherical coordinates, in particular $\{\Lambda_\Gamma, \Phi_\Gamma\}$ are its direction parameters. Here we solve the problem of determining;

(a) the 3×3 rotation matrix $\mathbf{R}$ and,
(b) the triplet $\{\Lambda_\Gamma, \Phi_\Gamma, \Sigma_\Gamma\}$ of orientation parameters from GPS/LPS measurements by means of the partial Procrustes algorithm.

Procrustes Solution of the Orientation Problem

Consider coordinates to be given in two configurations with the same three-dimensional space in the local level reference frame $\mathbb{F}^*$ and global reference frame $\mathbb{F}^\bullet$. For such a three-dimensional space, where $i = 3$ (i.e., 3 target points), the relationship in (8.20) between the two systems is expressed as

$$\begin{bmatrix} x_1 - x & x_2 - x & x_3 - x \\ y_1 - y & y_2 - y & y_3 - y \\ z_1 - z & z_2 - z & z_3 - z \end{bmatrix}_{\mathbb{F}^*} = \mathbf{R} \begin{bmatrix} X_1 - X & X_2 - X & X_3 - X \\ Y_1 - Y & Y_2 - Y & Y_3 - Y \\ Z_1 - Z & Z_2 - Z & Z_3 - Z \end{bmatrix}_{\mathbb{F}^\bullet} . \quad (8.21)$$

For n target points, (8.21) becomes

$$\begin{bmatrix} x_1 - x & x_2 - x & \dots & x_n - x \\ y_1 - y & y_2 - y & \dots & y_n - y \\ z_1 - z & z_2 - z & \dots & z_n - z \end{bmatrix}_{\mathbb{F}^*} = \mathbf{R} \begin{bmatrix} X_1 - X & X_2 - X & \dots & X_n - X \\ Y_1 - Y & Y_2 - Y & \dots & Y_n - Y \\ Z_1 - Z & Z_2 - Z & \dots & Z_n - Z \end{bmatrix}_{\mathbb{F}^\bullet}$$
$$3 \times n \qquad\qquad 3 \times 3 \qquad\qquad 3 \times n,$$
$$(8.22)$$

with their respective dimensions given below them. The transpose of (8.22) is expressed as

$$
\begin{bmatrix}
x_1 - x & y_1 - y & z_1 - z \\
x_2 - x & y_2 - y & z_2 - z \\
\cdot & \cdot & \cdot \\
\cdot & \cdot & \cdot \\
\cdot & \cdot & \cdot \\
x_n - x & y_n - y & z_n - z
\end{bmatrix}_{\mathbb{F}*}
=
\begin{bmatrix}
X_1 - X & Y_1 - Y & Z_1 - Z \\
X_2 - X & Y_2 - Y & Z_2 - Z \\
\cdot & \cdot & \cdot \\
\cdot & \cdot & \cdot \\
\cdot & \cdot & \cdot \\
X_n - X & Y_n - Y & Z_n - Z
\end{bmatrix}_{\mathbb{F}\bullet}
\mathbf{R}'.
$$

$$ n \times 3 \qquad\qquad n \times 3 \qquad\qquad 3 \times 3 \tag{8.23} $$

Equation (8.23) contains the relative position vectors of corresponding points in two reference frames. Let us indicate the matrix on the left-hand-side by $\mathbf{A}$, the one on the right-hand-side by $\mathbf{B}$, and denote the rotation matrix $\mathbf{R}'$ by $\mathbf{T}$. The partial Procrustes problem is now concerned with fitting the configuration of $\mathbf{B}$ into $\mathbf{A}$ as close as possible. The problem reduces to that of determination of the rotation matrix $\mathbf{T}$. The operations involved in the solution of the orientation problem, therefore, are:

- Solution of $\mathbf{T}^* = \mathbf{V}\mathbf{U}'$.
- Obtaining the rotation elements from $\mathbf{R} = (\mathbf{T}^*)'$.

The rotation matrix $\mathbf{T}^*$ is the best possible matrix out of the set of all orthogonal matrices $\mathbf{T}$ which are obtained by imposing the restriction $\mathbf{T}\mathbf{T}' = \mathbf{T}'\mathbf{T} = \mathbf{I}$. The matrix $\mathbf{T}$ could otherwise be any matrix, which means, geometrically, that $\mathbf{T}$ is some linear transformation which in general may not preserve the shape of $\mathbf{B}$. A summary of the computational procedure for the three-dimensional orientation parameters based on Example 8.1 is given in Fig. 8.3.

Example 8.1 (Computation of the three-dimensional orientation problem). The partial Procrustes approach discussed in Sect. 8.2 is applied to the Test network of Stuttgart Central presented in Sect. 7.4. 8 GPS stations are used to determine the three-dimensional orientation parameters $\{\Lambda_\Gamma, \Phi_\Gamma, \Sigma_\Gamma\}$. From the observations of Table 7.3 on p. 88, the matrix $\mathbf{A}$ in (8.1) is computed in terms of the spherical coordinates using (8.20). The Matrix $\mathbf{B}$ is obtained by subtracting the coordinates of station $K1$ from those of other stations in Table 7.1. The rotation matrix $\mathbf{T}$ is then computed using partial Procrustes algorithm, i.e., (8.1) to (8.10). For this network, the computed three-dimensional orientation parameters $\{\Lambda_\Gamma, \phi_\Gamma, \Sigma_\Gamma\}$ gave the values $\phi_\Gamma = 48^\circ 46' 54''.3$

and $\Lambda_\Gamma = 9^o10'30''.1$, which when compared to $\phi_\Gamma = 48^o46'54''.9$ and $\Lambda_\Gamma = 9^o10'29''.8$ in [212, p. 46] deviates by $\Delta\Lambda_\Gamma = -0''.3$ and $\Delta\Phi_\Gamma = 0''.6$.

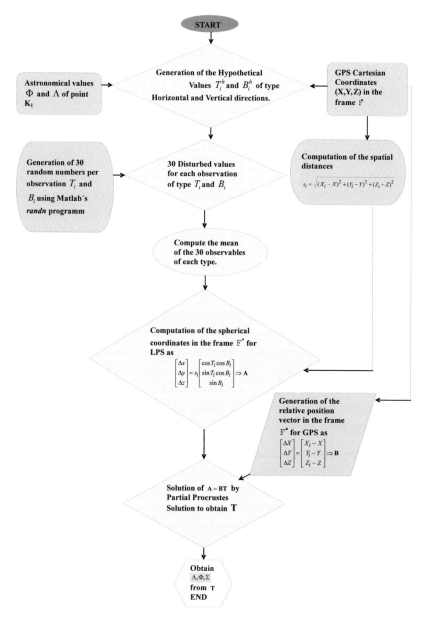

Fig. 8.3. Flow chart for computing three-dimensional orientation parameters

8.4.2 Determination of Vertical Deflection

As soon as we have determined the astronomical longitude Λ_Γ and astronomical latitude Φ_Γ, the deflection of the vertical can be computed with respect to a well chosen reference frame, e.g., the ellipsoidal normal vector field. Traditionally, orientation parameters $\{\Lambda_\Gamma, \Phi_\Gamma\}$ have been obtained from stellar observations and related to geodetic coordinates $\{\lambda, \phi\}$ to obtain the deflection of the vertical. Through the integration of GPS and LPS systems however, the astronomical observations of type $\{\Lambda_\Gamma, \Phi_\Gamma\}$ are obtained from the three-dimensional orientation solutions as discussed in Sect. 8.4.1. The approach alleviates the tiresome and expensive night stellar observation. Such pioneering approach in geodesy can be traced to the works of [13, 46, 137, 156].

To determine the vertical deflection, the reference direction is parameterized in terms of "surface normal"; ellipsoidal longitude λ and ellipsoidal latitude ϕ. These are then subtracted from the local vertical parameterized in terms of astronomical longitude Λ_Γ and astronomical latitude Φ_Γ as

$$\begin{aligned} \Lambda_\Gamma - \lambda \\ \Phi_\Gamma - \phi, \end{aligned} \tag{8.24}$$

to access the vertical deflections. In such a procedure, the topographical surface which is embedded into a three-dimensional Euclidean space $\mathbb{R}^3$ is mapped point-wise into a reference ellipsoid of revolution through the procedure discussed in Chap. 10. Indeed as outlined in Solution 10.5 on p. 159 for instance, those direction parameters $\{\Lambda, \Phi\}$ are conveniently computed from GPS Cartesian coordinates $\{X, Y, Z\}$ of the station point with respect to the global reference frame $\{\mathbb{F}_{1\bullet}, \mathbb{F}_{2\bullet}, \mathbb{F}_{3\bullet}\}$. The deflection of the vertical is then computed from (8.24) as

$$\begin{aligned} \delta\Lambda_\Gamma := \Lambda_\Gamma - \lambda, \delta\Phi_\Gamma := \Phi_\Gamma - \phi \\ \eta := \delta\Lambda_\Gamma \cos\Phi, \xi := \delta\Phi_\Gamma. \end{aligned} \tag{8.25}$$

Equation (8.25) are simple representation of the east vertical deflection η and the north vertical deflection ξ. The results in Table (3.1) of [137] document the precise determination of the orientation parameters of type astronomic longitude Λ_Γ, astronomic latitude Φ_Γ, horizontal orientation unknown Σ_Γ in the range of fraction of seconds of arc as well as vertical deflection $\{\xi, \eta\}$ in the same range exclusively from GPS-LPS observations.

8.5 Concluding Remarks

The partial Procrustes algorithm presented in this chapter provides a powerful tool for solving rotation and orientation related problems in general. The approach is straight forward and does not require linearization, which bog down least squares and other techniques commonly used. In Chap. 15, it shall be demonstrated how the general Procrustes approach determines scale and translation parameters of transformation, in addition to the rotation elements. For complete exposition of Procrustes approach, we refer to the works of [65, 69, 70, 76, 89, 90, 93, 97, 98, 105, 121, 122, 126, 160, 165, 166, 242, 243, 268, 288, 289, 307, 323].

9

Positioning by Ranging

9.1 Applications of Distances

Throughout history, position determination has been one of the fundamental task undertaken by man on daily basis. Each day, one has to know where one is, and where one is going. To mountaineers, pilots, sailors etc., the knowledge of position is of great importance. The traditional way of locating one's position has been the use of maps or campus to determine directions. In modern times, the entry into the game by Global Navigation Satellite Systems GNSS that comprise the Global Positioning System (GPS), Russian based GLONASS and the proposed European's GALILEO have revolutionized the art of positioning.

In the new field of *GPS meteorology* for example, as well as geodesy, robotics and geoinformatics etc., distances (ranges) play a key role in determining unknown parameters. In the recently developed *Spatial Reference System*[1] designed to check and control the accuracy of three-dimensional coordinate measuring machines and tooling equipments, coordinates of the edges of the instrument are computed from distances of the bars. This signifies that industrial application of distances is fast gaining momentum just as in geosciences. In GPS meteorology that we will discuss in Chap. 13 for example, distances traveled by GPS satellites signals through the atmosphere are measured and related do the would be distances in vacuo (i.e., in the absence of the atmosphere). Since these signals traverse the atmosphere, they enable *accurate global remote sensing* of the atmosphere to retrieve vertical profiles of *temperature*, *pressure* and *water vapour*.

---

[1]Metronom US., Inc., Ann Arbor: http://www.metronomus.com

Apart from distances being used to determine the user's position and its application in GPS meteorology, they find use in quick station search in engineering and cadastral surveying operations. Ranging, together with resection and intersection techniques (see e.g., Chaps. 11 and 12) are useful in densifying geodetic networks as illustrated by Fig. 9.1. Densification is vital for extending network control in areas where GPS receivers fail, e.g., in tunnels and forests (see Fig. 9.1). Distances are also used in photogrammetry to determine the perspective center coordinates from measured photo and the ground coordinates. Another area of application is in robotics.

Measured distances (ranges) are normally related to the desired parameters via nonlinear systems of equations that require explicit/exact solutions. Approximate numerical procedures used for solving such nonlinear distance equations are normally iterative in nature, and often require linearization of the nonlinear equations. Where closed form solutions exist, they require differencing and substitution steps which are laborious and time consuming. The desire therefore is to have procedures that can offer direct solutions without linearization, iterations or substitutional steps.

In this chapter, direct procedures for solving nonlinear systems of equations for distances without linearization, iteration, forward and backward substitutions are presented. In particular, the advantages of fast computers with large storage capacities, and computer algebraic software of Mathematica, Maple and Matlab are exploited by the algebraic based approaches. These methods which were presented in Chaps. 4, 5 and 6 directly deliver the position of unknown station from distance measurements. They do so by eliminating variables appearing in the nonlinear systems of equations resulting in univariate polynomials that are solvable using Matlab's *"roots"* command.

The improvements made on measuring instruments has led to Electromagnetic Distance Measuring (EDM) equipments that measure distances to higher accuracies. By measuring distances from an unknown station to two known stations, two nonlinear distance equations, whose geometrical properties have been studied by [147, 148] are formed. They have to be solved for the planar position of the unknown station. If distances are measured from an unknown station to three known stations instead, three nonlinear distance equations have to be solved for the unknown position. In Chaps. 4 and 6, planar distances were encountered in Figs. 4.1 and 6.1 respectively, where they were used to illustrate the concepts that were discussed. The position $\{x_0, y_0\}$ of the unknown station P_0 was related to the measured distances by (4.1) and (4.2) on p. 30.

The term ranging is broadly used in this chapter to incorporate the GPS pseudo-range measurements. For Local Positioning Systems (e.g., using EDMs), distances can be measured directly. For Global Positioning System (GPS) however, distances are not directly measured owing to satellites and receivers' clock uncertainties.

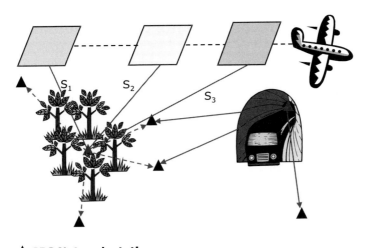

▲ **GPS Network stations**

▲ **New stations to be positioned**

Fig. 9.1. Point densification in forest and inside a tunnel

9.2 Ranging by Global Positioning System (GPS)

9.2.1 The Pseudo-ranging Four-Points Problem

If one has access to a hand held GPS receiver, a mobile phone or a watch fitted with a GPS receiver, one needs only to press the button to know the position where one is standing. Basically, the operations involve distance measurements to GPS satellites whose properties were discussed in Sect. 7.2. The receiver measures the travel time of the signal transmitted from the satellites. This distance is calculated from the relationship

$$distance = velocity \times time,$$

where velocity is given by the speed of light in vacuum. The distances S_i are then related to the position of the unknown station $\{X_0, Y_0, Z_0\}$ by

$$S_i = \sqrt{(X^i - X_0)^2 + (Y^i - Y_0)^2 + (Z^i - Z_0)^2}, \qquad (9.1)$$

where $\{X^i, Y^i, Z^i\}$ are the position of the satellite i. Geometrically,

Distance measurements to three satellites

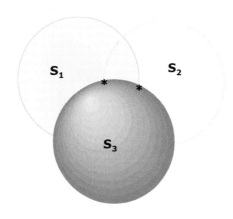

Fig. 9.2. Pseudo-ranging geometry

the three unknowns $\{X_0, Y_0, Z_0\}$ are obtained from the intersection of three spherical cones given by the pseudo-ranging equations. Distance measurements to only one satellite puts the user's position anywhere within the sphere formed by distance S_1 in Fig. 9.2. Measurements to two satellites narrow the position to the intersection of the two spheres S_1 and S_2. 9.2. A third satellite is therefore required to definitely fix the user's position. This is achieved by the intersection of the third sphere

S_3 with the other two. If direct distance measurements to the satellites were possible, (9.1) would have sufficed to provide the user's location. Distance measurements to satellites as already stated are however not direct owing to the satellites and receivers' clock biases. Satellites' clock biases can be modelled while the receivers' clock biases have to be determined as an unknowns. For GPS positioning therefore, in addition to position determination from measured distances, the receiver's clock bias has to be added in the observation equations as unknown. Since distances to the satellites in (9.1) are derived from the transmitted signals that are affected by both satellites and receivers' clock uncertainties, they are normally referred to as pseudo-ranges. What one measures therefore are not the actual distances (ranges) but pseudo-ranges. Pseudo-range measurements lead to GPS pseudo-ranging four-points problem ("pseudo 4P4"), which is the problem of determining the four unknowns. The unknowns comprise the three components of receiver position $\{X_0, Y_0, Z_0\}$ and the stationary receiver *range bias*. Minimum observations required to obtain receiver position and range bias are pseudo-range observations to four satellites as depicted in Fig. 9.3. Besides pseudo-range observations, phase measurements are often

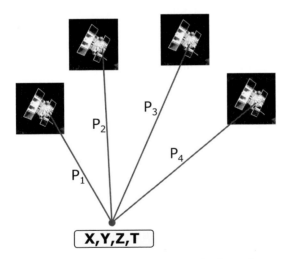

Fig. 9.3. Point positioning using GPS satellites

used where accurate results are desired.

Four pseudo-range equations are formed from (9.1) and expressed algebraically as

$$
\begin{bmatrix}
(x_1 - a_0)^2 + (x_2 - b_0)^2 + (x_3 - c_0)^2 - (x_4 - d_0)^2 = 0 \\
(x_1 - a_1)^2 + (x_2 - b_1)^2 + (x_3 - c_1)^2 - (x_4 - d_1)^2 = 0 \\
(x_1 - a_2)^2 + (x_2 - b_2)^2 + (x_3 - c_2)^2 - (x_4 - d_2)^2 = 0 \\
(x_1 - a_3)^2 + (x_2 - b_3)^2 + (x_3 - c_3)^2 - (x_4 - d_3)^2 = 0 \\
\qquad where \quad x_1, x_2, x_3, x_4 \in \\
(a_0, b_0, c_0) = (x^0, y^0, z^0) \sim P^0 \\
(a_1, b_1, c_1) = (x^1, y^1, z^1) \sim P^1 \\
(a_2, b_2, c_2) = (x^2, y^2, z^2) \sim P^2 \\
(a_3, b_3, c_3) = (x^3, y^3, z^3) \sim P^3.
\end{bmatrix}
\tag{9.2}
$$

In (9.2), $\{P^0, P^1, P^2, P^3\}$ are the positions of the four GPS satellites whose signals are tracked by the receiver at an unknown station P_0. The satellites' positions are given by the coordinates $\{x^i, y^i, z^i | i = 0, 1, 2, 3\}$, where i indicating a particular satellite number. The measured pseudo-ranges to these satellites from a stationary receiver at P_0 are given by $\{d_0, d_1, d_2, d_3\}$. The parameters $\{a_0, b_0, c_0\}$, $\{a_1, b_1, c_1\}$, $\{a_2, b_2, c_2\}$, $\{a_3, b_3, c_3\}$, $\{d_0, d_1, d_2, d_3\}$ are known elements of the spherical cone that intersect at P_0 to give the unknown coordinates $\{x_1, x_2, x_3\}$ of the receiver and the stationary receiver range bias x_4. Several procedures have been put forward to obtain exact solution of (9.2), e.g., [47, 150, 198, 199, 221, 296]. In what follows, we present alternative solutions to (9.2) based on algebraic approaches of Groebner bases and polynomial resultants discussed in Chaps. 4 and 5 respectively. Equation (9.2) is expanded and arranged in the lexicographic order $\{x_1 > x_2 > x_3 > x_4\}$ as

$$
\begin{bmatrix}
x_1^2 - 2a_0 x_1 + x_2^2 - 2b_0 x_2 + x_3^2 - 2c_0 x_3 - x_4^2 + 2d_0 x_4 + a_0^2 + b_0^2 + c_0^2 - d_0^2 = 0 \\
x_1^2 - 2a_1 x_1 + x_2^2 - 2b_1 x_2 + x_3^2 - 2c_1 x_3 - x_4^2 + 2d_1 x_4 + a_1^2 + b_1^2 + c_1^2 - d_1^2 = 0 \\
x_1^2 - 2a_2 x_1 + x_2^2 - 2b_2 x_2 + x_3^2 - 2c_2 x_3 - x_4^2 + 2d_2 x_4 + a_2^2 + b_2^2 + c_2^2 - d_2^2 = 0 \\
x_1^2 - 2a_3 x_1 + x_2^2 - 2b_3 x_2 + x_3^2 - 2c_3 x_3 - x_4^2 + 2d_3 x_4 + a_3^2 + b_3^2 + c_3^2 - d_3^2 = 0,
\end{bmatrix}
\tag{9.3}
$$

where the unknown variables to be determined are $\{x_1, x_2, x_3, x_4\}$. The other terms are known constants. Equation (9.3) is written with the linear terms on the right-hand-side and the nonlinear terms on the left-hand-side as

$$
\begin{bmatrix}
x_1^2 + x_2^2 + x_3^2 - x_4^2 = 2a_0 x_1 + 2b_0 x_2 + 2c_0 x_3 - 2d_0 x_4 + d_0^2 - a_0^2 - b_0^2 - c_0^2 \\
x_1^2 + x_2^2 + x_3^2 - x_4^2 = 2a_1 x_1 + 2b_1 x_2 + 2c_1 x_3 - 2d_1 x_4 + d_1^2 - a_1^2 - b_1^2 - c_1^2 \\
x_1^2 + x_2^2 + x_3^2 - x_4^2 = 2a_2 x_1 + 2b_2 x_2 + 2c_2 x_3 - 2d_2 x_4 + d_2^2 - a_2^2 - b_2^2 - c_2^2 \\
x_1^2 + x_2^2 + x_3^2 - x_4^2 = 2a_3 x_1 + 2b_3 x_2 + 2c_3 x_3 - 2d_3 x_4 + d_3^2 - a_3^2 - b_3^2 - c_3^2.
\end{bmatrix}
\tag{9.4}
$$

Subtracting the last expression (9.4iv) from the first three expressions (9.4i), (9.4ii), and (9.4iii) leads to

$$\left[\begin{array}{l} a_{03}x_1 + b_{03}x_2 + c_{03}x_3 + d_{30}x_4 + e_{03} = 0 \\ a_{13}x_1 + b_{13}x_2 + c_{13}x_3 + d_{31}x_4 + e_{13} = 0 \\ a_{23}x_1 + b_{23}x_2 + c_3x_3 + d_{32}x_4 + e_{23} = 0, \end{array}\right. \qquad (9.5)$$

where

$$\left[\begin{array}{l} a_{03} = 2(a_0 - a_3),\ b_{03} = 2(b_0 - b_3),\ c_{03} = 2(c_0 - c_3),\ d_{30} = 2(d_3 - d_0), \\ a_{13} = 2(a_1 - a_3),\ b_{13} = 2(b_1 - b_3),\ c_{13} = 2(c_1 - c_3),\ d_{31} = 2(d_3 - d_1), \\ a_{23} = 2(a_2 - a_3),\ b_{23} = 2(b_2 - b_3),\ c_{23} = 2(c_2 - c_3),\ d_{32} = 2(d_3 - d_2), \\ e_{03} = (d_0^2 - a_0^2 - b_0^2 - c_0^2) - (d_3^2 - a_3^2 - b_3^2 - c_3^2), \\ e_{13} = (d_1^2 - a_1^2 - b_1^2 - c_1^2) - (d_3^2 - a_3^2 - b_3^2 - c_3^2), \\ e_{23} = (d_2^2 - a_2^2 - b_2^2 - c_2^2) - (d_3^2 - a_3^2 - b_3^2 - c_3^2). \end{array}\right.$$

We note immediately that (9.5) comprises three equations which are linear with four unknowns leading to an underdetermined system of equations. This is circumvented by treating one variable, say x_4, as a constant thereby leading to a system of three equations in three unknowns. We then apply either Groebner basis or polynomial resultants techniques to solve the linear system of equation for $x_1 = g(x_4), x_2 = g(x_4), x_3 = g(x_4)$, where $g(x_4)$ is a linear function.

Sturmfels' Approach

The Sturmfels' [305] approach discussed in Sect. 5.3.2 is applied to solve (9.5). Depending on which variable one wants, (9.5) is rewritten such that this particular variable is hidden (i.e., is treated as a constant). If our interest is to solve $x_1 = g(x_4)$ for instance, (9.5) is first homogenized using x_5 (see Definition 5.1 on p. 48) and then written by hiding x_1 as

$$\left[\begin{array}{l} f_1 := (a_{03}x_1 + d_{30}x_4 + e_{03})x_5 + b_{03}x_2 + c_{03}x_3 \\ f_2 := (a_{13}x_1 + d_{31}x_4 + e_{13})x_5 + b_{13}x_2 + c_{13}x_3 \\ f_3 := (a_{23}x_1 + d_{32}x_4 + e_{13})x_5 + b_{23}x_2 + c_{23}x_3. \end{array}\right. \qquad (9.6)$$

The Jacobian determinant of (9.6) then becomes

$$J_{x_1} = \det \begin{bmatrix} \dfrac{\partial f_1}{\partial x_2} & \dfrac{\partial f_1}{\partial x_3} & \dfrac{\partial f_1}{\partial x_5} \\[2ex] \dfrac{\partial f_2}{\partial x_2} & \dfrac{\partial f_2}{\partial x_3} & \dfrac{\partial f_2}{\partial x_5} \\[2ex] \dfrac{\partial f_3}{\partial x_2} & \dfrac{\partial f_3}{\partial x_3} & \dfrac{\partial f_3}{\partial x_5} \end{bmatrix} = \det \begin{bmatrix} b_{03} & c_{03} & (a_{03}x_1 + d_{30}x_4 + e_{03}) \\ b_{13} & c_{13} & (a_{13}x_1 + d_{31}x_4 + e_{13}) \\ b_{23} & c_{23} & (a_{23}x_1 + d_{32}x_4 + e_{23}) \end{bmatrix}.$$

$$(9.7)$$

The determinant obtained in (9.7) gives the expression for $x_1 = g(x_4)$ as

$$x_1 = -(e_{03}b_{13}c_{23} + d_{32}x_4 b_{03}c_{13} + d_{30}x_4 b_{13}c_{23} - d_{30}x_4 c_{13}b_{23} - \\ d_{31}x_4 b_{03}c_{23} - e_{03}c_{13}b_{23} - e_{13}b_{03}c_{23} + e_{13}c_{03}b_{23} + e_{23}b_{03}c_{13} + \\ d_{31}x_4 c_{03}b_{23} - d_{32}x_4 c_{03}b_{13} - e_{23}c_{03}b_{13})/(a_{23}c_{13}b_{03} + a_{13}b_{23}c_{03} - \\ a_{13}c_{23}b_{03} - a_{23}b_{13}c_{03} - a_{03}c_{13}b_{23} + a_{03}c_{23}b_{13}).$$

For $x_2 = g(x_4)$, we have

$$\begin{bmatrix} f_4 := (b_{03}x_2 + d_{30}x_4 + e_{03})x_5 + a_{03}x_1 + c_{03}x_3 \\ f_5 := (b_{13}x_2 + d_{31}x_4 + e_{13})x_5 + a_{13}x_1 + c_{13}x_3 \\ f_6 := (b_{23}x_2 + d_{32}x_4 + e_{23})x_5 + a_{23}x_1 + c_{23}x_3, \end{bmatrix} \qquad (9.8)$$

whose Jacobian determinant is given by

$$J_{x_2} = \det \begin{bmatrix} \dfrac{\partial f_4}{\partial x_1} & \dfrac{\partial f_4}{\partial x_3} & \dfrac{\partial f_4}{\partial x_5} \\[2ex] \dfrac{\partial f_5}{\partial x_1} & \dfrac{\partial f_5}{\partial x_3} & \dfrac{\partial f_5}{\partial x_5} \\[2ex] \dfrac{\partial f_6}{\partial x_1} & \dfrac{\partial f_6}{\partial x_3} & \dfrac{\partial f_6}{\partial x_5} \end{bmatrix} = \det \begin{bmatrix} a_{03} & c_{03} & (b_{03}x_2 + d_{30}x_4 + e_{03}) \\ a_{13} & c_{13} & (b_{13}x_2 + d_{31}x_4 + e_{13}) \\ a_{23} & c_{23} & (b_{23}x_2 + d_{32}x_4 + e_{23}) \end{bmatrix}.$$

$$(9.9)$$

The determinant obtained in (9.9) gives the expression for $x_2 = g(x_4)$ as

$$x_2 = -(a_{23}c_{13}d_{30}x_4 + a_{03}c_{23}d_{31}x_4 + a_{03}c_{23}e_{13} - a_{23}c_{03}d_{31}x_4 - \\ a_{03}c_{13}d_{32}x_4 - a_{03}c_{13}e_{23} + a_{13}c_{03}d_{32}x_4 - a_{13}c_{23}d_{30}x_4 - \\ a_{13}c_{23}e_{03} - a_{23}c_{03}e_{13} + a_{23}c_{13}e_{03} + a_{13}c_{03}e_{23})/(a_{23}c_{13}b_{03} + \\ a_{13}b_{23}c_{03} - a_{13}c_{23}b_{03} - a_{23}b_{13}c_{03} - a_{03}c_{13}b_{23} + a_{03}c_{23}b_{13}).$$

Finally $x_3 = g(x_4)$ leads to

$$\begin{bmatrix} f_7 := (c_{03}x_3 + d_{30}x_4 + e_{03})x_5 + a_{03}x_1 + b_{03}x_2 \\ f_8 := (c_{13}x_3 + d_{31}x_4 + e_{13})x_5 + a_{13}x_1 + b_{13}x_2 \\ f_9 := (c_{23}x_3 + d_{32}x_4 + e_{23})x_5 + a_{23}x_1 + b_{23}x_2, \end{bmatrix} \quad (9.10)$$

whose Jacobian determinant is given by

$$J_{x_3} = det \begin{bmatrix} \dfrac{\partial f_7}{\partial x_1} & \dfrac{\partial f_7}{\partial x_2} & \dfrac{\partial f_7}{\partial x_5} \\[2mm] \dfrac{\partial f_8}{\partial x_1} & \dfrac{\partial f_8}{\partial x_2} & \dfrac{\partial f_8}{\partial x_5} \\[2mm] \dfrac{\partial f_9}{\partial x_1} & \dfrac{\partial f_9}{\partial x_2} & \dfrac{\partial f_9}{\partial x_5} \end{bmatrix} = det \begin{bmatrix} a_{03} & b_{03} & (c_{03}x_3 + d_{30}x_4 + e_{03}) \\ a_{13} & b_{13} & (c_{13}x_3 + d_{31}x_4 + e_{13}) \\ a_{23} & b_{23} & (c_{23}x_3 + d_{32}x_4 + e_{23}) \end{bmatrix}.$$

$$(9.11)$$

The determinant obtained in (9.7) gives the expression for $x_3 = g(x_4)$ as

$$x_3 = -(a_{23}b_{03}d_{31}x_4 + a_{03}b_{13}d_{32}x_4 + a_{03}b_{13}e_{23} - a_{23}b_{13}d_{30}x_4 - a_{03}b_{23}d_{31}x_4 - a_{03}b_{23}e_{13} + a_{13}b_{23}d_{30}x_4 - a_{13}b_{03}d_{32}x_4 - a_{13}b_{03}e_{23} - a_{23}b_{13}e_{03} + a_{23}b_{03}e_{13} + a_{13}b_{23}e_{03})/(a_{23}b_{03}c_{13} + a_{13}b_{23}c_{03} - a_{13}b_{03}c_{23} - a_{23}b_{13}c_{03} - a_{03}b_{23}c_{13} + a_{03}b_{13}c_{23}).$$

On substituting the obtained expressions of $x_1 = g(x_4), x_2 = g(x_4)$ and $x_3 = g(x_4)$ in (9.3i), we obtain a quadratic function in x_4. The structure of the quadratic equation is given in [11, Box 3-12, p. 54].

Groebner Basis Approach

Using (4.36) on p. 43, the Groebner basis of (9.5) is computed as

$$Groebner\,Basis \begin{bmatrix} \{a_{03}x_1 + b_{03}x_2 + c_{03}x_3 + d_{30}x_4 + e_{03}, \\ a_{13}x_1 + b_{13}x_2 + c_{13}x_3 + d_{31}x_4 + e_{13}, \\ a_{23}x_1 + b_{23}x_2 + c_3x_3 + d_{32}x_4 + e_{23}\}, \{x_1, x_2, x_3, x_4\} \end{bmatrix},$$

$$(9.12)$$

leading to Solution 9.1.

Solution 9.1. [Computed Groebner basis for GPS pseudo-ranging equations.]

$g_1 := (-a_{23})b_{13}e_{03} + a_{13}b_{23}e_{03} + a_{23}b_{03}e_{13} - a_{03}b_{23}e_{13} - a_{13}b_{03}e_{23} + a_{03}b_{13}e_{23} - a_{23}b_{13}c_{03}x_3 + a_{13}b_{23}c_{03}x_3 + a_{23}b_{03}c_{13}x_3 - a_{03}b_{23}c_{13}x_3 - a_{13}b_{03}c_{23}x_3 + a_{03}b_{13}c_{23}x_3 - a_{23}b_{13}d_{30}x_4 + a_{13}b_{23}d_{30}x_4 + a_{23}b_{03}d_{31}x_4 - $

$a_{03}b_{23}d_{31}x_4 - a_{13}b_{03}d_{32}x_4 + a_{03}b_{13}d_{32}x_4.$

$g_2 := (-a_{23})e_{13} + a_{13}e_{23} - a_{23}b_{13}x_2 + a_{13}b_{23}x_2 - a_{23}c_{13}x_3 + a_{13}c_{23}x_3 - a_{23}d_{31}x_4 + a_{13}d_{32}x_4.$

$g_3 := (-a_{23})e_{03} + a_{03}e_{23} - a_{23}b_{03}x_2 + a_{03}b_{23}x_2 - a_{23}c_{03}x_3 + a_{03}c_{23}x_3 - a_{23}d_{30}x_4 + a_{03}d_{32}x_4.$

$g_4 := (-a_{13})e_{03} + a_{03}e_{13} - a_{13}b_{03}x_2 + a_{03}b_{13}x_2 - a_{13}c_{03}x_3 + a_{03}c_{13}x_3 - a_{13}d_{30}x_4 + a_{03}d_{31}x_4.$

$g_5 := e_{23} + a_{23}x_1 + b_{23}x_2 + c_{23}x_3 + d_{32}x_4.$

$g_6 := e_{13} + a_{13}x_1 + b_{13}x_2 + c_{13}x_3 + d_{31}x_4.$

$g_7 := e_{03} + a_{03}x_1 + b_{03}x_2 + c_{03}x_3 + d_{30}x_4.$

From Solution 9.1, one notes that g_1 is a polynomial in the variables x_3 and x_4. With g_1 expressed as $x_3 = g(x_4)$, it is substituted in g_2 to obtain $x_2 = g(x_4)$, which together with $x_3 = g(x_4)$ are substituted in g_5 to give $x_1 = g(x_4)$. On substituting the obtained expressions of $x_1 = g(x_4), x_2 = g(x_4)$ and $x_3 = g(x_4)$ in (9.3i), a quadratic equation in x_4 (i.e., $h_2x_4^2 + h_1x_4 + h_0 = 0$) is obtained. The coefficients are as given in [11, Box 3-14, p. 55]. The desired variables $x_1 = g(x_4)$, $x_2 = g(x_4)$ and $x_3 = g(x_4)$ could also be obtained directly using the reduced Groebner basis (4.38) on p. 44. If one desired $x_3 = g(x_4)$ for example, (9.12) could be formulated as

$$Groebner\,Basis \begin{bmatrix} \{a_{03}x_1 + b_{03}x_2 + c_{03}x_3 + d_{30}x_4 + e_{03}, \\ a_{13}x_1 + b_{13}x_2 + c_{13}x_3 + d_{31}x_4 + e_{13}, \\ a_{23}x_1 + b_{23}x_2 + c_3x_3 + d_{32}x_4 + e_{23}\}, \{x_1, x_2, x_3, x_4\}, \\ \{x_1, x_2, x_4\} \end{bmatrix},$$

(9.13)

giving only the value of g_1 in Solution 9.1. This is repeated for $x_1 = g(x_4)$ and $x_2 = g(x_4)$. The algorithms for solving the unknown value x_4 of the receiver range bias from the quadratic equation $\{h_2x_4^2 + h_1x_4 + h_0 = 0\}$ and the respective stationary receiver coordinates are;

- Awange-Grafarend Groebner basis algorithm and,
- Awange-Grafarend Multipolynomial resultants algorithm.

They can be accessed in the GPS toolbox[2] and are discussed in detail in [21]. The distinction between the polynomial resultants method and the approach proposed by [150] is that the former does not have to

---

[2]http://www.ngs.noaa.gov/gps-toolbox/awange.htm

invert the coefficient matrix. It instead uses the *necessary* and *sufficient* conditions requiring the determinant to vanish if the four equations have a nontrivial solution. With the coefficients h_1, h_2 and h_3, the value of x_4 could also be solved from (3.8) or (3.9) on p. 25. Let us consider the example in [150, 198].

Example 9.1 (Ranging to four satellites). From the coordinates of four GPS satellites given in Table 9.1, we apply the Awange-Grafarend algorithms listed above to compute coordinates of a stationary GPS receiver and the receiver range bias term. The computed coefficients using either

Table 9.1. Geocentric coordinates of four GPS satellites and the pseudo-range observations

i	$x^i = a_i$	$y^i = b_i$	$z^i = c_i$	d_i
0	1.483230866e+7	-2.046671589e+7	-7.42863475e+6	2.4310764064e+7
1	-1.579985405e+7	-1.330112917e+7	1.713383824e+7	2.2914600784e+7
2	1.98481891e+6	-1.186767296e+7	2.371692013e+7	2.0628809405e+7
3	-1.248027319e+7	-2.338256053e+7	3.27847268e+6	2.3422377972e+7

of the algorithms are:

$$\begin{bmatrix} h_2 = -9.104704113943708e - 1 \\ h_1 = 5.233385578536521e7 \\ h_0 = -5.233405293375e9. \end{bmatrix}$$

Once these coefficients have been computed, the algorithms proceed to solve the roots x_4 of the quadratic equation $\{h_2 x_4^2 + h_1 x_4 + h_0 = 0\}$ giving the stationary receiver range bias term. The admissible value of the stationary receiver range bias term is then substituted in the expressions $x_1 = g(x_4), x_2 = g(x_4), x_3 = g(x_4)$ in Solution 9.1 to give the values of stationary receiver coordinates $\{x_1 = X, x_2 = Y, x_3 = Z\}$ respectively. With $x_4^- =$-57479918.164 m or $x_4^+ =$-100.0006 m, the complete pair of solutions with units in meters are

$$\left(\begin{array}{l} X = -2892123.412, Y = 7568784.349, Z = -7209505.102 \, \big| x_4^- \\ \text{or} \quad X = 1111590.460, Y = -4348258.631, Z = 4527351.820 \, \big| x_4^+ \end{array} \right)$$

The results indicate that the solution space is non unique. In-order to decide on the admissible solution from the pair above, we compute the *norm* (radial distance from the center of the Earth) of the positional vector $\{X, Y, Z\} \, \big| x_4^-$ and $\{X, Y, Z\} \, \big| x_4^+$ using

$$norm = \sqrt{(X^2 + Y^2 + Z^2)}.$$

If the receiver coordinates are in the global reference frame (see Sect. 7.2), the norm of the positional vector of the receiver station will approximate the value of the Earth's radius. The norm of the other solution pair will be in space. The computed norms are

$$\begin{bmatrix} \{X,Y,Z\} \, | x_4^- = 10845636.826\,m \\ \{X,Y,Z\} \, | x_4^+ = 6374943.214\,m, \end{bmatrix}$$

thus clearly giving the second solution $\{X, Y, Z\} \, | x_4^+$ as the admissible solution of the receiver position.

9.2.2 Ranging to more than Four GPS Satellites

In Sect. 9.2.1, we have looked at the case where ranging can be performed to only four satellites (minimum case). In this section, we will extend the concept to the case where more than four GPS satellites are in view as is usually the case in practice. Using Gauss-Jacobi combinatorial approach, it is demonstrated how one can obtain the stationary receiver position and range bias without reverting to iterative and linearization procedures such as Newton's or least squares approach.

The common features with the non-algebraic approaches in solving nonlinear problems are that they all have to do with some starting values, linearization of the observation equations and iterations as we have pointed out before. Although the issue of approximate starting values has been addressed in the works of [345, 346], the algebraic approach of Gauss-Jacobi combinatorial enjoys the advantage that all the requirements of non-algebraic approaches listed above are immaterial. The nonlinear problem is solved in an exact form with linearization permitted only during the formation of the variance-covariance matrix to generate the weight matrix of the pseudo-observations (see also [22]). The fact to note is that one has to be able to solve in a closed (exact) form nonlinear systems of equations, a condition already presented in Sect. 9.2.

Let us consider next the example of [302]. The algorithm is used to solve without linearization or iteration the overdetermined pseudo-range problem. The results are then compared to those of linearized least squares solutions.

Example 9.2 (Ranging to more than four satellites). Pseudo-ranges d_i are measured to six satellites whose coordinates $\{x^i, y^i, z^i\}$ are given in Table 9.2. From the data in Table 9.2 and using (6.26) on p. 69, 15 possible combinations listed in Table 9.3 are obtained. The Position Dilution of Precision (PDOP) are computed as suggested in [186] and presented in Table 9.3. From the computed PDOP, it is noticed that the 10th combination had a poor geometry, a fact validated by the plot of the PDOP values versus the combination numbers in Fig. 9.4. Using Gauss-Jacobi combinatorial algorithm, this weaker geometry is accounted for during the adjustment process. Variance-covariance matrix computed through nonlinear error propagation for that respective set is used. Groebner basis or polynomial resultants are used as computing engine (see Fig. 6.4 on p. 75) to compute the minimal combinatorial set as discussed in Sect. 6.5. The computed coefficients are presented in Table 9.4.

From the computed coefficients in Table 9.4, the 10th combination is once again identified as having significantly different values from the rest. This fact highlights the power of the Gauss-Jacobi combinatorial algorithm in identifying poor geometry. Using the coefficients of Table 9.4, the solution of receiver position $\{X, Y, Z\}$ and the range bias $\{cdt\}$ for each minimal combinatorial set is carried out as discussed in Sect. 9.2. The results are presented in Table 9.5. The final adjusted position is obtained using linear Gauss-Markov model (6.10) on p. 62. The random pseudo-observation values of Table 9.5 are placed in the vector of observation $\mathbf{y}$ and the dispersion matrix $\mathbf{\Sigma}$ obtained by nonlinear error propagation using (6.31) on p. 71. The coefficients of the unknowns $\{X, Y, Z, cdt\}$ form the design matrix $\mathbf{A}$. The dispersion of the estimated parameters are then obtained from (6.11).

Figure 9.8 gives the plot of the scatter of the 15 combinatorial solutions (shown by points) around the adjusted value (indicated by a star). Figure 9.9 is a magnification of Fig. 9.8 for the scatter of 14 solutions (shown by points) that are very close to the adjusted value (indicated by a star). The outlying point in Fig. 9.8 is ignored.

Table 9.2. Geocentric coordinates of six GPS satellites and pseudo-range observations

PRN	$x^i = a_i$	$y^i = b_i$	$z^i = c_i$	d_i
23	14177553.47	-18814768.09	12243866.38	21119278.32
9	15097199.81	-4636088.67	21326706.55	22527064.18
5	23460342.33	-9433518.58	8174941.25	23674159.88
1	-8206488.95	-18217989.14	17605231.99	20951647.38
21	1399988.07	-17563734.90	19705591.18	20155401.42
17	6995655.48	-23537808.26	-9927906.48	24222110.91

Table 9.3. Possible combinations and the computed PDOP

Combination Number	Combination	Computed PDOP
1	23-9-5-1	4.8
2	23-9-5-21	8.6
3	23-9-5-17	4.0
4	23-9-1-21	6.5
5	23-9-1-17	3.3
6	23-9-21-17	3.6
7	23-5-1-21	6.6
8	23-5-1-17	6.6
9	23-5-21-17	4.8
10	23-1-21-17	137.8
11	9-5-1-21	5.6
12	9-5-1-17	14.0
13	9-5-21-17	6.6
14	9-1-21-17	5.2
15	5-1-21-17	6.6

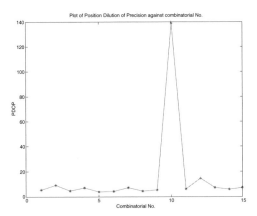

Fig. 9.4. A plot of PDOP for respective combinations

Table 9.4. Computed coefficients of the combinations

C/No.	c_2	c_1	c_0
1	-0.914220949236445	52374122.9848733	49022682.3125
2	-0.934176403102736	50396827.4998945	7915541824.84375
3	-0.921130625833683	51741826.0147786	343282824.25
4	-0.865060899130107	54950460.2842167	-10201105114.5
5	-0.922335616484969	51877166.0451888	280298481.625
6	-0.919296962706157	51562232.9601199	1354267366.4375
7	-0.894980063579044	53302005.6927825	-3642644147.5625
8	-0.917233949644576	52194946.1124139	132408747.46875
9	-0.925853049262193	51140847.6331213	3726719112.1875
10	3369.83293928593	-1792713339.80277	6251615074927.06
11	-0.877892756651551	54023883.5656926	-6514735288.13762
12	-0.942581538318523	50793361.5303674	784684294.241371
13	-0.908215141659006	52246642.0794924	-2499054749.05572
14	-0.883364070549387	53566554.3869961	-5481411035.37882
15	-0.866750765656126	54380648.2092251	-7320871488.80859

Table 9.5. Computed combinatorial solution points in a polyhedron

C/No.	$X(m)$	$Y(m)$	$Z(m)$	$cdt(m)$
1	596925.3485	-4847817.3618	4088206.7822	-0.9360
2	596790.3124	-4847765.7637	4088115.7092	-157.0638
3	596920.4198	-4847815.4785	4088203.4581	-6.6345
4	596972.8261	-4847933.4365	4088412.0909	185.6424
5	596924.2118	-4847814.5827	4088201.8667	-5.4031
6	596859.9715	-4847829.7585	4088228.8277	-26.2647
7	596973.5779	-4847762.4719	4088399.8670	68.3398
8	596924.2341	-4847818.6302	4088202.3205	-2.5368
9	596858.7650	-4847764.5341	4088221.8468	-72.8716
10	596951.5275	-4852779.5675	4088758.6420	3510.4002
11	597004.7562	-4847965.2225	4088300.6135	120.5901
12	596915.8657	-4847799.7045	4088195.5770	-15.4486
13	596948.5619	-4847912.9549	4088252.1599	47.8319
14	597013.7194	-4847974.1452	4088269.3206	102.3292
15	597013.1300	-4848019.6766	4088273.9565	134.6230

9.2.3 Least Squares versus Gauss-Jacobi Combinatorial

Let us now compare the least squares solution and the Gauss-Jacobi combinatorial approach. Using the combinatorial approach, the stationary receiver position and range bias are computed as discussed in Sect. 9.2.2. For the least squares approach, the nonlinear observation equations (9.2) are first linearized using Taylor series expansion for the

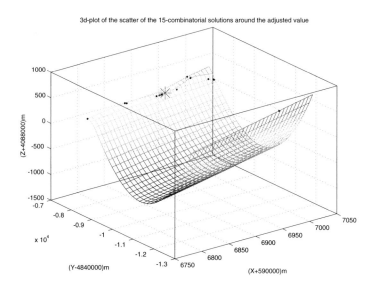

Fig. 9.5. Scatter of the *15* combinatorial solutions (•) around the adjusted value (⋆)

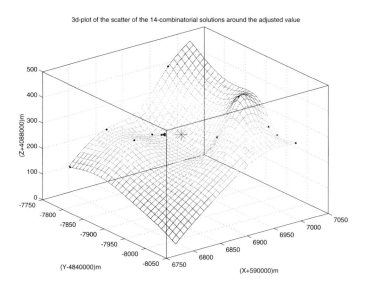

Fig. 9.6. Magnification of the scatter of *14* solutions (•) around the adjusted value (⋆) of Fig. 9.8

6 satellites in Table 9.2. This linearization process generates the Jacobi matrix required by the approach. After linearization, the desired values are estimated iteratively using linear models. As approximate starting values for the iterations, let us assign the stationary receiver position and the stationary receiver range bias zero values. Let us also set a convergence limit of 1×10^{-8}, as the difference between values of two successive iterations. With these settings, 6 iterations are required for the threshold condition above to be achieved. In the second case, the values of the combinatorial algorithm are used as approximate starting values for least squares solution. This time round, only two iterations were required to achieve convergence. For users who prefer least squares approach, Gauss-Jacobi combinatorial algorithm can therefore be used to offer quick approximate starting values that lead to faster convergence.

From (9.2) on p. 110, and the results of both procedures, residuals are computed, squared and used to compute the error norm from

$$
norm = \sqrt{\left\{ \sum_{i=1}^{6} \left(d_i - [\sqrt{(\hat{X} - a_i)^2 + (\hat{Y} - b_i)^2 + (\hat{Z} - c_i)^2} - \hat{x}_4] \right)^2 \right\}}.
$$

$$(9.14)$$

In (9.14), $\{\hat{X}, \hat{Y}, \hat{Z}, \hat{x}_4\}$ are the computed values of the stationary receiver position and range bias. The entities $\{a_i, b_i, c_i\} \mid \forall i = \{1, \ldots, 6\}$ are the coordinates of the six satellites in Table 9.2 and $\{d_i\} \mid \forall i = \{1, \ldots, 6\}$ the measured pseudo-ranges.

Table 9.6 compares the results from the Gauss-Jacobi combinatorial algorithm and those obtained from least squares approach. Table 9.7 presents the *root-mean-square-errors*. In Table 9.8, we present the computed residuals, their sum of squares and the computed error norm from (9.14). The computed error norm are identical for both procedures. Further comparison of the two procedures will be given in Chap. 15 where they are used to compute the 7-parameter datum transformation problem.

Table 9.6. Computed stationary receiver position and range bias

	X (m)	Y (m)	Z (m)	$cdt(m)$
Combinatorial approach	596929.6542	-4847851.5021	4088226.7858	-15.5098
Least squares	596929.6535	-4847851.5526	4088226.7957	-15.5181
Difference	0.0007	0.0505	-0.0098	0.0083

Table 9.7. Computed root-mean-square errors

	σ_X (m)	σ_Y (m)	σ_Z (m)	σ_{cdt} (m)
Combinatorial Approach	6.4968	11.0141	5.4789	8.8071
Least Squares	34.3769	58.2787	28.9909	46.6018

Table 9.8. Computed residuals, squares of residuals and error norm

PRN	Combinatorial approach (m)	Least squares (m)
23	-16.6260	-16.6545
9	-1.3122	-1.3106
5	2.2215	2.2189
1	-16.4369	-16.4675
21	26.8623	26.8311
17	5.4074	5.3825
Sum of squares	1304.0713	1304.0680
Error norm	36.1119	36.1119

9.3 Ranging by Local Positioning Systems (LPS)

As opposed to GPS ranging where the targets being observed are satellites in space and in motion, Local Positioning Systems' targets are fixed on the surface of the Earth as illustrated in Fig. 9.1 on p. 107. We present both planar and three-dimensional ranging within the LPS system. Planar ranging can be used for quick point search during engineering and cadastral surveying.

9.3.1 Planar Ranging

Conventional Approach

Consider two distances $\{S_1, S_2\}$ measured from an unknown station $P_0 \in \mathbb{E}^2$ to two known stations $P_1 \in \mathbb{E}^2$ and $P_2 \in \mathbb{E}^2$ as shown in Fig. 9.7. The two dimensional distance ranging problem involves the determination of the planar coordinates $\{X_0, Y_0\}_{P_0}$ of the unknown station $P_0 \in \mathbb{E}^2$ given;

- the observed distances $\{S_1, S_2\}$,
- the planar coordinates $\{X_1, Y_1\}_{P_1}$ of station $P_1 \in \mathbb{E}^2$ and $\{X_2, Y_2\}_{P_2}$ of stations $P_2 \in \mathbb{E}^2$.

The nonlinear distance equations relating the given values above with the coordinates of unknown station are expressed (see e.g., (4.1) and (4.2) on p. 30) as

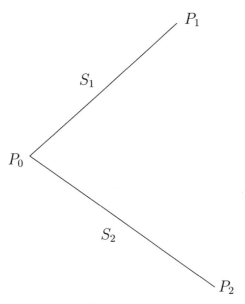

Fig. 9.7. Distance observations

$$\left[\begin{array}{l} (X_1 - X_0)^2 + (Y_1 - Y_0)^2 = S_1^2 \\ (X_2 - X_0)^2 + (Y_2 - Y_0)^2 = S_2^2, \end{array} \right. \tag{9.15}$$

which on expanding leads to

$$\left[\begin{array}{l} X_1^2 + Y_1^2 - 2X_1X_0 - 2Y_1Y_0 + X_0^2 + Y_0^2 = S_1^2 \\ \\ X_2^2 + Y_2^2 - 2X_2X_0 - 2Y_2Y_0 + X_0^2 + Y_0^2 = S_2^2. \end{array} \right. \tag{9.16}$$

The conventional analytic approach solves (9.16) by subtracting the first expression, i.e., (9.16i) from the second one, and expressing one unknown in terms of the other. This leads to

$$Y_0 = -\left\{ \frac{X_1 - X_2}{Y_1 - Y_2} \right\} X_0 + \frac{S_2^2 - S_1^2 + X_1^2 - X_2^2 + Y_1^2 - Y_2^2}{2(Y_1 - Y_2)}, \tag{9.17}$$

which is substituted for Y_0 in the first expression of (9.16) to give

$$\left[\begin{array}{l} X_1^2 + Y_1^2 - 2X_1X_0 - 2Y_1 \left\{ -\left\{ \frac{X_1 - X_2}{Y_1 - Y_2} \right\} X_0 + \frac{S_2^2 - S_1^2 + X_1^2 - X_2^2 + Y_1^2 - Y_2^2}{2(Y_1 - Y_2)} \right\} \\ \\ + X_0^2 + \left\{ -\left\{ \frac{X_1 - X_2}{Y_1 - Y_2} \right\} X_0 + \frac{S_2^2 - S_1^2 + X_1^2 - X_2^2 + Y_1^2 - Y_2^2}{2(Y_1 - Y_2)} \right\}^2 - S_1^2 = 0. \end{array} \right. \tag{9.18}$$

On expanding and factorizing (9.18) leads to

$$\left[\begin{array}{l} (1 + a^2)X_0^2 + (2ab - 2X_1 - 2Y_1 a)X_0 + b^2 - 2Y_1 b + X_1^2 + Y_1^2 - S_1^2 = 0, \\[1em] \qquad\qquad\qquad\qquad with \\[1em] \qquad\qquad a = -\left\{ \dfrac{X_1 - X_2}{Y_1 - Y_2} \right\} \\[1em] \qquad\qquad\qquad\qquad and \\[1em] \qquad\quad b = \dfrac{S_2^2 - S_1^2 + X_1^2 - X_2^2 + Y_1^2 - Y_2^2}{2(Y_1 - Y_2)}. \end{array} \right.$$

$$(9.19)$$

The quadratic equation (9.19) is solved for X_0 using the quadratic formulae (3.8) or (3.9) on p. 25 and substituted back in (9.17) to give the values of Y_0.

Sylvester Resultants Approach

Whereas the conventional analytical approach presented above involves differencing, e.g., (9.17), and substitution as in (9.18), the Sylvester resultants technique discussed in Sect. 5.2 solves (9.16) directly. In-order to achieve this, (9.16) is first expressed in algebraic form as

$$\left[\begin{array}{l} g_1 := X_1^2 + Y_1^2 - 2X_1 X_0 - 2Y_1 Y_0 + X_0^2 + Y_0^2 - S_1^2 = 0 \\[1em] g_2 := X_2^2 + Y_2^2 - 2X_2 X_0 - 2Y_2 Y_0 + X_0^2 + Y_0^2 - S_2^2 = 0. \end{array} \right.$$

$$(9.20)$$

Next, the *hide variable* technique is applied. By hiding the variable Y_0 (i.e., considering it as a constant), the coefficient matrix of the variable X_0 is formed as shown in Example 5.2 on p. 49. In (9.20), we note that the polynomials g_1 and g_2 are both of degree 2 and thus both i and j (e.g., (5.1) on p. 48) are equal to 2 resulting into a (4×4) matrix. The coefficient matrix of the variable X_0 formed by hiding the variable Y_0 (i.e., considering the coefficients of X_0 to be polynomials in Y_0) is

$$\mathbf{A}_X = \begin{bmatrix} 1 & -2X_1 & (Y_0^2 - 2Y_1Y_0 + X_1^2 + Y_1^2 - S_1^2) & 0 \\ 0 & 1 & -2X_1 & (Y_0^2 - 2Y_1Y_0 + X_1^2 + Y_1^2 - S_1^2) \\ 1 & -2X_2 & (Y_0^2 - 2Y_2Y_0 + X_2^2 + Y_2^2 - S_2^2) & 0 \\ 0 & 1 & -2X_2 & (Y_0^2 - 2Y_2Y_0 + X_2^2 + Y_2^2 - S_2^2) \end{bmatrix}, \tag{9.21}$$

while that of the variable Y_0 formed by hiding X_0 (i.e., considering the coefficients of Y_0 to be polynomials in X_0) is

$$\mathbf{A}_Y = \begin{bmatrix} 1 & -2Y_1 & (X_0^2 - 2X_1X_0 + X_1^2 + Y_1^2 - S_1^2) & 0 \\ 0 & 1 & -2Y_1 & (X_0^2 - 2X_1X_0 + X_1^2 + Y_1^2 - S_1^2) \\ 1 & -2Y_2 & (X_0^2 - 2X_2X_0 + X_2^2 + Y_2^2 - S_2^2) & 0 \\ 0 & 1 & -2Y_2 & (X_0^2 - 2X_2X_0 + X_2^2 + Y_2^2 - S_2^2) \end{bmatrix}. \tag{9.22}$$

Sylvester resultants are now obtained from the determinants of the coefficient matrices (9.21) and (9.22) respectively as

$$\begin{bmatrix} Res(g_1, g_2, X_0) = det(\mathbf{A}_X) \\ \\ Res(g_1, g_2, Y_0) = det(\mathbf{A}_Y), \end{bmatrix} \tag{9.23}$$

where $Res(g_1, g_2, X_0)$ and $Res(g_1, g_2, Y_0)$ are the Sylvester resultants of algebraic equations in (9.20), with respect to the variables X_0 and Y_0 as in (9.21) and (9.22) respectively. From (9.23) we obtain two quadratic equations (9.24) for solving the variables X_0 and Y_0 which are the planar coordinates of the unknown station P_0. The coefficients of the quadratic equations are given in Solution 9.2.

Solution 9.2 (Sylvester resultants solution of planar coordinates $\{X_0, Y_0\}$).

$$\left(\begin{array}{l} a_2 Y_0^2 + a_1 Y_0 + a_0 = 0 \\ \\ b_2 X_0^2 + b_1 X_0 + b_0 = 0 \end{array} \right) \tag{9.24}$$

with the coefficients:

$$a_2 = (4Y_2^2 + 4X_1^2 - 8Y_1Y_2 + 4X_2^2 + 4Y_1^2 - 8X_2X_1).$$

$$a_1 = (-4X_2^2Y_1 - 4S_1^2Y_2 - 4X_1^2Y_2 + 8X_1X_2Y_1 + 4Y_1S_1^2 + 4Y_1Y_2^2 + 8X_2X_1Y_2 - 4Y_2^3 + 4Y_1^2Y_2 - 4Y_2X_2^2 - 4Y_1S_2^2 - 4Y_1^3 - 4Y_1X_1^2 + 4Y_2S_2^2).$$

$$a_0 = (X_2^4 + Y_2^4 + S_2^4 - 4X_2X_1Y_2^2 + 4X_2X_1S_2^2 - 4X_1X_2Y_1^2 + 4X_1X_2S_1^2 + 2X_2^2Y_2^2 - 2X_2^2S_2^2 - 2Y_2^2S_2^2 - 4X_1X_2^3 + 6X_2^2X_1^2 + 2X_2^2Y_1^2 - 2X_2^2S_1^2 + 2X_1^2Y_2^2 - 2X_1^2S_2^2 - 2Y_1^2Y_2^2 + 2Y_1^2S_2^2 + 2S_1^2Y_2^2 - 2S_1^2S_2^2 - 4X_2X_1^3 + 2X_1^2Y_1^2 - 2X_1^2S_1^2 - 2Y_1^2S_1^2 + X_1^4 + Y_1^4 + S_1^4).$$

$$b_2 = (-8Y_1Y_2 + 4X_1^2 + 4Y_2^2 - 8X_2X_1 + 4X_2^2 + 4Y_1^2).$$

$$b_1 = (-4X_1^3 + 4X_2S_2^2 + 8Y_2Y_1X_2 - 4X_2Y_2^2 - 4X_1Y_2^2 - 4X_1S_2^2 + 4X_1X_2^2 - 4X_2S_1^2 + 8Y_1Y_2X_1 - 4X_2Y_1^2 + 4X_2X_1^2 - 4X_1Y_1^2 + 4X_1S_1^2 - 4X_2^3).$$

$$b_0 = (4Y_2Y_1S_2^2 - 4Y_2X_2^2Y_1 - 2Y_1^2S_1^2 + X_2^4 + Y_2^4 + S_2^4 + X_1^4 + Y_1^4 + S_1^4 - 4Y_1Y_2^3 + 4Y_1S_1^2Y_2 - 4Y_1X_1^2Y_2 + 2X_2^2Y_2^2 - 2X_2^2S_2^2 - 2Y_2^2S_2^2 - 2X_2^2X_1^2 + 2X_2^2Y_1^2 + 2X_2^2S_1^2 + 2X_1^2Y_2^2 + 2X_1^2S_2^2 + 6Y_1^2Y_2^2 - 2Y_1^2S_2^2 - 2S_1^2Y_2^2 - 2S_1^2S_2^2 + 2X_1^2Y_1^2 - 2X_1^2S_1^2 - 4Y_1^3Y_2).$$

Reduced Groebner Basis Approach

Reduced Groebner basis (4.38) on p. 44 solves (9.20) directly through

$$\begin{bmatrix} GroebnerBasis[\{g_1, g_2\}, \{X_0, Y_0\}, \{X_0\}] \\ GroebnerBasis[\{g_1, g_2\}, \{X_0, Y_0\}, \{Y_0\}]. \end{bmatrix} \qquad (9.25)$$

The first expression of (9.25) ensures that one gets a quadratic equation only in Y_0 with X_0 eliminated, while the second expression ensures a quadratic equation only in X_0 with Y_0 eliminated. Solution 9.3 presents the results of (9.25).

Solution 9.3 (Reduced Groebner basis solution of planar coordinates $\{X_0, Y_0\}$).

$$\left(\begin{array}{c} e_2Y_0^2 + e_1Y_0 + e_0 = 0 \\ \\ f_2X_0^2 + f_1X_0 + f_0 = 0 \end{array} \right) \qquad (9.26)$$

with the coefficients: $e_2 = (4X_1^2 - 8X_1X_2 - 8Y_1Y_2 + 4X_2^2 + 4Y_2^2 + 4Y_1^2).$

$$e_1 = (-4X_1^2Y_1 + 4S_1^2Y_1 - 4Y_1^3 - 4X_2^2Y_2 - 4X_1^2Y_2 + 4Y_1^2Y_2 + 4Y_1Y_2^2 - 4Y_2^3 + 4S_2^2Y_2 - 4S_2^2Y_1 - 4S_1^2Y_2 - 4X_2^2Y_1 + 8X_1X_2Y_1 + 8X_1X_2Y_2).$$

$$e_0 = (S_2^4 + 2X_1^2Y_2^2 + 4S_1^2X_1X_2 + 4S_2^2X_1X_2 - 2S_2^2X_2^2 - 2Y_1^2Y_2^2 + S_1^4 - 2S_1^2X_1^2 + X_2^4 + 2S_1^2Y_2^2 - 2S_1^2S_2^2 + X_1^4 + Y_1^4 + 2X_2^2Y_1^2 - 4X_1X_2Y_2^2 + 6X_1^2X_2^2 - 2S_2^2X_1^2 - 2S_2^2Y_2^2 + 2X_1^2Y_1^2 - 2S_1^2X_2^2 + 2S_2^2Y_1^2 - 2S_1^2Y_1^2 + 2X_2^2Y_2^2 - 4X_1X_2^3 - 4X_1X_2Y_1^2 + Y_2^4 - 4X_1^3X_2).$$

$$f_2 = (4X_1^2 - 8X_1X_2 - 8Y_1Y_2 + 4X_2^2 + 4Y_2^2 + 4Y_1^2).$$

$$f_1 = (-4X_2Y_1^2 - 4X_1Y_1^2 + 4X_1^2X_2 - 4S_2^2X_1 - 4X_2Y_2^2 - 4X_1^3 + 8X_1Y_1Y_2 + 4S_1^2X_1 + 8X_2Y_1Y_2 + 4X_1X_2^2 - 4X_2^3 - 4X_1Y_2^2 + 4S_2^2X_2 - 4S_1^2X_2).$$

$$f_0 = (S_2^4 + 2X_1^2Y_2^2 - 4X_2^2Y_1Y_2 - 2S_2^2X_2^2 + 6Y_1^2Y_2^2 + 4S_1^2Y_1Y_2 + S_1^4 - 4X_2^2Y_1Y_2 - 2S_1^2X_1^2 + \ldots X_2^4 - 2S_1^2Y_2^2 - 2S_1^2S_2^2 + 4S_2^2Y_1Y_2 + X_1^4 + Y_1^4 + 2X_2^2Y_1^2 - 2X_1^2X_2^2 + 2S_2^2X_1^2 - 4Y_1^3Y_2 - 2S_2^2Y_2^2 + 2X_1^2Y_1^2 + 2S_1^2X_2^2 - 2S_2^2Y_1^2 - 2S_1^2Y_1^2 + 2X_2^2Y_2^2 - 4Y_1Y_2^3 + Y_2^4).$$

With the given values of known stations and measured distances as listed on p. 122, all that is required of the practitioner, therefore, is to compute the coefficients $\{a_2, a_1, a_0, b_2, b_1, b_0\}$ using the Sylvester resultants Solution 9.2 or $\{e_2, e_1, e_0, f_2, f_1, f_0\}$ using the reduced Groebner basis Solution 9.3. Once the coefficients have been computed, the Matlab's roots command is applied to solve the univariate polynomials (9.26) or (9.26) for the position of the unknown station. The admissible position from the computed pair of solution is chosen with the help of prior information e.g., from existing maps.

Example 9.3 (Ranging to two known planar stations). Consider the Example of [195, p. 240] where two distances $\{S_1 = 294.330\,m, S_2 = 506.420\,m\}$ have been measured from an unknown station $P_0 \in \mathbb{E}^2$ to two known stations $P_1 \in \mathbb{E}^2$ and $P_2 \in \mathbb{E}^2$ (e.g., Fig. 9.7). The Cartesian planar coordinates of the two known stations P_1 and P_2 are given as $\{X_1 = 328.760\,m, Y_1 = 1207.850\,m\}_{P_1}$ and $\{X_2 = 925.040\,m, Y_2 = 954.330\,m\}_{P_2}$ respectively. The planar ranging problem now involves determining the planar coordinates $\{X_0, Y_0\}_{P_0}$ of the unknown station $P_0 \in \mathbb{E}^2$. Using the given values of known stations and measured distances in either Solution 9.2 or 9.3, the coefficients $\{a_2, a_1, a_0, b_2, b_1, b_0\}$ of the quadratic equation (9.24) or $\{e_2, e_1, e_0, f_2, f_1, f_0\}$ of (9.26) are computed. Using these coefficients and applying Matlab's roots command (see e.g., (4.40) in Example 4.11 on p. 44) leads to

$$X_0 = \{1336.940, \ 927.797\}\, m$$
$$Y_0 = \{593.271, \ 419.316\}\, m$$

In a four step procedure, [195, p. 240] obtained the values $\{X_0(m) = 927.90\}$ and $\{Y_0(m) = 419.42\}$. The algebraic approaches are however direct and fast (i.e., avoids forward and backwards substitutions).

Geometrically, the algebraic curves given by (9.15) would result in a conic intersection of two circles with the centers $\{X_1, Y_1\}$ and $\{X_2, Y_2\}$ and radiuses S_1 and S_2 respectively. The applied polynomial approaches decompose these complicated geometries to those of Figs. 9.8 and 9.9 which represent univariate polynomials and are simpler to solve. Figs. 9.8 and 9.9 indicate the solutions of (9.26) for the Example presented above. The intersection of the quadratic curves with the zero line are the solution points. In Solution 9.4, we present the critical configuration of the planar ranging problem. The computed determinants, (9.32) and (9.33) indicate the critical configuration (where solution ceases to exist) to be cases when points $P_0(X, Y)$, $P_1(X_1, Y_1)$ and $P_2(X_2, Y_2)$ all lie on a straight line with gradient $-\dfrac{c}{b}$ and intercept $-\dfrac{a}{b}$.

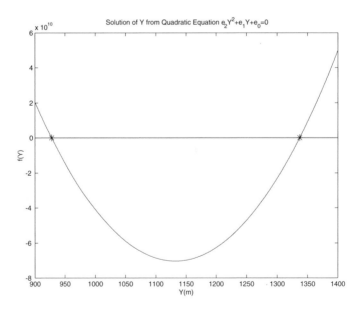

Fig. 9.8. Solution of the Y coordinates

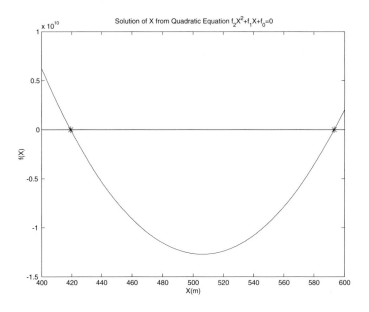

Fig. 9.9. Solution of the X coordinates

Solution 9.4 (Critical configuration of the planar ranging problem).

$$\begin{bmatrix} f_1(X,Y;X_1,Y_1,S_1) = (X_1 - X)^2 + (Y_1 - Y)^2 - S_1^2 \\ f_2(X,Y;X_2,Y_2,S_2) = (X_2 - X)^2 + (Y_2 - Y)^2 - S_2^2 \end{bmatrix} \tag{9.27}$$

$$\begin{bmatrix} \dfrac{\partial f_1}{\partial X} = -2(X_1 - X), \ \dfrac{\partial f_2}{\partial X} = -2(X_2 - X) \\[2mm] \dfrac{\partial f_1}{\partial Y} = -2(Y_1 - Y), \ \dfrac{\partial f_2}{\partial Y} = -2(Y_2 - Y), \end{bmatrix} \tag{9.28}$$

$$\begin{bmatrix} D = \left| \dfrac{\partial f_i}{\partial X_j} \right| = 4 \begin{vmatrix} X_1 - X & X_2 - X \\ Y_1 - Y & Y_2 - Y \end{vmatrix} \\[3mm] D \Leftrightarrow \begin{vmatrix} X_1 - X & X_2 - X \\ Y_1 - Y & Y_2 - Y \end{vmatrix} = \begin{vmatrix} X & Y & 1 \\ X_1 & Y_1 & 1 \\ X_2 & Y_2 & 1 \end{vmatrix} = 0, \end{bmatrix} \tag{9.29}$$

$$\begin{bmatrix} \tfrac{1}{4}D = (X_1 - X)(Y_2 - Y) - (X_2 - X)(Y_1 - Y) \\ = X_1 Y_2 - X_1 Y - X Y_2 + XY - X_2 Y_1 + X_2 Y + X Y_1 - XY \\ = X(Y_1 - Y_2) + Y(X_2 - X_1) + X_1 Y_2 - X_2 Y_1, \end{bmatrix} \tag{9.30}$$

thus

$$\begin{vmatrix} X & Y & 1 \\ X_1 & Y_1 & 1 \\ X_2 & Y_2 & 1 \end{vmatrix} = 2 \times \text{Area of triangle } P(X,Y), P_1(X_1,Y_1), \text{ and } P_2(X_2,Y_2)$$

(9.31)

$$D = \begin{vmatrix} X & Y & 1 \\ X_1 & Y_1 & 1 \\ X_2 & Y_2 & 1 \end{vmatrix} \begin{bmatrix} a \\ b \\ c \end{bmatrix} = 0, \qquad (9.32)$$

results in a system of homogeneous equations

$$\begin{cases} aX + bY + c = 0 \\ aX_1 + bY_1 + c = 0 \\ aX_2 + bY_2 + c = 0. \end{cases} \qquad (9.33)$$

Planar Ranging to more than Two Known Stations

In-order to solve the overdetermined two-dimensional ranging problem, the combinatorial algorithm is applied. In the *first step*, combinatorials are formed using (6.26) on p. 69 and solved in a closed form using either (9.24) or (9.26). In the *second step*, the dispersion matrix $\boldsymbol{\Sigma}$ is obtained from (6.31) on p. 71. Finally the pseudo-observations are adjusted using linear Gauss-Markov model (see e.g., Definition 6.1 on p. 61) in the *third step*, with the unknown parameters estimated via **B**est **L**inear **U**niformly **U**nbiased **E**stimator **BLUUE** (6.10). The dispersion of the estimated parameters are then obtained using (6.11) on p. 62.

Example 9.4 (Planar ranging to more than two known stations). Let us consider the example of [195, pp. 240–241] which is also solved in [29]. In this example, the coordinates of station N are to be determined from distance observations to four stations P_1, P_2, P_3 and P_4 [195, Fig. 6.4.4, p. 229]. In preparation for adjustment, the distances are corrected and reduced geometrically to Gauss-Krueger projection and are as given in Table 9.9. Using Gauss-Jacobi combinatorial algorithm, the coordinates of station N are computed and compared to those of least squares in [195, p. 242]. From (6.26), six combinations in the minimal sense are formed and solved for $\{x, y\}_N$ for position of station N using either (9.24) or (9.26). The combinatorial solutions are presented in Table 9.10.

The adjusted position of the unknown station N is now obtained either by;

Table 9.9. Distance observations to unknown station N

Pt. No.	Easting $x[m]$	Northing $y[m]$	s_i $[m]$
1	48177.62	6531.28	611.023
2	49600.15	7185.19	1529.482
3	49830.93	5670.69	1323.884
4	47863.91	5077.24	1206.524

Table 9.10. Position of station N computed for various combinatorials

Combinatorial No.	combinatorial points	x $[m]$	y $[m]$
1	1-2	48565.2783	6058.9770
2	1-3	48565.2636	6058.9649
3	1-4	48565.2701	6058.9702
4	2-3	48565.2697	6058.9849
5	2-4	48565.3402	6058.9201
6	2-5	48565.2661	6058.9731

(a) simply taking the arithmetic mean of the combinatorial solutions in columns 3 and 4 of Table 9.10 (an approach which does not take into account full information in terms of the variance-covariance matrix) or,

(b) using *special linear Gauss-Markov model* through the estimation by the **Best Linear Uniformly Unbiased Estimator BLUUE** in (6.10). The dispersion of the estimated parameters are subsequently obtained using (6.11).

The results are presented in Table 9.11 and plotted in Fig. 9.10. In Table 9.11, we present the coordinates $\{x, y\}$ of station N obtained using the least squares approach in [195], Gauss-Jacobi combinatorial (**BLUUE**) and the Gauss-Jacobi combinatorial (arithmetic mean) in columns 2 and 3, with their respective standard deviations $\{\sigma_x, \sigma_y\}$ in columns 4 and 5. In columns 6 and 7, the deviations $\{\Delta_x, \Delta_y\}$ of the computed coordinates of station N using Gauss-Jacobi combinatorial from the least squares' values of [195] are presented. The deviations of the exact solutions of each combination (columns 3 and 4 of Table 9.10) from the adjusted values of **Best Linear Uniformly Unbiased Estimator BLUUE** (i.e., second and third columns of Table 9.11) are plotted in Fig. 9.11.

Table 9.11. Position of station N after adjustments

Approach	$x(m)$	$y(m)$	$\sigma_x(m)$	$\sigma_y(m)$	$\Delta_x(m)$	$\Delta_y(m)$
Least Squares	48565.2700	6058.9750	0.006	0.006	-	-
Gauss-Jacobi (BLUUE)	48565.2709	6058.9750	0.0032	0.0034	-0.0009	0.0000
Gauss-Jacobi (Mean)	48565.2813	6058.9650	-	-	-0.01133	0.0100

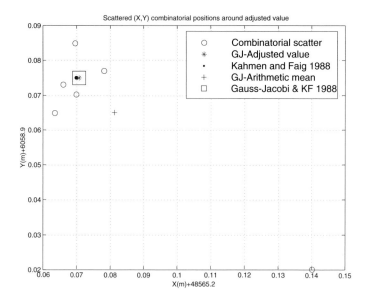

Fig. 9.10. Plot of the position of N from various approaches

From the results in Table 9.11 and Fig. 9.10, we note that when full information of the observations is taken into account via the *nonlinear error/variance-covariance* propagation, the results of Gauss-Jacobi combinatorial algorithm and least squares from [195] are in the same range. In addition to giving the adjusted position, Gauss-Jacobi algorithm can accurately pinpoint a poor combinatorial geometry (e.g., combination 5). This is taken care of through weighting. Fig. 9.10 shows the combinatorial scatter denoted by {∘} and the Gauss-Jacobi combinatorial adjusted value by {∗}. Least squares estimation from [195] is denoted by {●} and the arithmetic mean by {+}. One notes that the estimates from Gauss-Jacobi's BLUUE {∗} and least squares solution almost coincide. In the Figure, both estimates are enclosed by {□} for clarity purpose. Figure 9.11 indicates the deviations of the combinatorial scatter from the BLUUE adjusted position of N. These results

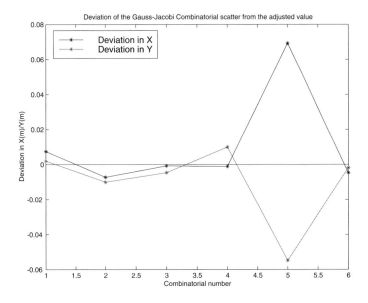

Fig. 9.11. Deviations of the combinatorial scatter from the BLUUE adjusted position of N

indicate the capability of the Gauss-Jacobi combinatorial algorithm to solve in a closed form the overdetermined planar ranging problems.

9.3.2 Three-dimensional Ranging

Closed Form Three-dimensional Ranging

Three-dimensional ranging problem differs from the planar ranging in terms of the number of unknowns to be determined. In the planar case, the interest is to obtain from the measured distances the two-dimensional coordinates $\{x_0, y_0\}$ of the unknown station P. For the three-dimensional ranging, the coordinates $\{X, Y, Z\}$ have to be derived from the measured distances. Since three coordinates are involved, distances *must* be measured to at least three known stations for the solution to be determined. If the stations observed are more than three, the case is an overdetermined one. The main task involved is the determination of the unknown position of a station given distance measurements from unknown station $P \in \mathbb{E}^3$, to three known stations $P_i \in \mathbb{E}^3 \mid i = 1, 2, 3$. In general, the three-dimensional closed form ranging problem can be formulated as follows: Given distance measurements from an unknown station $P \in \mathbb{E}^3$ to a minimum of three known stations $P_i \in \mathbb{E}^3 \mid i = 1, 2, 3$, determine the position $\{X, Y, Z\}$ of the unknown station $P \in \mathbb{E}^3$ (see e.g., Fig. 9.12).

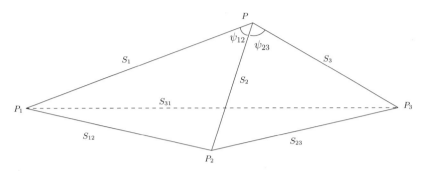

Fig. 9.12. Tetrahedron: three-dimensional distance and space angle observations

From the three nonlinear Pythagoras distance observation equations (9.34) in Solution 9.5, two equations with three unknowns are derived. Equation (9.34) is expanded in the form given by (9.35) and differenced to give (9.36) with the quadratic terms $\{X^2, Y^2, Z^2\}$ eliminated. Collecting all the known terms of (9.36) to the right-hand-side and those relating to the unknowns (i.e., a and b) on the left-hand-side leads to (9.38). The solution of the unknown terms $\{X, Y, Z\}$ now involves solving (9.37), which has two equations with three unknowns. Equation (9.37) is similar to (9.4) on p. 110 which was considered in the case of GPS pseudo-range. Four approaches are considered for solving (9.37), where more unknowns than equations are solved. Similar to the case of GPS pseudo-ranging that we considered, the underdetermined system (9.37) is overcome by determining two of the unknowns in terms of the third unknown (e.g., $X = g(Z), Y = g(Z)$).

Solution 9.5 (Differencing of the nonlinear distance equations).

$$\begin{bmatrix} S_1^2 = (X_1 - X)^2 + (Y_1 - Y)^2 + (Z_1 - Z)^2 \\ S_2^2 = (X_2 - X)^2 + (Y_2 - Y)^2 + (Z_2 - Z)^2 \\ S_3^2 = (X_3 - X)^2 + (Y_3 - Y)^2 + (Z_3 - Z)^2 \end{bmatrix} \qquad (9.34)$$

$$\begin{bmatrix} S_1^2 = X_1^2 + Y_1^2 + Z_1^2 + X^2 + Y^2 + Z^2 - 2X_1 X - 2Y_1 Y - 2Z_1 Z \\ S_2^2 = X_2^2 + Y_2^2 + Z_2^2 + X^2 + Y^2 + Z^2 - 2X_2 X - 2Y_2 Y - 2Z_2 Z \\ S_3^2 = X_3^2 + Y_3^2 + Z_3^2 + X^2 + Y^2 + Z^2 - 2X_3 X - 2Y_3 Y - 2Z_3 Z \end{bmatrix}$$
$$(9.35)$$

differencing above

$$\begin{bmatrix} S_1^2 - S_2^2 = X_1^2 - X_2^2 + Y_1^2 - Y_2^2 + Z_1^2 - Z_2^2 + a \\ S_2^2 - S_3^2 = X_2^2 - X_3^2 + Y_2^2 - Y_3^2 + Z_2^2 - Z_3^2 + b, \end{bmatrix} \tag{9.36}$$

where

$$\begin{bmatrix} a = 2X(X_2 - X_1) + 2Y(Y_2 - Y_1) + 2Z(Z_2 - Z_1) \\ b = 2X(X_3 - X_2) + 2Y(Y_3 - Y_2) + 2Z(Z_3 - Z_2). \end{bmatrix} \tag{9.37}$$

Making a and b the subject of the formula in (9.36) leads to

$$\begin{bmatrix} a = S_1^2 - S_2^2 - X_1^2 + X_2^2 - Y_1^2 + Y_2^2 - Z_1^2 + Z_2^2 \\ b = S_2^2 - S_3^2 - X_2^2 + X_3^2 - Y_2^2 + Y_3^2 - Z_2^2 + Z_3^2. \end{bmatrix} \tag{9.38}$$

Conventional Approaches

Solution by Elimination Approach-1

In the elimination approach presented in Solution 9.6, (9.37) is expressed in the form (9.39); with two equations and two unknowns $\{X, Y\}$. In this equation, Z is treated as a constant. By first eliminating Y, X is obtained in terms of Z and substituted in either of the two expressions of (9.39) to give the value of Y. The values of $\{X, Y\}$ are depicted in (9.40) with the coefficients $\{c, d, e, f\}$ given by (9.41). The values of $\{X, Y\}$ in (9.40) are substituted in the first expression of (9.34) to give the quadratic equation (9.42) in terms of Z as the unknown. The quadratic formula (3.8) on p. 25 is then applied to obtain the two solutions of Z (see the second expression of (9.42)). The coefficients $\{g, h, i\}$ are given in (9.43). Once we solve (9.42) for Z, we substitute in (9.40) to obtain the corresponding pair of solutions for $\{X, Y\}$.

Solution 9.6 (Solution by elimination).

$$\begin{bmatrix} 2X(X_2 - X_1) + 2Y(Y_2 - Y_1) = a - 2Z(Z_2 - Z_1) \\ 2X(X_3 - X_2) + 2Y(Y_3 - Y_2) = b - 2Z(Z_3 - Z_2) \end{bmatrix} \tag{9.39}$$

$$\begin{bmatrix} X = c - dZ \\ Y = e - fZ \end{bmatrix} \tag{9.40}$$

$$
\left[
\begin{array}{l}
c = \dfrac{a(Y_3 - Y_2) - b(Y_2 - Y_1)}{2\left\{(X_2 - X_1)(Y_3 - Y_2) - (X_3 - X_2)(Y_2 - Y_1)\right\}} \\[4mm]
d = \dfrac{\left\{(Z_2 - Z_1)(Y_3 - Y_2) - (Z_3 - Z_2)(Y_2 - Y_1)\right\} Z}{\left\{(X_2 - X_1)(Y_3 - Y_2) - (X_3 - X_2)(Y_2 - Y_1)\right\}} \\[4mm]
e = \dfrac{a(X_3 - X_2) - b(X_2 - X_1)}{2\left\{(Y_2 - Y_1)(X_3 - X_2) - (Y_3 - Y_2)(X_2 - X_1)\right\}} \\[4mm]
f = \dfrac{\left\{(Z_2 - Z_1)(X_3 - X_2) - (Z_3 - Z_2)(X_2 - X_1)\right\} Z}{\left\{(Y_2 - Y_1)(X_3 - X_2) - (Y_3 - Y_2)(X_2 - X_1)\right\}}
\end{array}
\right. \tag{9.41}
$$

substituting 9.40 in 9.34i

$$
\left[
\begin{array}{l}
gZ^2 + hZ + i = 0 \\[3mm]
Z_{1,2} = \dfrac{-h \pm \sqrt{h^2 - 4gi}}{2g},
\end{array}
\right. \tag{9.42}
$$

where

$$
\left[
\begin{array}{l}
g = d^2 + f^2 + 1 \\
h = 2(dX_1 + fY_1 - Z_1 - cd - ef) \\
i = X_1^2 + Y_1^2 + Z_1^2 - 2X_1c - 2Y_1e - S_1^2 + c^2 + e^2.
\end{array}
\right. \tag{9.43}
$$

Solution by Elimination Approach-2

The second approach presented in Solution 9.7 involves first expressing (9.37) in the form (9.44) which can also be expressed in matrix form as in (9.45). We now seek the matrix solution of $\{Y, Z\}$ in terms of the unknown element X as expressed by (9.46), which is written in a simpler form in (9.47). The elements of (9.47) are as given by (9.48). The solution of (9.46) for $\{Y, Z\}$ in terms of X is given by (9.49), (9.50) and (9.51). The coefficients of (9.51) are given by (9.52). Substituting the obtained values of $\{Y, Z\}$ in terms of X in the first expression of (9.34) leads to quadratic equation (9.53) in terms of X as an unknown. Applying the quadratic formula (3.8) on p. 25, two solutions for X are obtained as in the second expression of (9.53). These are then substituted back in (9.51) to obtain the values of $\{Y, Z\}$. The coefficients $\{l, m, n\}$ in (9.53) are given by (9.54).

A pair of solutions $\{X_1, Y_1, Z_1\}$ and $\{X_2, Y_2, Z_2\}$ are obtained. For GPS pseudo-ranging in Sect. 9.2, we saw that the admissible solution could easily be chosen from the pair of solutions. The desired solution was easily chosen as one set of solution was in space while the other set was on the Earth's surface. The solution could therefore be distinguished by computing the radial distances (positional norms). The admissible solution from the pair of the three-dimensional LPS ranging techniques is however difficult to isolate and must be obtained with the help of prior information, e.g., from an existing map.

Solution 9.7 (Solution by matrix approach).

$$\left[\begin{matrix} 2Y(Y_2 - Y_1) + 2Z(Z_2 - Z_1) = a - 2X(X_2 - X_1) \\ 2Y(Y_3 - Y_2) + 2Z(Z_3 - Z_2) = b - 2X(X_3 - X_2) \end{matrix} \right. \tag{9.44}$$

$$\left[\begin{matrix} Y_2 - Y_1 & Z_2 - Z_1 \\ Y_3 - Y_2 & Z_3 - Z_2 \end{matrix} \right] \left[\begin{matrix} Y \\ Z \end{matrix} \right] = \frac{1}{2} \left\{ \left[\begin{matrix} a \\ b \end{matrix} \right] - 2 \left[\begin{matrix} X_2 - X_1 \\ X_3 - X_2 \end{matrix} \right] X \right\} \tag{9.45}$$

$$\left[\begin{matrix} Y \\ Z \end{matrix} \right] = \frac{1}{2} d \left[\begin{matrix} Z_3 - Z_2 & -(Z_2 - Z_1) \\ -(Y_3 - Y_2) & (Y_2 - Y_1) \end{matrix} \right] \left\{ \left[\begin{matrix} a \\ b \end{matrix} \right] - 2 \left[\begin{matrix} X_2 - X_1 \\ X_3 - X_2 \end{matrix} \right] X, \right\} \tag{9.46}$$

with

$$d = \{(Y_2 - Y_1)(Z_3 - Z_2) - (Y_3 - Y_2)(Z_2 - Z_1)\}^{-1}.$$

$$\left[\begin{matrix} Y \\ Z \end{matrix} \right] = \{a_{11}a_{22} - a_{12}a_{21}\}^{-1} \left[\begin{matrix} a_{22} & -a_{12} \\ -a_{21} & a_{11} \end{matrix} \right] \left\{ \left[\begin{matrix} b_1 \\ b_2 \end{matrix} \right] + \left[\begin{matrix} c_1 \\ c_2 \end{matrix} \right] X, \right\} \tag{9.47}$$

where

$$\left[\begin{matrix} a_{11} = Y_2 - Y_1, \ a_{12} = Z_2 - Z_1, \ a_{21} = Y_3 - Y_2, \ a_{22} = Z_3 - Z_2 \\ c_1 = -(X_2 - X_1), \ c_2 = -(X_3 - X_2), \ b_1 = \frac{1}{2}a, \ b_2 = \frac{1}{2}b. \end{matrix} \right. \tag{9.48}$$

$$\left[\begin{matrix} Y = \{a_{11}a_{22} - a_{12}a_{21}\}^{-1} \{a_{22}(b_1 + c_1 X) - a_{12}(b_2 + c_2 X)\} \\ Z = \{a_{11}a_{22} - a_{12}a_{21}\}^{-1} \{a_{11}(b_2 + c_2 X) - a_{21}(b_1 + c_1 X)\} \end{matrix} \right. \tag{9.49}$$

$$\left[\begin{matrix} Y = e\left[\{a_{22}b_1 - a_{12}b_2\} + \{a_{22}c_1 - a_{12}c_2\} X\right] \\ Z = e\left[\{a_{11}b_2 - a_{21}b_1\} + \{a_{11}c_2 - a_{21}c_1\} X\right] \end{matrix} \right. \tag{9.50}$$

$$\left[\begin{matrix} Y = e(f + gX) \\ Z = e(h + iX) \end{matrix} \right. \tag{9.51}$$

$$\left[\begin{matrix} e = (a_{11}a_{22} - a_{12}a_{21})^{-1}, \ f = a_{22}b_1 - a_{12}b_2, \ g = a_{22}c_1 - a_{12}c_2 \\ h = a_{11}b_2 - a_{21}b_1, \ i = a_{11}c_2 - a_{21}c_1, \ k = X_1^2 + Y_1^2 + Z_1^2. \end{matrix} \right. \tag{9.52}$$

substituting (9.51) in (9.34i)

$$
\left[
\begin{array}{l}
lX^2 + mX + n = 0 \\
\\
X_{1,2} = \dfrac{-m \pm \sqrt{m^2 - 4ln}}{2l},
\end{array}
\right.
\tag{9.53}
$$

where

$$
\left[
\begin{array}{l}
l = e^2 i^2 + e^2 g^2 + 1 \\
m = 2(e^2 fg + e^2 hi - X_1 - egY_1 - eiZ_1) \\
n = k - S_1^2 - 2Y_1 ef + e^2 f^2 - 2Z_1 eh + e^2 h^2.
\end{array}
\right.
\tag{9.54}
$$

Groebner Basis Approach

Equation (9.37) is expressed in algebraic form (9.55) in Solution 9.8 with the coefficients as in (9.56). Groebner basis of (9.55) is then obtained in (9.57) using (4.36) on p. 43. The obtain Groebner basis solution of the three-dimensional ranging problem is presented in (9.58). The first expression of (9.58) is solved for $Y = g_1(Z)$, and the output presented in (9.59). This value is substituted in the second expression of (9.58) to give $X = g_2(Z)$ in (9.60). The obtained values of Y and X are substituted in the first expression of (9.34) to give a quadratic equation in Z. Once this quadratic equation has been solved for Z using (3.8) on p. 25, the values Y and X are obtained from (9.59) and (9.60) respectively. Instead of solving for $Y = g_1(Z)$ and substituting in the second expression of (9.58) to give $X = g_2(Z)$, direct solution of $X = g(Z)$ in (9.61) could be obtained by computing the reduced Groebner basis (4.38) on p. 44. Similarly we could obtain $Y = g(Z)$ alone by replacing Y with X in the option part of the reduced Groebner basis.

Solution 9.8 (Groebner basis solution).

$$
\begin{array}{l}
a_{02}X + b_{02}Y + c_{02}Z + f_{02} = 0 \\
a_{12}X + b_{12}Y + c_{12}Z + f_{12} = 0
\end{array}
\tag{9.55}
$$

$$\begin{bmatrix} a_{02} = 2(X_1 - X_2), b_{02} = 2(Y_1 - Y_2), c_{02} = 2(Z_1 - Z_2) \\[2mm] a_{12} = 2(X_2 - X_3), b_{12} = 2(Y_2 - Y_3), c_{12} = 2(Z_2 - Z_3) \\[2mm] f_{02} = (S_1^2 - X_1^2 - Y_1^2 - Z_1^2) - (S_2^2 - X_2^2 - Y_2^2 - Z_2^2) \\[2mm] f_{12} = (S_2^2 - X_2^2 - Y_2^2 - Z_2^2) - (S_3^2 - X_3^2 - Y_3^2 - Z_3^2). \end{bmatrix} \quad (9.56)$$

$$Groebner\,Basis[\{a_{02}X + b_{02}Y + c_{02}Z + f_{02}, a_{12}X + b_{12}Y + c_{12}Z + f_{12}\}, \{X, Y\}] \quad (9.57)$$

$$\begin{bmatrix} g_1 = a_{02}b_{12}Y - a_{12}b_{02}Y - a_{12}c_{02}Z + a_{02}c_{12}Z + a_{02}f_{12} - a_{12}f_{02} \\[2mm] g_2 = a_{12}X + b_{12}Y + c_{12}Z + f_{12} \\[2mm] g_3 = a_{02}X + b_{02}Y + c_{02}Z + f_{02}. \end{bmatrix}$$

$$(9.58)$$

$$Y = \frac{\{(a_{12}c_{02} - a_{02}c_{12})Z + a_{12}f_{02} - a_{02}f_{12}\}}{(a_{02}b_{12} - a_{12}b_{02})} \quad (9.59)$$

$$X = \frac{-(b_{12}Y + c_{12}Z + f_{12})}{a_{12}}, \quad (9.60)$$

or

$$X = \frac{\{(b_{02}c_{12} - b_{12}c_{02})Z + b_{02}f_{12} - b_{12}f_{02}\}}{(a_{02}b_{12} - a_{12}b_{02})}. \quad (9.61)$$

Polynomial Resultants Approach

The problem is solved in *four steps* as illustrated in Solution 9.9. In the *first step*, we solve for the first variable X in (9.55) by hiding it as a constant and homogenizing the equation using a variable W as in (9.62). In the *second step*, the Sylvester resultants discussed in Sect. 5.2 on p. 48 or the *Jacobian determinant* is obtained as in (9.63). The resulting determinant (9.64) is solved for $X = g(Z)$ and presented in (9.65). The procedure is repeated in *steps three* and *four* from (9.66) to (9.69) to solve for $Y = g(Z)$. The obtained values of $X = g(Z)$ and $Y = g(Z)$ are substituted in the first expression of (9.34) to give a quadratic equation in Z. Once this quadratic has been solved for Z, the values of X and Y are then obtained from (9.65) and (9.69) respectively.

Solution 9.9 (Polynomial resultants solution). <u>Step 1</u>: Solve for X in terms of Z

$$\begin{aligned}
f_1 &:= (a_{02}X + c_{02}Z + f_{02})W + b_{02}Y \\
f_2 &:= (a_{12}X + c_{12}Z + f_{12})W + b_{12}Y
\end{aligned} \tag{9.62}$$

<u>Step 2</u>: Obtain the Sylvester resultant

$$J_X = \det \begin{bmatrix} \dfrac{\partial f_1}{\partial Y} & \dfrac{\partial f_1}{\partial W} \\[2mm] \dfrac{\partial f_2}{\partial Y} & \dfrac{\partial f_2}{\partial W} \end{bmatrix} = \det \begin{bmatrix} b_{02} & (a_{02}X + c_{02}Z + f_{02}) \\ b_{12} & (a_{12}X + c_{12}Z + f_{12}) \end{bmatrix} \tag{9.63}$$

$$J_X = b_{02}a_{12}X + b_{02}c_{12}Z + b_{02}f_{12} - b_{12}a_{02}X - b_{12}c_{02}Z - b_{12}f_{02} \tag{9.64}$$

from (9.64)

$$X = \frac{\{(b_{12}c_{02} - b_{02}c_{12})Z + b_{12}f_{02} - b_{02}f_{12}\}}{(b_{02}a_{12} - b_{12}a_{02})} \tag{9.65}$$

<u>Step 3</u>: Solve for Y in terms of Z

$$\begin{aligned}
f_3 &:= (b_{02}Y + c_{02}Z + f_{02})W + b_{02}X \\
f_4 &:= (b_{12}Y + c_{12}Z + f_{12})W + a_{12}X
\end{aligned} \tag{9.66}$$

<u>Step 4</u>: Obtain the Sylvester resultant

$$J_Y = \det \begin{bmatrix} \dfrac{\partial f_3}{\partial X} & \dfrac{\partial f_3}{\partial W} \\[2mm] \dfrac{\partial f_4}{\partial X} & \dfrac{\partial f_4}{\partial W} \end{bmatrix} = \det \begin{bmatrix} a_{02} & (b_{02}Y + c_{02}Z + f_{02}) \\ a_{12} & (b_{12}Y + c_{12}Z + f_{12}) \end{bmatrix} \tag{9.67}$$

$$J_Y = a_{02}b_{12}Y + a_{02}c_{12}Z + a_{02}f_{12} - a_{12}b_{02}Y - a_{12}c_{02}Z - a_{12}f_{02} \tag{9.68}$$

from (9.68)

$$Y = \frac{\{(a_{12}c_{02} - a_{02}c_{12})Z + a_{12}f_{02} - a_{02}f_{12}\}}{(a_{02}b_{12} - a_{12}b_{02})} \tag{9.69}$$

Example 9.5 (Three-dimensional ranging to three known stations). Consider distance measurements of Fig. 9.12 as $S_1 = 1324.2380\,m$, $S_2 =$

$542.2609\,m$ and $S_3 = 430.5286\,m$, the position of P is obtained using either of the procedures above as $X = 4157066.1116\,m$, $Y = 671429.6655\,m$ and $= 4774879.3704\,m$. Figures 9.13, 9.14 and 9.15 indicate the solutions of $\{X, Y, Z\}$ respectively. The stars (intersection of the quadratic curves with the zero line) are the solution points. The critical configuration of the three-dimensional ranging problem is presented in Solution 9.10. Equations (9.75) and (9.76) indicate the critical configuration to be the case where points $P(X, Y, Z)$, $P_1(X_1, Y_1, Z_1)$, $P_2(X_2, Y_2, Z_2)$, and $P_3(X_3, Y_3, Z_3)$ all lie on a plane.

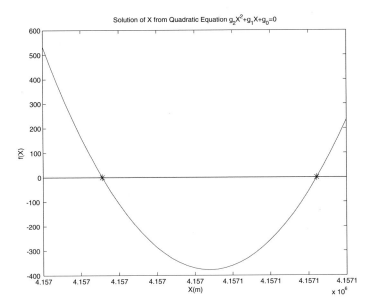

Fig. 9.13. Solution of the X coordinates

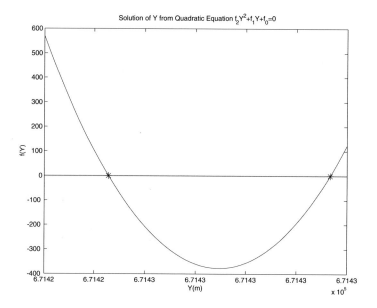

Fig. 9.14. Solution of the Y coordinates

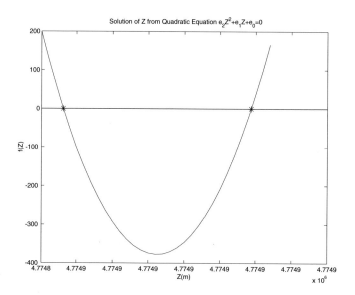

Fig. 9.15. Solution of the Z coordinates

Solution 9.10 (Critical configuration of three-dimensional ranging).

$$\begin{bmatrix} f_1(X,Y,Z;X_1,Y_1,Z_1,S_1) = (X_1 - X)^2 + (Y_1 - Y)^2 + (Z_1 - Z)^2 - S_1^2 \\ f_2(X,Y,Z;X_2,Y_2,Z_2,S_2) = (X_2 - X)^2 + (Y_2 - Y)^2 + (Z_2 - Z)^2 - S_2^2 \\ f_3(X,Y,Z;X_3,Y_3,Z_3,S_3) = (X_3 - X)^2 + (Y_3 - Y)^2 + (Z_3 - Z)^2 - S_3^2. \end{bmatrix}$$
$$(9.70)$$

$$\begin{bmatrix} \dfrac{\partial f_1}{\partial X} = -2(X_1 - X), & \dfrac{\partial f_2}{\partial X} = -2(X_2 - X), & \dfrac{\partial f_3}{\partial X} = -2(X_3 - X) \\[2mm] \dfrac{\partial f_1}{\partial Y} = -2(Y_1 - Y), & \dfrac{\partial f_2}{\partial Y} = -2(Y_2 - Y), & \dfrac{\partial f_3}{\partial Y} = -2(Y_3 - Y) \\[2mm] \dfrac{\partial f_1}{\partial Z} = -2(Z_1 - Z), & \dfrac{\partial f_2}{\partial Z} = -2(Z_2 - Z), & \dfrac{\partial f_3}{\partial Zv} = -2(Z_3 - Z). \end{bmatrix}$$
$$(9.71)$$

$$\begin{bmatrix} D = \left| \dfrac{\partial f_i}{\partial X_j} \right| = -8 \begin{vmatrix} X_1 - X & Y_1 - Y & Z_1 - Z \\ X_2 - X & Y_2 - Y & Z_1 - Z \\ X_3 - X & Y_3 - Y & Z_1 - Z \end{vmatrix} \\[4mm] D \Leftrightarrow \begin{vmatrix} X_1 - X & Y_1 - Y & Z_1 - Z \\ X_2 - X & Y_2 - Y & Z_1 - Z \\ X_3 - X & Y_3 - Y & Z_1 - Z \end{vmatrix} = \begin{vmatrix} X & Y & Z & 1 \\ X_1 & Y_1 & Z_1 & 1 \\ X_2 & Y_2 & Z_2 & 1 \\ X_3 & Y_3 & Z_3 & 1 \end{vmatrix} = 0. \end{bmatrix}$$
$$(9.72)$$

$$\begin{bmatrix} -\tfrac{1}{8}D = \{-Z_1Y_3 + Y_1Z_3 - Y_2Z_3 + Y_3Z_2 - Y_1Z_2 + Y_2Z_1\} X \\[2mm] + \{-Z_1X_2 - X_1Z_3 + Z_1X_3 + X_1Z_2 - X_3Z_2 + X_2Z_3\} Y \\[2mm] + \{Y_1X_2 - Y_1X_3 + Y_3X_1 - X_2Y_3 - X_1Y_2 + Y_2X_3\} Z \\[2mm] +X_1Y_2Z_3 - X_1Y_3Z_2 - X_3Y_2Z_1 + X_2Y_3Z_1 - X_2Y_1Z_3 + X_3Y_1Z_2, \end{bmatrix}$$
$$(9.73)$$

thus

$$\begin{vmatrix} X & Y & Z & 1 \\ X_1 & Y_1 & Z_1 & 1 \\ X_2 & Y_2 & Z_2 & 1 \\ X_3 & Y_3 & Z_3 & 1 \end{vmatrix},$$
$$(9.74)$$

describes six times volume of the tetrahedron formed by the points $P(X,Y,Z)$, $P_1(X_1,Y_1,Z_1)$, $P_2(X_2,Y_2,Z_2)$, and $P_3(X_3,Y_3,Z_3)$. Therefore

$$D = \begin{vmatrix} X & Y & Z & 1 \\ X_1 & Y_1 & Z_1 & 1 \\ X_2 & Y_2 & Z_2 & 1 \\ X_3 & Y_3 & Z_3 & 1 \end{vmatrix} \begin{bmatrix} a \\ b \\ c \\ d \end{bmatrix} = 0, \tag{9.75}$$

results in a system of homogeneous equations

$$\begin{bmatrix} aX + bY + cZ + d = 0 \\ aX_1 + bY_1 + cY_1 + d = 0 \\ aX_2 + bY_2 + cZ_2 + d = 0 \\ aX_3 + bY_3 + cZ_3 + d = 0. \end{bmatrix} \tag{9.76}$$

Three-dimensional Ranging to more than Three Known Stations

The Gauss-Jacobi combinatorial algorithm is here applied to solve the overdetermined three-dimensional ranging problem. An example based on the test network Stuttgart Central in Fig. 7.1 is considered.

Example 9.6 (Three-dimensional ranging to more than three known stations). From the test network Stuttgart Central in Fig. 7.1 of Sect. 7.4, the three-dimensional coordinates $\{X, Y, Z\}$ of the unknown station $K1$ are desired. One proceeds in three steps as follows:

Step 1 (combinatorial solution):
From Fig. 7.1 on p. 86 and using (6.26) on p. 69, 35 combinatorial subsets are formed whose systems of *nonlinear distance equations* are solved for the position $\{X, Y, Z\}$ of the unknown station $K1$ in closed form. Use is made of either Groebner basis derived equations (9.59) and (9.60) or polynomial resultants derived (9.65) and (9.69). 35 different positions $X, Y, Z|_{K1}$ of the same station $K1$, totalling to 105 (35×3) values of X, Y, Z are obtained and treated as pseudo-observations.

Step 2 (determination of the dispersion matrix Σ):
The variance-covariance matrix is computed for each of the combinatorial set $j = 1, \ldots, 35$ using error propagation. The closed form observational equations are written algebraically as

$$\begin{bmatrix} f_1 := (X_1 - X)^2 + (Y_1 - Y)^2 + (Z_1 - Z)^2 - S_1^2 \\ f_2 := (X_2 - X)^2 + (Y_2 - Y)^2 + (Z_2 - Z)^2 - S_2^2 \\ f_3 := (X_3 - X)^2 + (Y_3 - Y)^2 + (Z_3 - Z)^2 - S_3^2, \end{bmatrix} \tag{9.77}$$

where $S_i^j | i \in \{1, 2, 3\} | j = 1$ are the distances between known GPS stations $P_i \in \mathbb{E}^3 | i \in \{1, 2, 3\}$ and the unknown station $K1 \in \mathbb{E}^3$ for first combination set $j = 1$. Equation (9.77) is used to obtain the dispersion matrix Σ in (6.31) as discussed in Example 6.4 on p. 72.

Step 3 (rigorous adjustment of the combinatorial solution points in a polyhedron):

For each of the 35 computed coordinates of point $K1$ in step 2, we write the observation equations as

$$
\begin{bmatrix}
X^j = X + \varepsilon_X^j |, j \in \{1, 2, 3, 4, 5, 6, 7, \ldots, 35\} \\
Y^j = Y + \varepsilon_Y^j | j \in \{1, 2, 3, 4, 5, 6, 7, \ldots, 35\} \\
Z^j = Z + \varepsilon_Z^j |, j \in \{1, 2, 3, 4, 5, 6, 7, \ldots, 35\}.
\end{bmatrix}
\tag{9.78}
$$

The values $\{X^j, Y^j, Z^j\}$ are treated as pseudo-observation and placed in the vector of observation $\mathbf{y}$, while the coefficients of the unknown positions $\{X, Y, Z\}$ are placed in the design matrix $\mathbf{A}$. The vector $\boldsymbol{\xi}$ comprise the unknowns $\{X, Y, Z\}$. The solutions are obtained via (6.10) and the root-mean-square errors of the estimated parameters through (6.11). In the experiment above, the computed position of station $K1$ is given in Table 9.12. The deviations of the combinatorial solutions from the *true (measured) GPS value* are given in Table 9.13. Figure 9.16 indicates the plot of the combinatorial scatter $\{\bullet\}$ around the adjusted values $\{*\}$.

Table 9.12. Position of station $K1$ computed by *Gauss-Jacobi combinatorial algorithm*

$X(m)$	$Y(m)$	$Z(m)$	σ_X	σ_Y	σ_Z
4157066.1121	671429.6694	4774879.3697	0.00005	0.00001	0.00005

Table 9.13. Deviation of the computed position of $K1$ in Table (9.12) from the real measured GPS values

$\Delta X(m)$	$\Delta Y(m)$	$\Delta Z(m)$
-0.0005	-0.0039	0.0007

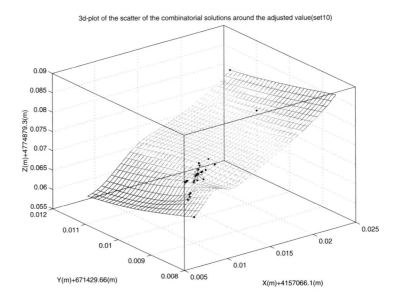

Fig. 9.16. Scatter of combinatorial solutions

9.4 Concluding Remarks

In cases where positions are required from distance measurements such as point location in engineering and cadastral surveying, the algorithms presented in this chapter are handy. Users need only to insert measured distances and the coordinates of known stations in these algorithms to obtain their positions. In essence, one does not need to re-invent the wheel by going back to the Mathematica software! Additional literature on the topic are [1, 48, 86, 171, 208, 279].

From Geocentric Cartesian to Ellipsoidal Coordinates

10.1 Mapping Topographical Points onto Reference Ellipsoid

The projection of points from the topographical surface to their equivalent on the reference ellipsoid remains one of the fundamental tasks undertaken in geodesy and geoinformatics. This is because the reference ellipsoid of revolution is the mathematical representation of the geoid. Geoid is the surface that approximates mean sea level, and provides vertical datum for heights. It is of prime importance in engineering and geosciences in general. From it, geophysicists can infer on processes taking place below and above the Earth such as earthquakes and rise in sea level. Hydrologists need it to infer on water table, while engineers need it for height determination during roads and structural constructions.

Measurements are normally related to the geoid for computation via its mathematical form, the reference ellipsoid of revolution. There exist two ways of projecting points from a topographical surface onto the reference ellipsoid of revolution. One approach projects a point P onto the geoid p_g and then finally onto the reference ellipsoid of revolution p. This type of projection is called the Pizetti's projection. The other approach directly projects a topographical point P through the ellipsoidal normal onto a point p on the reference ellipsoid of revolution. The distance between the topographical point P and the ellipsoidal point p gives the geometrical height H above the ellipsoid. The topographical position of point P would therefore be referred by the ellipsoidal height H and the geographical coordinates L, B. In this case, the geographical coordinate L is the longitude and B the latitude. The set of coordinates

$\{L, B, H\}$ defining the point P are called geodetic or ellipsoidal coordinates. This second projection is called the Helmert's projection which will be considered in this chapter. The two projections are discussed in detail in [184, pp. 178–184].

The *forward transformation* from ellipsoid to Cartesian coordinates, i.e., $\{L, B, H\} \rightarrow \{X, Y, Z\}$, is demonstrated by Solutions 10.1 and 10.2. The challenge is the inverse transformation which projects topographical points to the ellipsoid. One way of achieving this projection is by first converting topographical Cartesian coordinates into ellipsoidal cartesian coordinates. Once this is done, the ellipsoidal Cartesian coordinates are then converted to their equivalent geodetic coordinates. The problem is formulated as follows: Given topographical coordinates $\{X, Y, Z\}$ of a point P, obtain the geodetic coordinates $\{L, B, H\}$. This problem is a one-to-one mapping of

$$
\begin{bmatrix}
\{X, Y, Z\} & \longrightarrow & \{L, B, H\} \\
Topography & & Ellipsoid
\end{bmatrix}
\tag{10.1}
$$

Table 10.1 outlines the existing methods by other authors to convert Cartesian coordinates $\{X, Y, Z\}$ to Gauss ellipsoidal coordinates $\{L, B, H\}$ in (10.1). The target of this chapter is to invert algebraically $\{X, Y, Z\} \rightarrow \{L, B, H\}$ by means of *minimum distance mapping* through the map in (10.2) as

$$
\begin{bmatrix}
\{X, Y, Z\} & \longrightarrow & \{x_1, x_2, x_3\} & \longrightarrow & \{L, B, H\} \\
Topography & & Ellipsoid & & Ellipsoid.
\end{bmatrix}
\tag{10.2}
$$

Grafarend [134] already constructed surface normal coordinates with respect to the *international reference ellipsoid*. In this chapter, we will be interested with setting up an algebraic minimum distance mapping to relate a point on the Earth's topographical surface uniquely (one-to-one) to a point on the international reference ellipsoid. The solution to such an optimization problem generates *projective ellipsoidal heights* and the standard transformation of the Gauss ellipsoidal coordinates $\{L, B, H\}$ to geocentric Cartesian coordinates $\{X, Y, Z\}$. The *inverse transformation* of geocentric Cartesian coordinates $\{X, Y, Z\}$ to Gauss ellipsoidal coordinates $\{L, B, H\}$ is here solved algebraically and examples presented.

Table 10.1. Characteristics of inverse transformation Cartesian coordinates to Gauss ellipsoidal coordinates

Author	Publication year	Characteristic
Awange et al. [36]	in press	closed (similar to [11])
Bartelme N, Meissl P [51]	1975	iterative
Benning W [60]	1974	closed
Benning W [61]	1987	iterative first point curve
Borkowski KM [71]	1987	iterative
Borkowski KM [72]	1989	iterative
Bowring BR [73]	1976	approximate "closed"
Bowring BR [74]	1985	approximate
Croceto N [96]	1993	iterative
Fitzgibbon A et al. [110]	1999	iterative
Fotiou A [112]	1998	approximate "closed"
Fröhlich H, Hansen HH [115]	1976	closed
Fukushima T [116]	1999	"fast" iterative
Gander W et al. [117]	1994	iterative
Grafarend EW [135]	2001	closed
Grafarend EW, Lohse P [141]	1991	closed form 4th order equation reduced to 3rd order
Grafarend EW et al. [159]	1995	closed form
Heck B [181]	1987	iterative
Heikkinen M [182]	1982	closed
Heiskannen WA, Moritz H [184]	1976	iterative
Hirvonen R, Moritz H [185]	1963	iterative
Hofman-Wellenhof B et al. [186]	2001	Identical to Bowring [73]
Lapaine M [215]	1990	algebraic equations of higher order
Lin KC, Wang J [223]	1995	iterative
Loskowski P [227]	1991	"simply iterative"
Ozone MI [261]	1985	3rd order equation
Paul MK [262]	1973	iterative
Penev P [263]	1978	Angular variable 3rd order equation
Pick M [265]	1985	approximate "closed"
Sjöberg LE [297]	1999	iterative
Soler T, Hothem LD [298]	1989	iterative "closed" Jacobi ellipsoidal coordinates
Sünkel H [306]	1976	series expansion
Torge W [313]	1991	iterative
Vaniceck P, Krakiwski E [318]	1982	higher order algebraic equation
Vincenty T [321]	1978	Iterative
Vincenty T [322]	1980	approximate "closed"
You RJ [349]	2000	iterative

Solution 10.1 (Forward transformation of Gauss ellipsoidal coordinates).

$$\mathbf{X}(L, B, H) = \mathbf{e_1} \left[\frac{a}{\sqrt{1 - e^2 \sin^2 B}} + H(L, B) \right] \cos B \cos L +$$

$$\mathbf{e_2} \left[\frac{a}{\sqrt{1 - e^2 \sin^2 B}} + H(L, B) \right] \cos B \sin L +$$

$$\mathbf{e_2} \left[\frac{a(1 - e^2)}{\sqrt{1 - e^2 \sin^2 B}} + H(L, B) \right] \sin B, \qquad (10.3)$$

$$\begin{bmatrix} X \\ Y \\ Z \end{bmatrix} = \begin{bmatrix} \left[\frac{a}{\sqrt{1-e^2 \sin^2 B}} + H(L, B) \right] \cos B \cos L \\ \left[\frac{a}{\sqrt{1-e^2 \sin^2 B}} + H(L, B) \right] \cos B \sin L \\ \left[\frac{a(1-e^2)}{\sqrt{1-e^2 \sin^2 B}} + H(L, B) \right] \sin B \end{bmatrix}, \qquad (10.4)$$

with $\{(L(X, Y, Z), B(X, Y, Z), H(X, Y, Z)\}$ as unknowns.

Solution 10.2 (Forward transformation of Gauss complex ellipsoidal coordinates). Consider

$$X + iY = \left[\frac{a}{\sqrt{1 - e^2 \sin^2 B}} + H(L, B) \right] \cos B (\cos L + i \sin L), \quad (10.5)$$

and

$$Z = \left[\frac{a(1 - e^2)}{\sqrt{1 - e^2 \sin^2 B}} + H(L, B) \right] \sin B, \qquad (10.6)$$

then

$$\mathbf{X} = \begin{bmatrix} X + iY & Z \\ -Z & X - iY \end{bmatrix} \in \mathbb{C}^{2 \times 2} \qquad (10.7)$$

10.2 Mapping Geometry

In [134], Gauss surface normal coordinates with respect to the international reference ellipsoid $\mathbb{E}^2_{a,a,b}$ are introduced and called $\{l, b\}$. The Gauss surface normal longitude is represented by l (geodetic longitude) and the Gauss surface normal latitude by b (geodetic latitude). Such a coordinate system build up the proper platform for introducing surface normal coordinates $\{L, B, H\}$ for mapping the Earth's topographical surface $\mathbb{T}^2$ with respect to the international reference ellipsoid. In particular, the minimum distance mapping which maps a topographic point $P \in \mathbb{T}^2$ onto a *nearest point* $p \in \mathbb{E}^2_{a,a,b}$ on the international reference ellipsoid is implemented. Such mapping, initiated by *C. F. Gauss*, is *isozenithal* since $\{l = L, b = B\}$: The orthogonal projection of $P \in \mathbb{T}^2$ onto $p \in \mathbb{E}^2_{a,a,b}$ as the nearest point is along the *surface normal* of $\mathbb{E}^2_{a,a,b}$. The minimum distance from point p to P, i.e., $\overline{pP}$ is called accordingly ellipsoidal height H ("geodetic height") complemented by *surface normal longitude* $l = L$ and *surface normal latitude* $b = B$.

In-order to gain a unique solution of *minimum distance mapping*, the assumption that the Earth's topographical surface is *starshaped* has to be made. Figure 10.1 illustrates a topographical surface which is *starshaped*, while Fig. 10.2 illustrates that which is *not*. With respect to these figures, the notion of a *starshaped compact* (closed and bounded) *topographical surface* may be obvious.

Definition 10.1 (Starshaped surface). *A region* $\mathbb{M} \in \mathbb{R}^3$ *is starshaped with respect to a point* $P \in \mathbb{R}^3$, *if the straight line which connects an arbitrary point* $Q \in \mathbb{M}$ *with* P *lies in* $\mathbb{M}$. *We call a surface starshaped, if it forms the boundary of a starshaped region.*

It is understood that the shape of a star guarantees that the minimum distance mapping of topographical surfaces in $\mathbb{R}^3$ onto the average sphere $\mathbb{S}^2_R$ (*"Bjerhammer sphere"*) is *one-to-one*. If the minimum distance mapping would *not* be one-to-one, it might happen that a point on the *average sphere* $\mathbb{S}^2_R$ *has more than one image* on the topographical surface. Here the condition of *starshaped* has to be relaxed if the topographic surface $\mathbb{T}^2 \subset \mathbb{R}^3$ is mapped onto the international reference ellipsoid $\mathbb{E}^2_{a,a,b}$, an ellipsoid of revolution of semi-major axis a and semi-minor axis b. If any surface normal to the ellipsoid of revolution $\mathbb{E}^2_{a,a,b}$ intersects the topographical surface *only once*, the topographical surface is *ellipsoidal starshaped*. Indeed this condition is not met by any arbitrary topographical surface like the Earth's. Instead we shall assume that we have properly regularized the Earth's topographical surface to meet our requirement. Otherwise the Gauss ellipsoidal coordinates $\{L, B, H\}$ would break down! Figures 10.3 and 10.4 gives a better insight into the notion of *ellipsoidal starshaped* and *anti-ellipsoidal starshaped*.

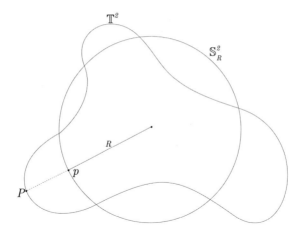

Fig. 10.1. Minimum distance mapping, starshaped topographic surface (orthogonal projection of $P \in \mathbb{T}^2$ onto $p \in \mathbb{S}_R^2$)

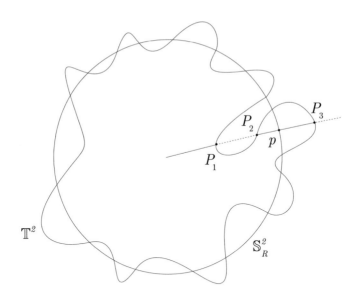

Fig. 10.2. Minimum distance mapping, non-starshaped topographic surface (orthogonal projection of $P \in \mathbb{T}^2$ onto $p \in \mathbb{S}_R^2$)

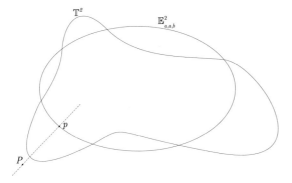

Fig. 10.3. Minimum distance mapping, ellipsoidal starshaped topographic surface (orthogonal projection of $P \in \mathbb{T}^2$ onto $p \in \mathbb{E}^2_{a,a,b}$)

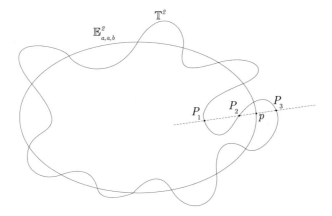

Fig. 10.4. Minimum distance mapping, a topographic surface $\mathbb{T}^2$ which is *not* ellipsoidal starshaped (orthogonal projection of $P \in \mathbb{T}^2$ onto $p \in \mathbb{E}^2_{a,a,b}$)

10.3 Minimum Distance Mapping

In-order to relate a point P on the Earth's topographic surface to a point on the international reference ellipsoid $\mathbb{E}^2_{a,a,b}$, a bundle of half straight lines so called *projection lines* which depart from P and intersect $\mathbb{E}^2_{a,a,b}$ either not at all or in two points are used. There is *one projection line* which is at minimum distance relating P to p. Figure 10.5 is an illustration of such a minimum distance mapping. Let us formulate such an optimization problem by means of the Lagrangean $\mathcal{L}(x_1, x_2, x_3, x_4)$ in Solution 10.3.

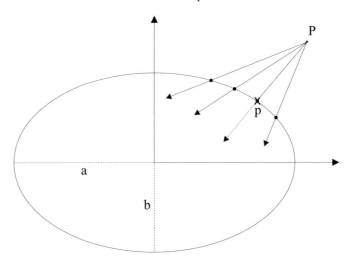

Fig. 10.5. Minimum distance mapping of a point P on the Earth's topographic surface to a point p on the *international reference ellipsoid* $\mathbb{E}^2_{a,a,b}$

Solution 10.3 (Constraint minimum distance mapping in terms of Cartesian coordinates).

$$\mathcal{L}(x_1, x_2, x_3, x_4) := \frac{1}{2}\|\mathbf{X} - \mathbf{x}\|^2 + \frac{1}{2}x_4\left[b^2(x_1^2 + x_2^2) + ax_3^2 - a^2b^2\right]$$

$$= \frac{1}{2}\{(X - x_1)^2 + (Y - x_2)^2 + (Z - x_3)^2$$

$$+ x_4\left[b^2(x_1^2 + x_2^2) + ax_3^2 - a^2b^2\right]\}$$

$$= \frac{min}{(x_1, x_2, x_3, x_4)} \tag{10.8}$$

$$\mathbf{x} \in \mathbb{X} := \{\mathbf{x} \in \mathbb{R}^3 | \frac{x_1^2 + x_2^2}{a^2} + \frac{x_3^2}{b^2} = 1\} =: \mathbb{E}^2_{a,a,b} \tag{10.9}$$

In the first case, the *Euclidean distance* between points P and p in terms of *Cartesian coordinates* of $P(X, Y, Z)$ and of $p(x_1, x_2, x_3)$ is represented. The *Cartesian coordinates* (x_1, x_2, x_3) of the projection point P are unknown. The constraint that the point p is an element of the ellipsoid of revolution

$$\mathbb{E}^2_{a,a,b} := \{\mathbf{x} \in \mathbb{R}^3 | b^2(x_1^2 + x_2^2) + a^2x_3^2 - a^2b^2 = 0, \mathbb{R}^+ \ni a > b \in \mathbb{R}^+\}$$

is substituted into the Lagrangean by means of the Lagrange multiplier x_4, which is unknown too. $\{(x_1^\wedge, x_2^\wedge, x_3^\wedge, x_4^\wedge) = arg\{£(x_1, x_2, x_3, x_4) = min\}$ is the argument of the minimum of the *constrained Lagrangean* $£(x_1, x_2, x_3, x_4)$. The result of the minimization procedure is presented by Lemma 10.1. Equation (10.10) provides the *necessary conditions* to constitute an extremum: The normal equations are of bilinear type. *Products* of the unknowns for instance $x_1 x_4, x_2 x_4, x_3 x_4$ and *squares* of the unknowns, for instance x_1^2, x_2^2, x_3^2 appear. Finally the matrix of second derivatives $\mathbf{H}_3$ in (10.12) which is *positive definite* constitutes the *sufficient condition* to obtain a minimum. Fortunately the matrix of second derivatives $\mathbf{H}_3$ is *diagonal*. Using (10.11i–10.11iv), together with (10.14) leads to (10.15), which are the eigenvalues of the Hesse matrix $\mathbf{H}_3$. These values are $\Lambda_1 = \Lambda_2 = X \backslash x_1^\wedge, \Lambda_3 = Z \backslash x_3^\wedge$ and must be positive.

Lemma 10.1 (Constrained minimum distance mapping). *The functional* $£(x_1, x_2, x_3, x_4)$ *is minimal, if the conditions (10.10) and (10.12) hold.*

$$\frac{\partial £}{\partial x_i}((x_1^\wedge, x_2^\wedge, x_3^\wedge, x_4^\wedge)) = 0 \quad \forall \quad i=1,2,3,4. \tag{10.10}$$

On taking partial derivatives with respect to x_i, *we have*

$$
\left[
\begin{array}{ll}
(i) & \dfrac{\partial £}{\partial(x_1^\wedge)} = -(X - x_1^\wedge) + b^2 x_1^\wedge x_4^\wedge = 0 \\[2ex]
(ii) & \dfrac{\partial £}{\partial(x_2^\wedge)} = -(Y - x_2^\wedge) + b^2 x_2^\wedge x_4^\wedge = 0 \\[2ex]
(iii) & \dfrac{\partial £}{\partial(x_3^\wedge)} = -(Z - x_3^\wedge) + a^2 x_3^\wedge x_4^\wedge = 0 \\[2ex]
(iv) & \dfrac{\partial £}{\partial(x_4^\wedge)} = \dfrac{1}{2}[b^2(x_1^{\wedge 2} + x_2^{\wedge 2})] + a^2 x_3^{\wedge 2} - a^2 b^2 = 0
\end{array}
\right. \tag{10.11}
$$

$$\frac{\partial^2 £}{\partial x_i \partial x_j}(x_1^\wedge, x_2^\wedge, x_3^\wedge, x_4^\wedge)) > 0 \quad \forall \quad i,j \in \{1,2,3\}. \tag{10.12}$$

$$\mathbf{H}_3 := \left[\frac{\partial^2 £}{\partial x_i \partial x_j}(\mathbf{x}^\wedge) \right]$$

$$= \begin{bmatrix} 1 + b^2 x_4^\wedge & 0 & \\ 0 & 1 + b^2 x_4^\wedge & 0 \\ 0 & 0 & 1 + a^2 x_4^\wedge \end{bmatrix} \in \mathbb{R}^{3\times 3} \qquad (10.13)$$

"eigenvalues"

$$|\mathbf{H}_3 - \Lambda \mathbf{I}_3| = 0 \quad \Longleftrightarrow \qquad (10.14)$$

$$\begin{bmatrix} \Lambda_1 = \Lambda_2 := 1 + b^2 x_4^\wedge = \dfrac{X}{x_1^\wedge} = \dfrac{Y}{x_2^\wedge} \\[2mm] \\ \Lambda_3 := 1 + a^2 x_4^\wedge = \dfrac{Z}{x_3^\wedge} \end{bmatrix} \qquad \begin{matrix} (10.15) \end{matrix}$$

In Sects. 10.3.1 and 10.3.2, we present algebraic solutions of the normal equations (10.11).

10.3.1 Grafarend-Lohse's Mapping of $\mathbb{T}^2 \longrightarrow \mathbb{E}^2_{a,a,b}$

Two approaches are proposed by [141] for mapping $\mathbb{T}^2 \longrightarrow \mathbb{E}^2_{a,a,b}$. The *first approach* which is presented in Solution 10.4 is based on substitution technique. The *second approach* is based on degenerate conics and will not be treated in this book. Instead, we refer the reader to [141]. Let us start with the algorithm that solves the normal equations (10.11) in a closed form. Solution 10.4 outlines the *first and second forward steps* of reduction which lead to a univariate polynomial equation (10.20) of *fourth order* (quartic polynomial) in terms of the *Lagrangean multiplier*. First, the solution of the quartic polynomial is implemented. One then continues to determine with the *backward step* the Cartesian coordinates (x_1, x_2, x_3) of the point $p \in \mathbb{E}^2_{a,a,b}$ by means of the minimum distance mapping of the point $P \in \mathbb{T}^2$ to $p \in \mathbb{E}^2_{a,a,b}$.

Solution 10.4 (Grafarend-Lohse MDM solution).

First forward step

Solve $(i), (ii), (iii)$ for x_1, x_2, x_3 respectively.

$$\begin{bmatrix} (i) & x_1^\wedge(1 + b^2 x_4^\wedge) = X \Rightarrow x_1^\wedge = \dfrac{X}{1 + b^2 x_4^\wedge} \\[2mm] (ii) & x_2^\wedge(1 + b^2 x_4^\wedge) = Y \Rightarrow x_2^\wedge = \dfrac{Y}{1 + b^2 x_4^\wedge} \\[2mm] (iii) & x_3^\wedge(1 + a^2 x_4^\wedge) = Z \Rightarrow x_3^\wedge = \dfrac{Z}{1 + a^2 x_4^\wedge}. \end{bmatrix} \qquad (10.16)$$

Second forward step

Substitute $(x_1^\wedge, x_2^\wedge, x_3^\wedge, x_4^\wedge)$ in (10.11iv)

$$x_1^{\wedge 2} + x_2^{\wedge 2} = \frac{1}{(1 + b^2 x_4^\wedge)^2}(X^2 + Y^2) \tag{10.17}$$

$$x_3^{\wedge 2} = \frac{1}{(1 + a^2 x_4^\wedge)^2} Z^2 \tag{10.18}$$

$$\left[\begin{array}{l} b^2(x_1^{\wedge 2} + x_2^{\wedge 2}) + a^2 x_3^{\wedge 2} - a^2 b^2 = 0 \Leftrightarrow \\[2mm] \Leftrightarrow b^2 \dfrac{X^2 + Y^2}{(1 + b^2 x_4^\wedge)^2} + a^2 \dfrac{Z^2}{(1 + a^2 x_4^\wedge)^2} - a^2 b^2 = 0. \end{array} \right. \tag{10.19}$$

Multiply (10.19) by $(1 + a^2 x_4)^2(1 + b^2 x_4)^2$ leads to the quartic polynomial (10.20).

$$(10.19i) \quad \left[\begin{array}{l} b^2(1 + a^2 x_4)^2(X^2 + Y^2) + a^2(1 + b^2 x_4)^2 Z^2 \\ -a^2 b^2(1 + a^2 x_4)^2(1 + b^2 x_4)^2 = 0 \end{array} \right.$$

$$\Longleftrightarrow (1 + 2a^2 x_4 + a^4 x_4^2)b^2(X^2 + Y^2) + (1 + 2b^2 x_4 + b^4 x_4^2)a^2 Z^2$$
$$-a^2 b^2(1 + 2a^2 x_4 + a^4 x_4^2)(1 + 2b^2 x_4 + b^4 x_4^2) = 0$$

$$(10.19ii) \quad \left[\begin{array}{l} -x_4^4 a^6 b^6 - 2x_4^3 a^4 b^4(a^2 + b^2) \\[2mm] +x_4^2 a^2 b^2[a^2(X^2 + Y^3) + b^2 Z^2 - 4a^2 b^2 - a^4 - b^4] + \\[2mm] 2x_4 a^2 b^2(X^2 + Y^2 + Z^2) + b^2(X^2 + Y^2) + a^2 Z^2 - a^2 b^2 = 0 \end{array} \right.$$

$$\left[\begin{array}{l} x_4^4 + 2x_4^3 \dfrac{a^2 + b^2}{a^2 b^2} + x_4^2 \dfrac{4a^2 b^2 + a^4 + b^4 - a^2(X^2 + Y^2) - b^2 Z^2}{a^4 b^4} \\[4mm] -2x_4 \dfrac{X^2 + Y^2 + Z^2}{a^4 b^4} - \dfrac{b^2(X^2 + Y^2) + a^2 Z^2 - a^2 b^2}{a^6 b^6} = 0 \end{array} \right. \tag{10.20}$$

Backward step

Substitute $x_4^\wedge$ into $x_1^\wedge(x_4^\wedge), x_2^\wedge(x_4^\wedge), x_3^\wedge(x_4^\wedge)$

$$x_1^\wedge = (1 + b^2 x_4^\wedge)^{-1} X, \quad x_2^\wedge = (1 + b^2 x_4^\wedge)^{-1} Y, \quad x_3^\wedge = (1 + a^2 x_4^\wedge)^{-1} Z \tag{10.21}$$

Test

$$\Lambda_1 = \Lambda_2 = 1 + b^2 x_4^\wedge > 0, \quad \Lambda_3 = 1 + a^2 x_4^\wedge > 0 \tag{10.22}$$

if $\Lambda_1 = \Lambda_2 > 0$ and $\Lambda_3 > 0$ then end.

10.3.2 Groebner Basis' Mapping of $\mathbb{T}^2 \longrightarrow \mathbb{E}^2_{a,a,b}$

Without the various forward and backward reduction steps, we could automatically generate an equivalent algorithm for solving the normal equations (10.11i)–(10.11iv) in a closed form by means of Groebner basis approach. Let us write the *Ideal* of the polynomials in *lexicographic order* "$x_1 > x_2 > x_3 > x_4$" (read: x_1 before x_2 before x_3 before x_4) as

$$Ideal\ I := \left\langle \left\{ \begin{array}{l} x_1 + b^2 x_1 x_4 - X, \\ x_2 + b^2 x_2 x_4 - Y, \\ x_3 + a^2 x_3 x_4 - Z, \\ b^2 x_1^2 + b^2 x_2^2 - a^2 x_3^2 - a^2 b^2 \end{array} \right\} \right\rangle. \tag{10.23}$$

Expressing the generators of Ideal (10.23) as

$$\left[\begin{array}{l} f_1 := x_1 + b^2 x_1 x_4 - X, \\ f_2 := x_2 + b^2 x_2 x_4 - Y, \\ f_3 := x_3 + a^2 x_3 x_4 - Z, \\ f_4 := b^2 x_1^2 + b^2 x_2^2 - a^2 x_3^2 - a^2 b^2, \end{array} \right. \tag{10.24}$$

the Groebner basis of these generators, characteristic for the minimum distance mapping problem, are computed using (4.36) on p. 43 as

$$Groebner\,Basis[\{f_1, f_2, f_3, f_4\}, \{x_1, x_2, x_3, x_4\}]. \tag{10.25}$$

Groebner basis computation (10.25) leads to 14 elements presented in Solution 10.5 interpreted as follows: The *first expression* is a univariate polynomial of order four (quartic) in the Lagrange multiplier, i.e.,

$$\left[\begin{array}{l} \boxed{c_4 x_4^4 + c_3 x_4^3 + c_2 x_4^2 + c_1 x_4 + c_o = 0} \\ c_4 = a^6 b^6 \\ c_3 = (2a^6 b^4 + 2a^4 b^6) \\ c_2 = (a^6 b^2 + 4a^4 b^4 + a^2 b^6 - a^4 b^2 X^2 - a^4 b^2 Y^2 - a^2 b^4 Z^2) \\ c_1 = (2a^4 b^2 + 2a^2 b^4 - 2a^2 b^2 X^2 - 2a^2 b^2 Y^2 - 2a^2 b^2 Z^2) \\ c_o = (a^2 b^2 - b^2 X^2 - b^2 Y^2 - a^2 Z^2), \end{array} \right. \tag{10.26}$$

and is identical to (10.19ii). With the admissible values x_4 substituted in linear equations (4),(8),(12) of the computed Groebner basis, i.e.,

$$\left[\begin{array}{l} (1 + a^2 x_4)x_3 - Z \\ (1 + b^2 x_4)x_2 - Y \\ (1 + b^2 x_4)x_1 - X, \end{array}\right. \tag{10.27}$$

the values $(x_1, x_2, x_3) = (x, y, z)$ are finally produced.

Solution 10.5 (Groebner basis MDM solution).

(1) $\left[\begin{array}{l} a^2 b^2 x_4^4 + (2a^6 b^4 + 2a^4 b^6)x_4^3 + (a^6 b^2 + 4a^4 b^4 + a^2 b^6 - a^4 b^2 X^2 - a^4 b^2 Y^2 - \\ a^2 b^4 Z^2)x_4^2 + +(2a^4 b^2 + 2a^2 b^4 - 2a^2 b^2 X^2 - 2a^2 b^2 Y^2 - 2a^2 b^2 Z^2)x_4 \\ +(a^2 b^2 - b^2 X^2 - b^2 Y^2 - a^2 Z^2). \end{array}\right.$

(2) $\left[\begin{array}{l} (a^4 Z - 2a^2 b^2 Z + b^4 Z)x_3 - a^6 b^6 x_4^3 - (2a^6 b^4 + a^4 b^6)x_4^2 \\ -(a^6 b^2 + 2a^4 b^4 - a^4 b^2 X^2 - a^4 b^2 Y^2 - a^2 b^4 Z^2)x_4 \\ -a^2 b^4 + a^2 b^2 X^2 + a^2 b^2 Y^2 + 2a^2 b^2 Z^2 - b^4 Z^2. \end{array}\right.$

(3) $\left[\begin{array}{l} (2b^2 Z + b^4 x_4 Z - a^2 Z)x_3 + a^4 b^6 x_4^3 + (2a^4 b^4 + a^2 b^6)x_4^2 \\ +(a^4 b^2 + 2a^2 b^4 - a^2 b^2 X^2 - a^2 b^2 Y^2 - b^4 Z^2)x_4 \\ +a^2 b^2 - b^2 X^2 - b^2 Y^2 - 2b^2 Z^2. \end{array}\right.$

(4) $\qquad\qquad (1 + a^2 x_4)x_3 - Z$

(5) $\left[\begin{array}{l} (a^4 - 2a^2 b^2 + b^4)x_3^2 + (2a^2 b^2 Z - 2b^4 Z)x_3 \\ -a^4 b^6 x_4^2 - 2a^4 b^4 x_4 - a^4 b^2 + a^2 b^2 X^2 + a^2 b^2 Y^2 + b^4 Z^2). \end{array}\right.$

(6) $\left[\begin{array}{l} (2b^2 - a^2 + b^4 x_4)x_3^2 - a^2 Z x_3 + a^4 b^6 x_4^3 + (2a^4 b^4 + 2a^2 b^6)x_4^2 \\ + + (a^4 b^2 + 4a^2 b^4 - a^2 b^2 X^2 - a^2 b^2 Y^2 - b^4 Z^2)x_4 \\ +2a^2 b^2 - 2b^2 X - 2bY^2 - 2b^2 Z^2. \end{array}\right.$

(7) $\left[\begin{array}{l} (X^2 + Y^2)x_2 + a^2 b^4 Y x_4^2 + Y(a^2 b^2 - b^2 x_3^2 - b^2 Z x_3)x_4 \\ +Y x_3^2 - Y^3 - Y Z x_3 - Y X^2. \end{array}\right.$

(8) $\qquad\qquad (1 + b^2 x_4)x_2 - Y$

(9) $\qquad\qquad a^2 x_3 - b^2 x_3 + b^2 Z)x_2 - a^2 x_3 Y$

(10) $\qquad\qquad Y x_1 - X x_2$

(11) $\quad X x_1 + a^2 b^4 x_4^2 + (a^2 b^2 + b^2 x_3^2 - b^2 Z x_3)x_4 + x_3^2 - Z x_3 + Y x_2 - X^2 - Y^2.$

(12) $\qquad\qquad (1 + b^2 x_4)x_1 - X$

(13) $\qquad\qquad (a^2 x_3 - b^2 x_3 + b^2 Z)x_1 - a^2 X x_3$

(14) $\quad x_1^2 + a^2 b^4 x_4^2 + (2a^2 b^2 + b^2 x_3^2 - b^2 Z x_3)x_4 + 2x_3^2 - 2Z x_3 + x_2^2 - X^2 - Y^2.$

Once the ellipsoidal Cartesian coordinates $\{x_1, x_2, x_3\}$ have been computed using either Solutions 10.4 or 10.5, they are transformed into their equivalent Gauss ellipsoidal coordinates $\{L, B, H\}$ using (10.28), (10.29) and (10.30) in Solution 10.6.

Solution 10.6 (Coordinates transformation from Cartesian to Gauss ellipsoidal).

$$\{X, Y, Z\} \in \mathbb{T}^2 \quad \{x_1, x_2, x_3\} \in \mathbb{E}^2_{a,a,b} \text{ to } \{L, B, H\}$$

"Pythagoras in three dimension"

$$H := \sqrt{(X - x_1)^2 + (Y - x_2)^2 + (Z - x_3)^2} \tag{10.28}$$

"convert $\{x_1, x_2, x_3\}$ and $\{X, Y, Z\}$
to $\{L, B\}$"

$$\tan L = \frac{Y - x_2}{X - x_1} = \frac{Y - y}{X - x} \tag{10.29}$$

$$\tan B = \frac{Z - x_3}{\sqrt{(X - x_1)^2 + (Y - x_2)^2}} = \frac{Z - x_3}{\sqrt{(X - x)^2 + (Y - y)^2}} \tag{10.30}$$

Example 10.1 (Example from [141]). Given are the geometric parameters of the ellipsoid of revolution; *semi-major axis* $a = 6378137.000$m and *first numerical eccentricity* $e^2 = 0.00669437999013$ from which the *semi-minor axis* b is to be computed. The input data are Cartesian coordinates of 8 points on the surface of the Earth presented in Table 10.2. Using these data, the coefficients of the univariate polynomial

Table 10.2. Cartesian coordinates of topographic points

Point	$X(m)$	$Y(m)$	$Z(m)$
1	3980192.960	0	4967325.285
2	0	0	6356852.314
3	0	0	-6357252.314
4	4423689.486	529842.355	4555616.169
5	4157619.145	664852.698	4775310.888
6	-2125699.324	6012793.226	-91773.648
7	5069470.828	3878707.846	-55331.828
8	213750.930	5641092.098	2977743.624

(10.26) are computed and used in the Matlab's roots command, e.g., (4.40) on p. 44 as $x_4 = roots\,[\,c_4\ c_3\ c_2\ c_1\ c_0\,]$. The obtained roots are then substituted in (10.27) to give the values of $\{x_3, x_2, x_1\}$ of the ellipsoidal Cartesian coordinates. The computed results presented in Table 10.3 are identical to those obtained by [141, Table 4, p. 108]. Once the ellipsoidal Cartesian coordinates have been derived, the ellipsoidal coordinates (ellipsoidal longitude L, ellipsoidal latitude B and height H) can be computed using (10.28), (10.29) and (10.30) in Solution 10.6.

Table 10.3. Computed ellipsoidal cartesian coordinates and the Lagrange factor

Point	$x_1(m)$	$x_2(m)$	$x_3(m)$	$x_4(m^{-2})$
1	3980099.549	0.000	4967207.921	5.808116e-019
2	0.000	0.000	6356752.314	3.867016e-019
3	0.000	0.000	-6356752.314	1.933512e-018
4	4420299.446	529436.317	4552101.519	1.897940e-017
5	4157391.441	664816.285	4775047.592	1.355437e-018
6	-2125695.991	6012783.798	-91773.503	3.880221e-020
7	5065341.132	3875548.170	-55286.450	2.017617e-017
8	213453.298	5633237.315	2973569.442	3.450687e-017

Example 10.2 (Case study: Baltic sea level project). Let us adopt the *world geodetic datum 2000* with the semi-major axis a=6378136.602 m and semi-minor axis b=6356751.860 m from [136]. Here we take advantage of given Cartesian coordinates of 21 points of the topographic surface of the Earth presented in Table 10.4. Using these data, the coefficients of (10.26) are computed and used to solve for x_4. With the admissible values of x_4 substituted in (10.27), the values of the ellipsoidal Cartesian coordinates $(x_1, x_2, x_3) = (x, y, z)$ are produced and are as presented in Table 10.5. They are finally converted by means of Solution 10.6 to (L, B, H) in Table 10.6. Figure 10.6 depicts the mapping of topographical points onto the reference ellipsoid.

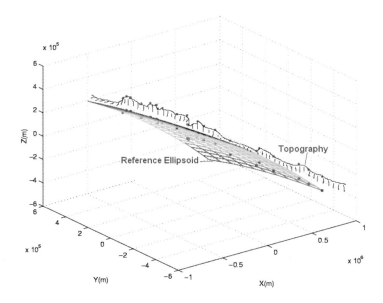

Fig. 10.6. Baltic sea level project topographic points mapped on to the *international reference ellipsoid* $\mathbb{E}^2_{a,a,b}$

Table 10.4. Baltic sea level project: Cartesian coordinates of topographic points

Station	$X(m)$	$Y(m)$	$Z(m)$
Borkum (Ger)	3770667.9989	446076.4896	5107686.2085
Degerby (Fin)	2994064.9360	1112559.0570	5502241.3760
Furuoegrund (Swe)	2527022.8721	981957.2890	5753940.9920
Hamina (Fin)	2795471.2067	1435427.7930	5531682.2031
Hanko (Fin)	2959210.9709	1254679.1202	5490594.4410
Helgoland (Ger)	3706044.9443	513713.2151	5148193.4472
Helsinki (Fin)	2885137.3909	1342710.2301	5509039.1190
Kemi (Fin)	2397071.5771	1093330.3129	5789108.4470
Klagshamn (Swe)	3527585.7675	807513.8946	5234549.7020
Klaipeda (Lit)	3353590.2428	1302063.0141	5249159.4123
List/Sylt (Ger)	3625339.9221	537853.8704	5202539.0255
Molas (Lit)	3358793.3811	1294907.4149	5247584.4010
Mntyluoto (Fin)	2831096.7193	1113102.7637	5587165.0458
Raahe (Fin)	2494035.0244	1131370.9936	5740955.4096
Ratan (Swe)	2620087.6160	1000008.2649	5709322.5771
Spikarna (Swe)	2828573.4638	893623.7288	5627447.0693
Stockholm (Swe)	3101008.8620	1013021.0372	5462373.3830
Ustka (Pol)	3545014.3300	1073939.7720	5174949.9470
Vaasa (Fin)	2691307.2541	1063691.5238	5664806.3799
Visby (Swe)	3249304.4375	1073624.8912	5364363.0732
OElands N. U. (Swe)	3295551.5710	1012564.9063	5348113.6687

Table 10.5. Computed ellipsoidal Cartesian coordinates $(x_1, x_2, x_3) = (x, y, z)$ and Lagrange multiplier x_4

Station	$x_1(m)$	$x_2(m)$	$x_3(m)$	$x_4(m^{-2})$
Borkum (Ger)	3770641.3815	446073.3407	5107649.9100	1.746947e-019
Degerby (Fin)	2994054.5862	1112555.2111	5502222.2279	8.554612e-020
Furuoegrund (Swe)	2527009.7166	981952.1770	5753910.8356	1.288336e-019
Hamina (Fin)	2795463.7019	1435423.9394	5531667.2524	6.643801e-020
Hanko (Fin)	2959199.2560	1254674.1532	5490572.5584	9.797001e-020
Helgoland (Ger)	3706019.4100	513709.6757	5148157.7376	1.705084e-019
Helsinki (Fin)	2885126.2764	1342705.0575	5509017.7534	9.533532e-020
Kemi (Fin)	2397061.6153	1093325.7692	5789084.2263	1.028464e-019
Klagshamn (Swe)	3527564.6083	807509.0510	5234518.0924	1.484413e-019
Klaipeda (Lit)	3353562.2593	1302052.1493	5249115.3164	2.065021e-019
List/Sylt (Ger)	3625314.3442	537850.0757	5202502.0726	1.746017e-019
Molas (Lit)	3358777.7367	1294901.3835	5247559.7944	1.152676e-019
Mntyluoto (Fin)	2831087.1439	1113098.9988	5587146.0214	8.370165e-020
Raahe (Fin)	2494026.5401	1131367.1449	5740935.7483	8.418639e-020
Ratan (Swe)	2620078.1000	1000004.6329	5709301.7015	8.988111e-020
Spikarna (Swe)	2828561.2473	893619.8693	5627422.6007	1.068837e-019
Stockholm (Swe)	3100991.6259	1013015.4066	5462342.8173	1.375524e-019
Ustka (Pol)	3544995.3045	1073934.0083	5174921.9867	1.328158e-019
Vaasa (Fin)	2691299.0138	1063688.2670	5664788.9183	7.577249e-020
Visby (Swe)	3249290.3945	1073620.2512	5364339.7330	1.069551e-019
OElands N. U. (Swe)	3295535.1675	1012559.8663	5348086.8692	1.231803e-019

Table 10.6. Baltic sea level project: Geodetic coordinates computed from ellipsoidal Cartesian coordinates in closed form

Station	Longitude L			Latitude B			Ellipsoidal height H
	°	′	″	°	′	″	m
Borkum (Ger)	6	44	48.5914	53	33	27.4808	45.122
Degerby (Fin)	20	23	4.0906	60	1	52.8558	22.103
Furuoegrund (Swe)	21	14	6.9490	64	55	10.2131	33.296
Hamina (Fin)	27	10	47.0690	60	33	52.9819	17.167
Hanko (Fin)	22	58	35.4445	59	49	21.6459	25.313
Helgoland (Ger)	7	53	30.3480	54	10	29.3979	44.042
Helsinki (Fin)	24	57	24.2446	60	9	13.2416	24.633
Kemi (Fin)	24	31	5.6737	65	40	27.7029	26.581
Klagshamn (Swe)	12	53	37.1597	55	31	20.3311	38.345
Klaipeda (Lit)	21	13	9.0156	55	45	16.5952	53.344
List/Sylt (Ger)	8	26	19.7594	55	1	3.0992	45.101
Molas (Lit)	21	4	58.8931	55	43	47.2453	29.776
Mntyluoto (Fin)	21	27	47.7777	61	35	39.3552	21.628
Raahe (Fin)	24	24	1.8197	64	38	46.8352	21.757
Ratan (Swe)	20	53	25.2392	63	59	29.5936	23.228
Spikarna (Swe)	17	31	57.9060	62	21	48.7645	27.620
Stockholm (Swe)	18	5	27.2528	59	19	20.4054	35.539
Ustka (Pol)	16	51	13.8751	54	35	15.6866	34.307
Vaasa (Fin)	21	33	55.9146	63	5	42.8394	19.581
Visby (Swe)	18	17	3.9292	57	38	21.3487	27.632
OElands N. U. (Swe)	17	4	46.8542	57	22	3.4508	31.823

10.4 Concluding Remarks

The chapter has presented a new and direct algebraic approach to the mapping problem that has attracted alot of research as evidenced in Table 10.1. All that is required is for the user to apply equations (10.26) and (10.27).

11

Positioning by Resection Methods

11.1 Resection Problem and its Importance

In Chap. 9, ranging method for positioning was presented where distances were measured to known targets. In this chapter, an alternative positioning technique which uses direction measurements as opposed to distances is presented. This positioning approach is known as the resection. Unlike in ranging where measured distances are affected by atmospheric refraction, resection methods have the advantage that the measurements are angles or directions which are not affected by refraction.

Resection methods find use in densification of GPS networks. In Fig. 9.1 for example, if the station inside the tunnel or forest is a GPS station, a GPS receiver can not be used due to signal blockage. In such a case, horizontal and vertical directions are measured to three known GPS stations using a theodolite or total station operating in the local positioning systems (LPS). These angular measurements are converted into global reference frame's equivalent using (7.18) and (7.19). The coordinates of the unknown tunnel or forest station is finally computed using resection techniques that we will discuss later in the chapter. A more recent application of resection is demonstrated by [123] who applies it to find the position and orientation of scanner head in object space (Fig. 11.1[1]). The scanner is then used to monitor deformation of a steep hillside in Fig. 11.2[1] which was inaccessible. The only permissible deformation monitoring method was through remote sensing scanning technique.

To understand the resection problem, consider Fig. 11.3.

---

[1]Courtesy of Survey Review: Gordon and Lichti (2004)

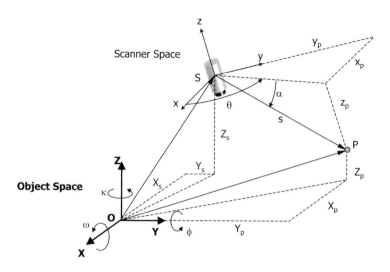

Fig. 11.1. Position and orientation of scanner head
©Survey Review: Gordon and Lichti (2004)

Fig. 11.2. Slope and Lower Walkway at Kings Park
©Survey Review: Gordon and Lichti (2004)

The planar (two-dimensional) resection problem is formulated as
follows: Given horizontal direction measurements T_i from unknown
station $P_0 \in \mathbb{E}^2$ to three known stations $P_i | i = 1, 2, 3 \in \mathbb{E}^2$ in Fig.
11.3, determine the position $\{x_0, y_0\}$ and orientation $\{\sigma\}$ of P_0. For
the three-dimensional resection, the unknown position $\{X_0, Y_0, Z_0\}$ of

point $P_0 \in \mathbb{E}^3$ and the orientation unknown Σ have to be determined. In this case therefore, in addition to horizontal directions T_i, vertical directions B_i have to be measured. In photogrammetry, image coordinates on the photographs are used instead of direction measurements.

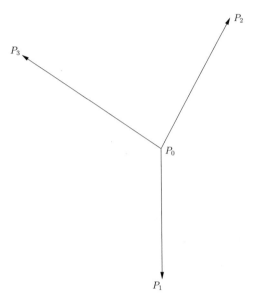

Fig. 11.3. Planar resection

Equations relating unknowns and the measurements are nonlinear and the solution has been by linearized numerical iterative techniques. This has been mainly due to the difficulty of solving in closed form the underlying nonlinear systems of equations. Procedures for solving planar nonlinear resection are reported by [68] to exceed 500! Several procedures put forward as early as 1900 concentrated on the solution of the overdetermined version as evidenced in the works of [173, 282, 331, 332, 333]. Most of these works were based on the graphical approaches. Procedures to solve *closed form planar resection* were put forward by [75] and later by [12, 32, 64, 156, 195].

The search towards the solution of the three-dimensional resection problem traces its origin to the work of a German mathematician J. A. Grunert [162] whose publication appeared in the year 1841. Grunert (1841) solved the three-dimensional resection problem – what was then known as the *"Pothenot's"* problem – in a closed form by solving an algebraic equation of degree four. The problem had hitherto been solved by iterative means mainly in photogrammetry and computer vision. Procedures developed later for solving the three-dimensional resection problem revolved around improvements of the approach of [162] with the aim of searching for optimal means of distances determination. Whereas [162] solved the problem by substitution approach in *three steps*, more recent desire has been to solve the distance equations in less steps as exemplified in the works of [107, 109, 156, 224, 225, 249]. In [176, 177, 255], extensive review of these procedures are presented. Other solutions of three-dimensional resection include the works of [11, 24, 25, 123] among others. The closed form solution of overdetermined three-dimensional resection is presented in [26] and elaborate literature on the subject presented in [11].

In this chapter, *Grunert's distance equations* for three-dimensional resection problem are solved using the algebraic techniques of Groebner basis and polynomial resultants. The resulting quartic polynomial is solved for the unknown distances and the admissible solution substituted in any equation of the original system of polynomial equations to determine the remaining two distances. Once distances have been obtained, the position $\{X_0, Y_0, Z_0\}$ are computed using the ranging techniques discussed in Chap. 9. The three-dimensional orientation unknown Σ is thereafter solved using partial Procrustes algorithm of Chap. 8.

11.2 Geodetic Resection

11.2.1 Planar Resection

For planar resection, if the horizontal directions are oriented arbitrarily, the unknown orientation in the horizontal plane σ has to be determined in addition to position $\{x, y\}$ of the observing unknown station. The coordinates $X_i, Y_i \mid i \in \{1, 2, 3\}$ of the known target stations $P_i \in \mathbb{E}^2 \mid i \in \{1, 2, 3\}$ are given in a particular reference frame. Horizontal directions $T_i \mid i \in \{1, 2, 3\}$ are observed from an unknown station to the three known target stations. The task at hand as already stated in Sect. 11.1 is to determine the unknowns $\{x, y, \sigma\}$. The observation equation is formulated as

$$tan(T_i + \sigma) = \frac{y_i - y}{x_i - x} \mid \forall_i = 1, 2, 3. \tag{11.1}$$

Next, we present three approaches which can be used to solve (11.1) namely; conventional analytical solution, Groebner basis and Sylvester resultants methods.

Conventional Analytical Solution

Using trigonometric additions theorem as suggested by [75], (11.1) is expressed as

$$\frac{\tan T_i + \tan \sigma}{1 - \tan T_i \tan \sigma} = \frac{y_i - y}{x_i - x} \mid \forall_i = 1, 2, 3, \tag{11.2}$$

leading to

$$(\tan T_i + \tan \sigma)(x_i - x) = (1 - \tan T_i \tan \sigma)(y_i - y). \tag{11.3}$$

Expanding (11.3) gives

$$y(\tan T_i \tan \sigma) - y_i(\tan T_i \tan \sigma) - y + y_i = x_i \tan T_i + x_i \tan \sigma - x\tan T_i - x\tan \sigma. \tag{11.4}$$

Equation (11.4) leads to a nonlinear system of equations in the unknowns $\{x, y, \sigma\}$ as

$$\begin{bmatrix} y(\tan T_1 \tan \sigma) - y_1(\tan T_1 \tan \sigma) - y + y_1 = x_1 \tan T_1 + x_1 \tan \sigma - x\tan T_1 - x\tan \sigma \\ y(\tan T_2 \tan \sigma) - y_2(\tan T_2 \tan \sigma) - y + y_2 = x_2 \tan T_2 + x_2 \tan \sigma - x\tan T_2 - x\tan \sigma \\ y(\tan T_3 \tan \sigma) - y_3(\tan T_3 \tan \sigma) - y + y_3 = x_3 \tan T_3 + x_3 \tan \sigma - x\tan T_3 - x\tan \sigma, \end{bmatrix} \tag{11.5}$$

which is solved in three steps for σ and then substituted in the first two equations of (11.5) to obtain the unknowns $\{x, y\}$. The procedure is performed stepwise as follows:

Step 1 (elimination): In this step, the variable y and the term $x\tan \sigma$ are eliminated from the three equations by subtracting the second and third expressions of (11.5) from the first. This results in

$$\begin{bmatrix} y\tan\sigma(\tan T_1 - \tan T_2) = (\tan T_2 - \tan T_1)x + (x_1 - x_2 + y_1\tan T_1 \\ \quad -y_2\tan T_2)\tan\sigma + x_1\tan T_1 - x_2\tan T_2 - y_1 + y_2 \\ \\ y\tan\sigma(\tan T_1 - \tan T_3) = (\tan T_3 - \tan T_1)x + (x_1 - x_3 + y_1\tan T_1 \\ \quad -y_3\tan T_3)\tan\sigma + x_1\tan T_1 - x_3\tan T_3 - y_1 + y_3. \end{bmatrix} \tag{11.6}$$

Step 2 (division): The first expression of (11.6) is divided by $(tanT_1 - tanT_2)$ and the second expression by $(tanT_1 - tanT_3)$. This is done in-order to make $ytan\,\sigma$ appearing on the left-hand-side of both equations the subject of the formula. The net results are:

$$\left[\begin{array}{l} ytan\sigma = -x + \dfrac{x_1 - x_2 + y_1 tanT_1 - y_2 tanT_2}{tanT_1 - tanT_2} tan\sigma \\[2em] \qquad + \dfrac{x_1 tanT_1 - x_2 tanT_2 - y_1 + y_2}{tanT_1 - tanT_2} \\[2em] ytan\sigma = -x + \dfrac{x_1 - x_3 + y_1 tanT_1 - y_3 tanT_3}{tanT_1 - tanT_3} tan\sigma \\[2em] \qquad + \dfrac{x_1 tanT_1 - x_3 tanT_3 - y_1 + y_3}{tanT_1 - tanT_3}. \end{array}\right. \tag{11.7}$$

Step 3 (elimination): In (11.7), we note that $ytan\,\sigma$ and x appear in both expressions. They are eliminated by subtracting the second expression from the first. On re-arranging the resulting expression leads to $tan\sigma$ on the left-hand-side as

$$\left[\begin{array}{l} \qquad\qquad tan\,\sigma = \dfrac{N}{D} \\[2em] N = \dfrac{x_1 tanT_1 - x_2 tanT_2 - y_1 + y_2}{tanT_1 - tanT_2} + \dfrac{x_3 tanT_3 - x_1 tanT_1 - y_3 + y_1}{tanT_1 - tanT_3} \\[2em] D = \dfrac{x_1 - x_3 + y_1 tanT_1 - y_3 tanT_3}{tanT_1 - tanT_3} + \dfrac{x_2 - x_1 + y_2 tanT_2 - y_1 tanT_1}{tanT_1 - tanT_2} \end{array}\right. \tag{11.8}$$

Step 4 (solution of $\{x, y\}$): Once we have solved for σ in (11.8), the first and the second expressions of (11.5) are re-written in the final step with x, y on the left-hand-side as

$$\left[\begin{array}{l} a_{11}y + a_{12}x = b_{11} \\ a_{21}y + a_{22}x = b_{22}, \end{array}\right. \tag{11.9}$$

where;

$$\left[\begin{array}{l} a_{11} = (tanT_1 tan\sigma - 1), \\ a_{12} = (tanT_1 + tan\sigma), \\ a_{21} = (tanT_2 tan\sigma - 1), \\ a_{22} = (tanT_2 + tan\sigma), \\ b_{11} = y_1 tanT_1 tan\sigma - y_1 + x_1 tanT_1 + x_1 tan\sigma, \\ b_{22} = y_2 tanT_2 tan\sigma - y_2 + x_2 tan\sigma + x_2 tanT_2. \end{array}\right.$$

In matrix form, (11.9) is expressed as

$$\begin{bmatrix} a_{11} & a_{12} \\ a_{21} & a_{22} \end{bmatrix} \begin{bmatrix} y \\ x \end{bmatrix} = \begin{bmatrix} b_{11} \\ b_{22} \end{bmatrix}, \tag{11.10}$$

giving the solutions as

$$\begin{bmatrix} y \\ x \end{bmatrix} = (a_{11}a_{22} - a_{21}a_{12})^{-1} \begin{bmatrix} a_{22} & -a_{12} \\ -a_{21} & a_{11} \end{bmatrix} \begin{bmatrix} b_{11} \\ b_{22} \end{bmatrix} \tag{11.11}$$

or

$$\begin{aligned} y &= (a_{11}a_{22} - a_{21}a_{12})^{-1}(a_{22}b_{11} - b_{22}a_{12}) \\ x &= (a_{11}a_{22} - a_{21}a_{12})^{-1}(a_{11}b_{22} - b_{11}a_{21}) \end{aligned}, \tag{11.12}$$

which completes the conventional analytic solution.

Groebner Basis Approach

Denoting $a = \tan T_1$, $b = \tan T_2$, $c = \tan T_3$, and $d = \tan \sigma$, (11.5) is simplified in lexicographic order $y > x > d$ as

$$\begin{bmatrix} f_1 := -y + ady + ax + xd - y_1ad - x_1d - x_1a + y_1 = 0 \\ f_2 := -y + bdy + bx + xd - y_2bd - x_2d - x_2b + y_2 = 0 \\ f_3 := -y + cdy + cx + xd - y_3cd - x_3d - x_3c + y_3 = 0. \end{bmatrix} \tag{11.13}$$

The reduced Groebner basis (4.38) on p. 44 is then computed as

$$\begin{bmatrix} GroebnerBasis\left[\{f_1,\ f_2,\ f_3\},\{x,\ y,\ d\},\{x,\ y\}\right] \\ GroebnerBasis\left[\{f_1,\ f_2,\ f_3\},\{x,\ y,\ d\},\{y\}\right] \\ GroebnerBasis\left[\{f_1,\ f_2,\ f_3\},\{x,\ y,\ d\},\{x\}\right]. \end{bmatrix} \tag{11.14}$$

The first expression of (11.14) gives a linear equation in the variable d allowing the computation of the unknown orientation parameter σ. The second and the third expressions respectively give linear equations in x and y in the variable d. The computed reduced Groebner basis rearranged with the unknown terms on the left-hand-side are presented in Solution 11.1.

Solution 11.1 (Reduced Groebner basis computation of planar resection).

$$\left[\begin{array}{l} d = \dfrac{N1}{D1} \\[4mm] x = -\dfrac{N2}{(-cd^2 + a - c + ad^2)} \\[4mm] y = -\dfrac{N3}{(b - c + bd^2 - cd^2)}, \end{array} \right. \qquad (11.15)$$

where

$$\left[\begin{array}{l} N1 = -(abX_1 - acX_1 + aY_2 - abX_2 + bcX_2 - aY_3 + acX_3 \\ \qquad\quad -bcX_3 - bY_1 + cY_1 - cY_2 + bY_3) \\[3mm] D1 = (bX_1 - aX_2 + aX_3 - bX_3 + abY_1 - acY_1 - cX_1 - abY_2 \\ \qquad\quad +bcY_2 + cX_2 + acY_3 - bcY_3), \end{array} \right.$$

$$N2 = \left[\begin{array}{l} (-adY_1 - cdY_1 + acd^2Y_1 - Y_3 - aX_1 - dX_1 + acdX_1 + cd^2X_1 \\ +cX_3 + dX_3 - acdX_3 - ad^2X_3 + Y_1 + adY_3 + cdY_3 - acd^2Y_3) \end{array} \right.$$

and

$$N3 = \left[\begin{array}{l} (-bcX_2 - bdX_2 - cdX_2 - d^2X_2 + bcX_3 + bdX_3 + cdX_3 + d^2X_3 \\ +cY_2 + dY_2 - bcdY_2 - bd^2Y_2 - bY_3 - dY_3 + bcdY_3 + cd^2Y_3). \end{array} \right.$$

Once d has been computed from the first expression of (11.15), it is inserted into the second and third expressions to solve the unknowns $\{x, y\}$ respectively. The unknown orientation in the horizontal plane is then computed via $\sigma = tan^{-1}d$.

Sturmfels' Resultant Approach

Let z be a homogenizing variable for (11.13). In-order to solve for the variable d in (11.13), we hide d by making it a polynomial of degree zero (i.e., treating it as a constant) as

$$\left[\begin{array}{l} g_1 := -y + ady + ax + dx + (-y_1 ad - x_1 d - x_1 a + y_1)z = 0 \\ g_2 := -y + bdy + bx + dx + (-y_2 bd - x_2 d - x_2 b + y_2)z = 0 \\ g_3 := -y + cdy + cx + dx + (-y_3 cd - x_3 d - x_3 c + y_3)z = 0, \end{array} \right. \tag{11.16}$$

which is expressed in the form (5.12) as

$$J_d = \det \begin{bmatrix} \dfrac{\partial g_1}{\partial x} & \dfrac{\partial g_1}{\partial y} & \dfrac{\partial g_1}{\partial z} \\[2mm] \dfrac{\partial g_2}{\partial x} & \dfrac{\partial g_2}{\partial y} & \dfrac{\partial g_2}{\partial z} \\[2mm] \dfrac{\partial g_3}{\partial x} & \dfrac{\partial g_3}{\partial y} & \dfrac{\partial g_3}{\partial z} \end{bmatrix} = \det \begin{bmatrix} (a+d) & (ad-1) & (y_1 - y_1 ad - x_1 d - x_1 a) \\ (b+d) & (bd-1) & (y_2 - y_2 bd - x_2 d - x_2 b) \\ (c+d) & (cd-1) & (y_3 - y_3 cd - x_3 d - x_3 c) \end{bmatrix}.$$

$$(11.17)$$

Equation (11.17) leads to

$$\boxed{e_3 d^3 + e_2 d^2 + e_1 d + e_0 = 0},$$

with

$$e_3 = ax_3 - bx_3 - ay_2 b - ax_2 + cx_2 - cx_1 + cy_2 b + ay_3 c - cy_1 a + bx_1 + by_1 a - by_3 c$$

$$e_2 = cy_1 - ax_2 b + ay_2 + ax_3 c - by_1 - ay_3 + bx_1 a + by_3 + cx_2 b - cy_2 - cx_1 a - bx_3 c$$

$$e_1 = ax_3 - bx_3 - ay_2 b - ax_2 + cx_2 - cx_1 + cy_2 b + ay_3 c - cy_1 a + bx_1 + by_1 a - by_3 c$$

$$e_0 = cy_1 - ax_2 b + ay_2 + ax_3 c - by_1 - ay_3 + bx_1 a + by_3 + cx_2 b - cy_2 - cx_1 a - bx_3 c.$$

$$(11.18)$$

The value of d is then solved from (11.18) using Matlab's roots command (see 4.40 on p. 44). Comparing the expressions for d in (11.15) and (11.18), we note that the reduced Groebner basis in Solution 11.1 gave a linear function while the Sturmfels' approach results in a cubic polynomial. Both expressions however lead to the same numerical results. The advantage of reduced Groebner basis over the Sturmfels' approach, however, is that the solution is uniquely determined. Sturmfels' approach requires prior information to choose the admissible value of d from the three solutions. Once this value has been selected, the coordinates $\{x, y\}$ are then solved in terms of d as follows:

1. Hiding x and solving in terms of d from (f_1, f_2) of (11.13) gives

$$\begin{bmatrix} h_1 := (ad-1)y + (ax + dx - y_1 ad - x_1 d - x_1 a + y_1)z = 0 \\ h_2 := (bd-1)y + (bx + dx - y_2 bd - x_2 d - x_2 b + y_2)z = 0. \end{bmatrix}$$

$$(11.19)$$

Applying (5.12) leads to

$$J_x = \det \begin{bmatrix} \dfrac{\partial h_1}{\partial y} & \dfrac{\partial h_1}{\partial z} \\[2mm] \dfrac{\partial h_2}{\partial y} & \dfrac{\partial h_2}{\partial z} \end{bmatrix} = \det \begin{bmatrix} (ad-1) & (ax+dx-y_1ad-x_1d-x_1a+y_1) \\ (bd-1) & (bx+dx-y_2bd-x_2d-x_2b+y_2) \end{bmatrix}.$$

$$(11.20)$$

The Jacobian determinant of (11.20) is

$$\begin{bmatrix} x = -(y_1+ady_2-ad^2y_2b-ad^2x_2-adx_2b+x_1abd-y_2+y_2bd+x_2d \\ +x_2b+y_1ad^2b-y_1ad-y_1bd-x_1d-x_1a+x_1d^2b)/(a+ad^2-b-d^2b). \end{bmatrix}$$

$$(11.21)$$

2. Hiding y and solving in terms of d from (f_2, f_3) of (11.13) gives

$$\begin{bmatrix} k_1 := (b+d)x + (bdy-y-y_2bd-x_2d-x_2b+y_2)z = 0 \\ k_2 := (c+d)x + (cdy-y-y_3cd-x_3d-x_3c+y_3)z = 0, \end{bmatrix}$$

$$(11.22)$$

whose Jacobian determinant

$$J_y = \det \begin{bmatrix} \dfrac{\partial k_1}{\partial x} & \dfrac{\partial k_1}{\partial z} \\[2mm] \dfrac{\partial k_2}{\partial x} & \dfrac{\partial k_2}{\partial z} \end{bmatrix} = \det \begin{bmatrix} (b+d) & (bdy-y-y_2bd-x_2d-x_2b+y_2) \\ (c+d) & (cdy-y-y_3cd-x_3d-x_3c+y_3) \end{bmatrix}.$$

$$(11.23)$$

leads to

$$\begin{bmatrix} y = -(-y_2d+by_3-by_3cd-bx_3d-bx_3c+x_2bc+dy_3-y_3cd^2-x_3d^2 \\ -dx_3c+y_2bdc+y_2bd^2-y^2c+x_2d^2+x_2bd+x_2dc)/(-bd^2-b+cd^2+c) \end{bmatrix}$$

$$(11.24)$$

Once d has been computed from ((11.18), it is used in (11.21) and (11.24) to obtain x and y respectively. The unknown orientation in the horizontal plane can now be computed via $\sigma = \tan^{-1}d$.

Example 11.1. Let us consider the Example given by [195, p. 234] with our axis defined such that the Easting refer to the $Y-$axis and the Northing refer to the $X-$axis. The input data are given in Table 11.1 for the coordinates of three known stations A, B and M which are denoted by P_1, P_2 and P_3 respectively in Fig. 11.3. Table 11.2 gives directional observations $T_i \mid i \in \{1,2,3\}$ from the observing unknown station $P \in \mathbb{E}^2$ (whose unknown x, y coordinates and orientation parameter σ are sought) to three known stations $P_i \in \mathbb{E}^2 \mid i \in \{1,2,3\}$ whose coordinates $X_i, Y_i \mid i \in \{1,2,3\}$ are given in Table 11.1. The obtained results from either reduced Groebner basis or Sturmfels' resultant algebraic approaches are presented in Table 11.3. They are identical to those of [195, p. 234] once we interchange the axes. If Sturmfels' solution is adopted in (11.18), two complex and one real values of d are obtained. The real value, which is identical to that obtained from reduced Groebner basis solution (11.15) is used to solve for σ from $\sigma = tan^{-1}d$.

Table 11.1. Coordinates of known stations $P_i \in \mathbb{E}^2 \mid i \in \{1, 2, 3\}$

Station	Easting $Y(m)$	Northing $X(m)$
P_1	46867.94	5537.00
P_2	51293.86	6365.89
P_3	49666.56	4448.58

Table 11.2. Directions measured from unknown station $P \in \mathbb{E}^2$ to known stations $P_i \in \mathbb{E}^2 \mid i \in \{1, 2, 3\}$

Station	Horizontal directions ° ′ ″		
P_1	60	07	50
P_2	265	18	22
P_3	326	33	59

Table 11.3. Position and orientation of station $P \in \mathbb{E}^2$.

Station	Easting $Y(m)$	Northing $X(m)$	Orientation unknown ° ′ ″		
P	48613.3384	6361.1690			
σ			4	35	34.7

11.2.2 Three-dimensional Resection

Exact Solution

Closed form solution of three-dimensional resection problem concerns itself with the determination of position and orientation of a point P connected by angular observations of type horizontal directions T_i and vertical directions B_i to three known stations P_1, P_2, P_3 (see e.g., Fig.

9.12 on p. 134). From these angular measurements, distances are derived by solving equations known as *Grunert's equations*. Once the distances have been established, the unknown position P is determined using ranging techniques that we discussed in Sect. 9.3.2 of Chap. 9. The closed form solution of the three-dimensional resection problem is completed by solving the unknown orientation parameters that relate the global reference frame $\mathbb{F}^\bullet$ to the local level reference frame of type $\mathbb{F}^*$. As we have already pointed out in Sect. 11.1, several procedures have been suggested for solving Grunert's equations. This section presents three alternative algebraic methods for solving explicitly the three-dimensional resection problem namely; Groebner basis, polynomial resultants and Grafarend-Lohse-Schaffrin methods.

Solution of Grunert's Distance Equations

We begin in Solution 11.2 by deriving Grunert's distance equations. These equations relate;

(i) known distances S_{ij}, $i, j = 1, 2, 3 \,|\, i \neq j$ computed from known stations,

(ii) unknown distances S_i, $i = 1, 2, 3$ between the unknown station $P \in \mathbb{E}^3$, and three known stations $P_i \in \mathbb{E}^3 \,|\, i \in \{1, 2, 3\}$ and

(iii) the spatial angles ψ_{ij}, $i, j = 1, 2, 3 \,|\, i \neq j$ derived from measured horizontal directions T_i and vertical directions B_i in the local level reference frame $\mathbb{F}.^*$

In Solution 11.2, multiplying (7.11) on p. 83 by (11.25) leads to (11.26). After manipulations of (11.27),(11.28) and (11.29), space angles ψ_{ij} can be written in terms of spherical coordinates $\{T_i, B_i\}$, $\{T_j, B_j\}$ of points P_i and P_j with respect to a theodolite orthogonal Euclidean frame $\mathbb{F}^*$ as in (11.30). The Grunert's equations for the three unknown distances S_1, S_2, S_3 are then written in terms of known distances S_{12}, S_{23}, S_{31} and space angles $\psi_{12}, \psi_{23}, \psi_{31}$ (illustrated in Fig. 9.12, p. 134) as in (11.32).

Solution of (11.32) was first proposed by J. A. Grunert [162]. Procedures that were later developed sought to optimize the solution of (11.32) in terms of computational steps. In particular, the interest was to reduce the order of the univariate polynomial that resulted following the solution of (11.32). Such procedures were encountered in Sect. 11.1. In what follows, we present algebraic solution of (11.32).

Solution 11.2 (Derivation of Grunert's distance equations).

$$(-2) \left[\cos T_j \cos B_j, \sin T_j \cos B_j, \sin B_j \right] S_j \tag{11.25}$$

$$(-2)[\cos T_j \cos B_j, \sin T_j \cos B_j, \sin B_j] S_i S_j \begin{bmatrix} \cos T_i \cos B_i \\ \sin T_i \cos B_i \\ \sin B_i \end{bmatrix} = \tag{11.26}$$

$$(-2)[(X_j - X), (Y_j - Y), (Z_j - Z)] \begin{bmatrix} X_i - X \\ Y_i - Y \\ Z_i - Z \end{bmatrix}$$

$$\begin{bmatrix} (X_j - X)(X_i - X) = X_j X_i - X_j X - X_i X + X^2 \\ (X_i - X_j)(X_i - X_j) = X_i^2 - 2X_i X_j + X_j^2 \\ (X_i - X)(X_i - X) = X_i^2 - 2X_i X + X^2 \\ (X_j - X)(X_j - X) = X_j^2 - 2X_j X + X^2 \end{bmatrix} \Rightarrow$$

$$\Rightarrow (X_i - X_j)^2 - (X_i - X)^2 - (X_j - X)^2 = -2(X_j - X)(X_i - X) \tag{11.27}$$

$$(-2)[\cos T_j \cos B_j, \sin T_j \cos B_j, \sin B_j] S_i S_j \begin{bmatrix} \cos T_i \cos B_i \\ \sin T_i \cos B_i \\ \sin B_i \end{bmatrix} = \tag{11.28}$$

$$\left\{ \begin{array}{c} (X_i - X_j)^2 + (Y_i - Y_j)^2 + (Z_i - Z_j)^2 - \\ -(X_i - X)^2 - (Y_i - Y)^2 - (Z_i - Z)^2 - \\ -(X_j - X)^2 - (Y_j - Y)^2 - (Z_j - Z)^2 \end{array} \right\}$$

$$-2 \left\{ \sin B_j \sin B_i + \cos B_j \cos B_i \cos(T_j - T_i) \right\} S_i S_j =$$
$$= \left\{ \begin{array}{c} (X_i - X_j)^2 + (Y_i - Y_j)^2 + (Z_i - Z_j)^2 - \\ -(X_i - X)^2 - (Y_i - Y)^2 - (Z_i - Z)^2 - \\ -(X_j - X)^2 - (Y_j - Y)^2 - (Z_j - Z)^2 \end{array} \right\} \tag{11.29}$$

$$\cos \psi_{ij} = \cos B_i \cos B_j \cos(T_j - T_i) + \sin B_i \sin B_j \tag{11.30}$$

$$\begin{bmatrix} -2 \cos \psi_{ij} S_i S_j = S_{ij}^2 - S_i^2 - S_j^2 \\ S_{ij}^2 = S_i^2 + S_j^2 - 2S_i S_j \cos \psi_{ij} \end{bmatrix} \tag{11.31}$$

$$\begin{bmatrix} S_{12}^2 = S_1^2 + S_2^2 - 2S_1 S_2 \cos \psi_{12} \\ S_{23}^2 = S_2^2 + S_3^2 - 2S_2 S_3 \cos \psi_{23} \\ S_{31}^2 = S_3^2 + S_1^2 - 2S_3 S_1 \cos \psi_{31} \end{bmatrix} \tag{11.32}$$

Groebner Basis Solution of Grunert's Equations

In-order to quicken our understanding of the application of Groebner basis to solve Grunert's distance equations (11.32), let us consider a simple case of a regular tetrahedron. A regular tetrahedron presents a unique case where all the distances and spatial angles of Fig. 9.12 on p. 134 are equal. Instead of computing Groebner basis using 4.36 on p. 43, we will demonstrate by a hand computation how Groebner basis can be computed. Later, we will apply (4.36) to solve the general case of (11.32). We begin by expressing (11.32) in algebraic form

$$\left[\begin{array}{l} x_1^2 - 2a_{12}x_1x_2 + x_2^2 - a_0 = 0 \\ x_2^2 - 2b_{23}x_2x_3 + x_3^2 - b_0 = 0 \\ x_1^2 - 2c_{31}x_1x_3 + x_3^2 - c_0 = 0, \end{array} \right. \tag{11.33}$$

where the unknown distances $\{S_1, S_2, S_3\}$ that appear in (11.32) are denoted by $\{x_1, x_2, x_3\}$. The distances between known stations $\{S_{12}, S_{23}, S_{31}\}$ are denoted by $\{a_o, b_o, c_o\}$, while the constants $\{a_{12}, b_{23}, c_{31}\}$ represent $\{cos\,\psi_{12}, cos\,\psi_{23}, cos\,\psi_{31}\}$ respectively. Equation (11.33) therefore has only the distances $\{x_1, x_2, x_3\}$ as unknowns. These are the distances relating the unknown station P_0 to the known stations $P_i|_{\{i=1,2,3\}}$. Grafarend [156] demonstrated that for each of the quadratic equation in (11.33), there exists an elliptical cylinder in the planes $\{x_1, x_2\}$, $\{x_2, x_3\}$ and $\{x_3, x_1\}$ for the first, second and third equations respectively. These cylinders are constrained to their first quadrant since the distances are positive thus $\{x_1 \in \mathbb{R}^+\}$, $\{x_2 \in \mathbb{R}^+\}$ and $\{x_3 \in \mathbb{R}^+\}$. For a regular tetrahedron, the distances $x_1 = x_2 = x_3$ joining the unknown station $P \in \mathbb{E}^3$ to three known stations $P_i \in \mathbb{E}^3|_{\{i=1,2,3\}}$ are all equal to the distances $S_{12} = S_{23} = S_{31}$ between the known stations. Let us give these distances a value $+\sqrt{d}$. The spatial angles are also equal (i.e., $\psi_{12} = \psi_{23} = \psi_{31} = 60°$). In Solution 11.3, a hand computation of Groebner basis of (11.33) is carried out and used to find the Grunert's distances for the regular tetrahedron (i.e., show that the desired solutions for $\{x_1, x_2, x_3\} \in \mathbb{R}^+$ are $x_1 = x_2 = x_3 = +\sqrt{d}$.)

Solution 11.3 (Hand computation of Groebner basis of (11.32) for a regular tetrahedron). For a regular tetrahedron, where $\psi_{ij} = 60°$, and $a_0 = b_0 = c_0 = d$, (11.33) is re-written in lexicographic order $x_1 > x_2 > x_3$ as

$$\left[\begin{array}{l} x_1^2 - x_1x_2 + x_2^2 - d = 0 \\ x_2^2 - x_2x_3 + x_3^2 - d = 0 \\ x_1^2 - x_1x_3 + x_3^2 - d = 0, \end{array} \right. \tag{11.34}$$

giving rise to the Ideal I (e.g., 4.14 on p. 34) as

$$I = \langle x_1^2 - x_1 x_2 + x_2^2 - d, x_2^2 - x_2 x_3 + x_3^2 - d, x_1^2 - x_1 x_3 + x_3^2 - d \rangle$$
$$\subset \mathbb{R}[x_1, x_2, x_3],$$

$$(11.35)$$

whose generators G are

$$\begin{cases} g_1 = x_1^2 - x_1 x_2 + x_2^2 - d \\ g_2 = x_1^2 - x_1 x_3 + x_3^2 - d \\ g_3 = x_2^2 - x_2 x_3 + x_3^2 - d. \end{cases} \qquad (11.36)$$

Desired now are the Groebner basis (simplified structure) of the generators (11.36) of the Ideal I in (11.35). Using (4.23) on p. 39, the S pair polynomials $(g_1, g_2), (g_1, g_3), (g_2, g_3)$ are computed from the generators (11.36). From B. *Buchberger's* third criterion explained in Chap. 4, we notice that $\text{LM}(g_2) = x_1^2$ divides the $\text{LCM}(g_1, g_3) = x_1^2 x_2^2$. One therefore suppresses (g_1, g_3) and considers only $(g_1, g_2), (g_2, g_3)$ instead. $S(g_1, g_2)$ gives

$$S(g_1, g_2) = -x_1 x_2 + x_1 x_3 + x_2^2 - x_3^2, \qquad (11.37)$$

which is reduced with respect to G by subtracting g_3 to obtain

$$-x_1 x_2 + x_1 x_3 - 2x_3^2 + x_2 x_3 + d. \qquad (11.38)$$

Equation (11.38) does not reduce to zero and is added to the original list G of the generating set of the Ideal I as g_4. The S–polynomial pairs to be considered next are $S(g_2, g_3), S(g_2, g_4)$ and $S(g_3, g_4)$ from the new generating set $G = \{g_2, g_3, g_4\}$. Since $LM(g_2)$ and $LM(g_3)$ are relatively prime, $S(g_2, g_3)$ reduces to zero modulo G ($S(g_2, g_3) \rightarrow_G 0$). The S pair polynomials remaining for consideration are (g_2, g_4) and (g_3, g_4). $S(g_2, g_4)$ gives

$$S(g_2, g_4) = x_1^2 x_3 + x_1 d - 2x_1 x_3^2 + x_2 x_3^2 - x_2 d, \qquad (11.39)$$

which is reduced with respect to G by subtracting $x_3 g_2$ to give

$$x_1 d - x_1 x_3^2 + x_2 x_3^2 - x_2 d - x_3^3 + x_3 d, \qquad (11.40)$$

Equation (11.40) does not reduce to zero and is added to the list G of the generating set of the Ideal I as g_5. The S–polynomial pair to be considered next is $S(g_3, g_4)$ from the new generating set $G = \{g_2, g_3, g_4, g_5\}$. $S(g_3, g_4)$ gives

$$S(g_3, g_4) = -x_1 x_3^2 + x_1 d + 2 x_2 x_3^2 - x_2^2 x_3 - x_2 d, \tag{11.41}$$

which is reduced with respect to G by subtracting g_5 and adding $x_3 g_3$ to give

$$2 x_3^3 - 2 x_3 d. \tag{11.42}$$

Equation (11.42) is a univariate polynomial and completes the solution of the set G of Groebner basis summarized as

$$G := \begin{bmatrix} g_2 = x_1^2 - x_1 x_3 + x_3^2 - d \\ g_3 = x_2^2 - x_2 x_3 + x_3^2 - d \\ g_4 = -x_1 x_2 + x_1 x_3 - 2 x_3^2 + x_2 x_3 + d \\ g_5 = x_1 d - x_1 x_3^2 + x_2 x_3^2 - x_2 d - x_3^3 + x_3 d \\ g_6 = 2 x_3^3 - 2 x_3 d. \end{bmatrix} \tag{11.43}$$

From the computed Groebner basis in (11.43), one notes that the element $g_6 = 2 x_3^3 - 2 x_3 d$ is a *cubic polynomial* in x_3 and readily gives the values of $x_3 = \left\{0, \pm\sqrt{d}\right\}$. The solutions to the Grunert's distance equations (11.33) for a regular tetrahedron are then deduced as follows: Since $S_3 = x_3 \in \mathbb{R}^+$, the value of $S_3 = +\sqrt{d}$. This is substituted back in $g_3 = x_2^2 - x_2 x_3 + x_3^2 - d$ and $g_2 = x_1^2 - x_1 x_3 + x_3^2 - d$ to give $x_2 = \left\{0, +\sqrt{d}\right\}$ and $x_1 = \left\{0, +\sqrt{d}\right\}$ respectively. This completes the solution of Grunert's distance equations (11.33) for the unknown distances $x_1 = x_2 = x_3 = +\sqrt{d}$ as we had initially assumed.

Having demonstrated a hand computation of Groebner basis of the Grunert's distance equations (11.32) for a regular tetrahedron, let us consider next the general case. The geometry of the three-dimensional resection problem in practice is hardly a regular tetrahedron. Beginning by expressing (11.32) algebraically as

$$\begin{bmatrix} g_1 := x_1^2 + x_2^2 + a_{12} x_1 x_2 + a_0 = 0 \\ g_2 := x_2^2 + x_3^2 + b_{23} x_2 x_3 + b_0 = 0 \\ g_3 := x_3^2 + x_1^2 + c_{31} x_3 x_1 + c_0 = 0, \end{bmatrix} \tag{11.44}$$

where

$$\begin{bmatrix} S_1 = x_1 \in \mathbb{R}^+, S_2 = x_2 \in \mathbb{R}^+, S_3 = x_3 \in \mathbb{R}^+, \\ -2 \cos \psi_{12} = a_{12}, -2 \cos \psi_{23} = b_{23}, -2 \cos \psi_{31} = c_{31}, \\ -S_{12}^2 = a_0, -S_{23}^2 = b_0, -S_{31}^2 = c_0, \end{bmatrix} \tag{11.45}$$

one forms the Ideal

$$I = < x_1^2 + x_2^2 + a_{12}x_1x_2 + a_0, \; x_2^2 + x_3^2 + b_{23}x_2x_3 + b_0, \; x_3^2 + x_1^2 + c_{31}x_3x_1 + c_0 > .$$
$$(11.46)$$

We then seek the Groebner basis of the generators of the Ideal (11.46). Following lexicographic ordering $\{x_1 > x_2 > x_3\}$, (4.36) on p. 43 is applied as

$$Groebner Basis[\{g_1, g_2, g_3\}, \{x_1, x_2, x_3\}], \qquad (11.47)$$

giving the Groebner basis of the Ideal (11.46) expressed in [24, Boxes 3-3a and 3-3b]. Distances can also be derived from (11.44) using reduced Groebner basis. We leave it as an exercise for the reader to try and solve the unknown distances $\{x_1, x_2, x_3\}$ in (11.44) using reduced Groebner basis (4.38) on p. 44.

Polynomial resultants' solution of Grunert's distance equations

Besides the use of Groebner bases approach demonstrated above, polynomial resultants techniques can also be used to solve Grunert's equations for distances. We illustrate the solution of the problem using F. Macaulay formulation of Sect. 5.3.1 and B. Sturmfels' formulation presented in Sect. 5.3.2. We start by expressing (11.44) as

$$\begin{bmatrix} R_1 := x_1^2 + x_2^2 + a_{12}x_1x_2 + a_0 = 0 \\ R_2 := x_2^2 + x_3^2 + b_{23}x_2x_3 + b_0 = 0 \\ R_3 := x_1^2 + x_3^2 + c_{31}x_1x_3 + c_0 = 0. \end{bmatrix} \qquad (11.48)$$

Clearly, (11.48) is not homogeneous (see Definition 5.1 on p. 48). It is therefore homogenized by introducing the fourth variable x_4 and treating the variable which is to be solved, say x_1, as a constant (i.e., hiding it by giving it degree zero). The resulting homogenized polynomial is

$$\begin{bmatrix} R_{11} := x_2^2 + a_{12}x_1x_2x_4 + (a_0 + x_1^2)x_4^2 = 0 \\ R_{21} := x_2^2 + x_3^2 + b_{23}x_2x_3 + b_0x_4^2 = 0 \\ R_{31} := x_3^2 + c_{31}x_1x_3x_4 + (x_1^2 + c_0)x_4^2 = 0, \end{bmatrix} \qquad (11.49)$$

which is simplified as

$$\begin{bmatrix} R_{11} := x_2^2 + a_1x_2x_4 + a_2x_4^2 = 0 \\ R_{21} := x_2^2 + x_3^2 + b_1x_2x_3 + b_2x_4^2 = 0 \\ R_{31} := x_3^2 + c_1x_3x_4 + c_2x_4^2 = 0, \end{bmatrix} \qquad (11.50)$$

with the coefficients denoted as $a_1 = a_{12}x_1$, $a_2 = (a_0 + x_1^2)$, $b_1 = b_{23}$, $b_2 = b_0$, $c_1 = c_{31}x_1$, $c_2 = (c_0 + x_1^2)$.

Approach 1 (F. Macaulay Formulation):

The *first step* involves the determination of the total degree of (11.50) using (5.7) on p. 51 which gives $d = 4$. In the *second step*, one formulates the general set comprising the monomials of degree 4 in three variables by multiplying the monomials of (11.50) by each other. These monomials form the elements of the set X^d (e.g., 5.8 on p. 51) as

$$X^d = \left\{ \begin{array}{l} x_2^4,\ x_2^3 x_4,\ x_2^2 x_3^2,\ x_2^3 x_3,\ x_2^2 x_4^2,\ x_2^2 x_3 x_4,\ x_2 x_3^3 \\[2mm] x_2 x_4^3,\ x_2 x_3^2 x_4,\ x_2 x_3 x_4^2,\ x_3^2 x_4^2,\ x_3 x_4^3,\ x_4^4,\ x_3^4,\ x_3^3 x_4 \end{array} \right\}, \quad (11.51)$$

which is now partitioned in *step 3* according to (5.9) on p. 52 as

$$\left[\begin{array}{l} X_i^d = \{ x^\alpha \mid \alpha_i \geq d_i\ and\ \alpha_j < d_j, \forall_j < i \} \\[2mm] X_2^4 = \{ x_2^4,\ x_2^3 x_4,\ x_2^2 x_3^2,\ x_2^3 x_3,\ x_2^2 x_4^2,\ x_2^2 x_3 x_4 \} \\[2mm] X_3^4 = \{ x_2 x_3^2 x_4,\ x_3^2 x_4^2,\ x_2 x_3^3,\ x_3^4,\ x_3^3 x_4 \} \\[2mm] X_4^4 = \{ x_2 x_4^3,\ x_2 x_3 x_4^2,\ x_3 x_4^3,\ x_4^4 \}. \end{array} \right. \quad (11.52)$$

In the *fourth step*, the polynomials F_i are formed using the sets in (11.52) according to (5.10) on p. 52 giving rise to

$$\left[\begin{array}{l} F_1 := \dfrac{X_2^4}{x_2^2} f_1 = \{ x_2^2 f_1,\ x_2 x_4 f_1,\ x_3^2 f_1,\ x_2 x_3 f_1,\ x_4^2 f_1,\ x_3 x_4 f_1 \} \\[4mm] F_2 := \dfrac{X_3^4}{x_3^2} f_2 = \{ x_2 x_4 f_2,\ x_4^2 f_2,\ x_2 x_3 f_2,\ x_3^2 f_2,\ x_3 x_4 f_2 \} \quad (11.53) \\[4mm] F_3 := \dfrac{X_4^4}{x_4^2} f_3 = \{ x_2 x_4 f_3,\ x_2 x_3 f_3,\ x_3 x_4 f_3,\ x_4^2 f_3 \}. \end{array} \right.$$

Finally, the matrix $\mathbf{A}$ of dimension (15×15) is formed as discussed on p. 52. Its rows are the coefficients of the f_i in (11.53) and the columns are the monomials
$\{ c_1 = x_2^4,\ c_2 = x_2^3 x_3,\ c_3 = x_2^3 x_4,\ c_4 = x_2^2 x_3^2,\ c_5 = x_2^2 x_3 x_4,\ c_6 = x_2^2 x_4^2,$
$c_7 = x_2 x_3^3,\ c_8 = x_2 x_3^2 x_4,\ c_9 = x_2 x_3 x_4^2,\ c_{10} = x_2 x_4^3,\ c_{11} = x_3^4,\ c_{12} = x_3^3 x_4,$
$c_{13} = x_3^2 x_4^2,\ c_{14} = x_3 x_4^3$ and $c_{15} = x_4^4 \},$
elements of the sets formed in (11.52). The matrix $\mathbf{A}$ is

$$\mathbf{A} = $$

	c_1	c_2	c_3	c_4	c_5	c_6	c_7	c_8	c_9	c_{10}	c_{11}	c_{12}	c_{13}	c_{14}	c_{15}
$x_2^2 f_1$	1	0	a_1	0	0	a_2	0	0	0	0	0	0	0	0	0
$x_3^2 f_1$	0	0	0	1	0	0	0	a_1	0	0	0	0	a_2	0	0
$x_2 x_3 f_1$	0	1	0	0	a_1	0	0	0	a_2	0	0	0	0	0	0
$x_4^2 f_1$	0	0	0	0	0	1	0	0	0	a_1	0	0	0	0	a_2
$x_3 x_4 f_1$	0	0	0	0	1	0	0	0	0	a_1	0	0	0	a_2	0
$x_2 x_4 f_1$	0	0	1	0	0	a_1	0	0	0	a_2	0	0	0	0	0
$x_2 x_4 f_2$	0	0	1	0	b_1	0	0	1	0	b_2	0	0	0	0	0
$x_4^2 f_2$	0	0	0	0	0	1	0	0	b_1	0	0	0	1	0	b_2
$x_3^2 f_2$	0	0	0	1	0	0	b_1	0	0	0	1	0	b_2	0	0
$x_3 x_4 f_2$	0	0	0	0	1	0	0	b_1	0	0	0	1	0	b_2	0
$x_2 x_3 f_2$	0	1	0	b_1	0	0	1	0	b_2	0	0	0	0	0	0
$x_2 x_3 f_3$	0	0	0	0	0	0	1	c_1	c_2	0	0	0	0	0	0
$x_3 x_4 f_3$	0	0	0	0	0	0	0	0	0	0	0	1	c_1	c_2	0
$x_4^2 f_3$	0	0	0	0	0	0	0	0	0	0	0	0	1	c_1	c_2
$x_2 x_4 f_3$	0	0	0	0	0	0	0	1	c_1	c_2	0	0	0	0	0

The determinant of this matrix is a univariate polynomial of degree 8 in the variable x_1 given in [25, Box 3-1]. Its roots can be obtained using Matlab's *roots* command. Once these roots have been obtained, the admissible solution is substituted in the third expression of (11.48) on p. 181 to obtain the value of $x_3 \in \mathbb{R}^{.+}$ The obtained value of $x_3 \in \mathbb{R}^+$ is in turn substituted in the second expression of (11.48) to obtain the last variable $x_2 \in \mathbb{R}^{.+}$ The admissible values of distances are deduced with the help of prior information.

Approach 2 (B. Sturmfels' Formulation):

From (11.50) on p. 181, the determinant of the Jacobi matrix is computed as

$$J = \det \begin{bmatrix} \dfrac{\partial R_{11}}{\partial x_2} & \dfrac{\partial R_{11}}{\partial x_3} & \dfrac{\partial R_{11}}{\partial x_4} \\[2mm] \dfrac{\partial R_{21}}{\partial x_2} & \dfrac{\partial R_{21}}{\partial x_3} & \dfrac{\partial R_{21}}{\partial x_4} \\[2mm] \dfrac{\partial R_{31}}{\partial x_2} & \dfrac{\partial R_{31}}{\partial x_3} & \dfrac{\partial R_{31}}{\partial x_4} \end{bmatrix}, \tag{11.54}$$

respectively

$$J = \det \begin{bmatrix} 2x_2 + a_1 x_4 & 0 & 2a_2 x_4 + a_1 x_2 \\ 2x_2 + b_1 x_3 & 2x_3 + b_1 x_2 & 2b_2 x_4 \\ 0 & 2x_3 + c_1 x_4 & 2c_2 x_4 + c_1 x_3 \end{bmatrix}, \tag{11.55}$$

which gives a cubic polynomial in x_2, x_3, x_4 as

$$J = 8x_2x_3c_2x_4 + 4x_2c_1x_3^2 + 4b_1x_2^2c_2x_4 + 2b_1x_2^2c_1x_3 - 8x_2b2x_4x_3 - 4x_2b2x_4^2c_1 + 4a_1x_4^2x_3c_2 + 2a_1x_4c_1x_3^2 + 2a_1x_4^2b_1x_2c_2 + 2a_1x_4b_1x_2c_1x_3 - 4a_1x_4^2b2x_3 - 2a_1x_4^3b2c_1 + 8x_2a_2x_4x_3 + 4x_2a_2x_4^2c_1 + 4a_1x_3^2x_3 + 2a_1x_3^2c_1x_4 + 4b_1x_3^2a_2x_4 + 2b_1x_3a_2x_4^2c_1 + 2b_1x_3^2a_1x_2,$$

whose partial derivatives with respect to x_2, x_3, x_4 can be written in the form (5.14) on p. 54. The coefficients b_{ij} and a_{ij} are given as in [25]. The computation of the resultant of the matrix using (5.15) on p. 54 leads to a univariate polynomial in x_1 of degree eight, e.g., [25, Box 3-2].

Fischler and Bolles [109, pp. 386-387, Fig. 5] have demonstrated that because every term in (11.32) is either a constant or of degree 2, for every real positive solution, there exist a geometrically isomorphic negative solution. Thus there are at most four positive solutions to (11.32). This is because (11.32) has eight solutions according to [91, p. 415] who states that for n independent polynomial equations in n unknowns, there can be no more solution than the product of their respective degrees. Since each equation of (11.32) is of degree 2 there can only be up to eight solutions.

Finally, in comparing the polynomial resultants approach to Groebner basis method, the latter in most cases is slow and there is always a risk of the computer breaking down during computations. Besides, the Groebner basis approach computes unwanted intermediary elements which occupy more space and thus lead to storage problems. The overall speed of computation is said to be proportional to twice exponential the number of variables [235, 236, 237, 239]. This has led to various studies advocating for the use of the alternate method; the resultant and specifically multipolynomial resultant approach. Groebner bases can be made faster by computing the reduced Groebner bases as explained in Chap. 4. Polynomial resultants on the other hand involve computing with larger matrices which may require alot of work. For linear systems and ternary quadrics, Sturmfels' approach offers a remedy through the application of the Jacobi determinants. Once the distances have been computed, they are subjected to the ranging techniques (Chap. 9) to compute positions. Finally, the three-dimensional orientation parameters are computed from (8.10) on p. 97.

Grafarend-Lohse-Schaffrin Approach

In this approach, [156] begin by first setting up rigorous projective equations of resection problem in three-dimensional Euclidean space. They classify the equations as six dimensional algebraic system of nonlinear equations of cubic type. In the second part, a three step procedure is adopted for solving the Grunert's distance equations. The nonlinear system of distance equations are projected into linear equations by means of the technique of degenerate quadrics called the stencil method. The stencil method gives the solution of Grunert's equation (11.44) as

$$
\left[
\begin{array}{c}
x_1^2 = S_1^2 = -\dfrac{a_{00}}{1 + 2a_{12}p + p^2} \\[2em]
x_2^2 = S_2^2 = -\dfrac{b_{00}p^2}{p^2 + 2b_{23}pq + q^2} \\[2em]
x_3^2 = S_3^2 = -\dfrac{c_{00}}{1 + 2c_{31}q + q^2}.
\end{array}
\right. \tag{11.56}
$$

The solution for p and q are as discussed in [156]. Once the distances have been solved from (11.56), the three orientation parameters and the cartesian coordinates of the unknown stations are solved from a 6×6 system of linear equations. The linear system of equations are formed using the normalized Hamilton-quaternion (see e.g., p. 10). For a complete discussion on the approach and a numerical example, we refer to papers by [156]. Lohse [225] extends the approach by proposing an alternative solution of the Grunert's distance equations.

Example 11.2 (Three-dimensional resection given three known stations). In-order to position station $K1$ (see Fig. 7.1 on p. 86) by resection method, horizontal directions T_i and vertical directions B_i are measured to three known stations Haussmanstr., Eduardpfeiffer, and Liederhalle. The computation is performed in three steps as follows:

- In the *first* step, the spatial distances are computed. This involves solving the Grunert's distance equations.
- The *second step* is the computation of the GPS Cartesian coordinates $\{X, Y, Z\}$ of the unknown station $K1 \in \mathbb{E}^3$ in the global reference frame using the algebraic ranging techniques of Chap. 9.
- The three-dimensional orientation parameters are computed in the *final step* using the partial Procrustes approach (see Chap. 8).

Using the computed univariate polynomials in [24, Boxes 3-3a and 3-3b] or [25, Box 3-2], and the observations in Tables 7.1 and 7.3 on p. 88, the distances $S_i = x_i \in \mathbb{R},^+ \ i = \{1,2,3\} \in \mathbb{Z}_+^3$ between the unknown station $K1 \in \mathbb{E}^3$ and the known stations $P_i \in \mathbb{E}^3$ are determined. For control purposes, these distances are as expressed in Fig. 7.1. The unknown station $K1$ is located on top of one of the University's building at Kepler Strasse 11. Points $\{P_1, P_2, P_3\}$ of the tetrahedron $\{PP_1P_2P_3\}$ in Fig. 9.12 correspond to the chosen known GPS stations Haussmannstr., Eduardpfeiffer, and Liederhalle. The distance from $K1$ *to Haussmannstr.* is designated $S_1 = x_1 \in \mathbb{R},^+$ $K1$ *to Eduardpfeiffer* $S_2 = x_2 \in \mathbb{R},^+$ while that of $K1$ *to Liederhalle* is designated $S_3 = x_3 \in \mathbb{R}.^+$ The distances between the known stations $\{S_{12}, S_{23}, S_{31}\} \in \mathbb{R}^+$ are computed from their respective GPS coordinates as indicated in Solution 11.4. Their corresponding space angles $\psi_{12}, \psi_{23}, \psi_{31}$ are computed from (11.30) on p. 177.

In-order to control the computations, the Cartesian GPS coordinates of station $K1$ are also known. Solution 11.4 gives the unknowns distances $\{x_1, x_2, x_3\} \in \mathbb{R}^+$ computed using Groebner basis. The univariate polynomial in x_3 has eight roots, four of which are complex and four real. Of the four real roots two are positive and two are negative. The desired distance $x_3 \in \mathbb{R}^+$ is thus chosen amongst the two positive roots with the help of prior information and substituted in [24, g_{11} in Box 3-3b] to give two solutions of x_1, one of which is positive. Finally the obtained values of $\{x_1, x_3\} \in \mathbb{R}^+$ are substituted in [24, g_5 in Box 3-3b] to obtain the remaining indeterminate x_2. Using this procedure, we have in Solution 11.4 that $S_3 = \{430.5286, 153.7112\}$. Since $S_3 = x_3 \in \mathbb{R},^+$ from prior information (e.g., Fig. 7.1), we choose $S_3 = 430.5286$, leading to $S_1 = 1324.2381$, and $S_2 = 542.2608$. These values compare well with their real values depicted in Fig. 7.1 on p. 86.

Solution 11.4 (Computation of distances for test network Stuttgart Central).
Using the entries of Table 7.1 on p. 88, inter-station distances are computed by Pythagoras $S_{ij} = \sqrt{(X_j - X_i)^2 + (Y_j - Y_i)^2 + (Z_j - Z_i)^2}$, and spatial angles obtained from (11.30). The values are

$$
\begin{bmatrix} S_{12} = 1560.3302\,m \\ S_{23} = 755.8681\,m \\ S_{31} = 1718.1090\,m \end{bmatrix} and \begin{bmatrix} \psi_{12} = 1.843620 \\ \psi_{23} = 1.768989 \\ \psi_{31} = 2.664537 \end{bmatrix}
$$

and are substituted in (11.45) on p. 180 to compute the terms a_{12}, b_{23}, c_{31}, a_0, b_0, c_0 which are needed to determine the coefficients of the Groebner basis element g_1 in [24, Box 3-3a]. Expressing the univariate polynomial g_1 as $A_8 x_3^8 + A_6 x_3^6 + A_4 x_3^4 + A_2 x_3^2 + A_0 = 0$, the computed coefficients are:

$$\begin{bmatrix} A_0 = 4.833922266706213e + 023 \\ A_2 = -2.306847176510587e + 019 \\ A_4 = 1.104429253262719e + 014 \\ A_6 = -3.083017244255380e + 005 \\ A_8 = 4.323368172460818e - 004. \end{bmatrix}$$

The solutions to the univariate polynomial equation are then obtained using Matlab's roots command (e.g., 4.40 on p. 44) as

$$\begin{bmatrix} c = [A_8 \ A_7 \ A_6 \ A_5 \ A_4 \ A_3 \ A_2 \ A_1 \ A_0] \\ x_3 = roots(c), \end{bmatrix}$$

where A_7, A_5, A_3, A_1 are all zero. The obtained values of x_3 are:

$$x_3 = \begin{bmatrix} -20757.2530734872 + 8626.43262759353i \\ -20757.2530734872 - 8626.43262759353i \\ 20757.2530734872 + 8626.4326275935i \\ 20757.2530734872 - 8626.4326275935i \\ 430.528578109464 \\ -430.528578109464 \\ 153.711222705295 \\ -153.711222705295. \end{bmatrix}$$

Alternatively, the polynomial resultants techniques can be used to solve the Grunert's distance equations. They proceed as follows:

(a) The F. Macaulay formulation discussed in Sect. 5.3.1 solves for the determinant of the matrix $\mathbf{A}$ leading to a univariate polynomial in x_1. The solution of the obtained univariate polynomial equation expressed in [25, Box 3-1] leads to similar results as those of Groebner basis, i.e.,

$$
\begin{aligned}
det(A) &= A_8 x_1^8 + A_6 x_1^6 + A_4 x_1^4 + A_2 x_1^2 + A_0 \\
A_0 &= -4.8715498798062 2^{26}, \quad A_2 = 4.74815547158708^{20} \\
A_4 &= -113109755605017 \\
A_8 &= -0.000432336817247789, \quad A_6 = 435283.472057364
\end{aligned}
$$

$$
\begin{aligned}
x_1 = \ & -22456.4891074245 + 1735.29702574406i \\
& -22456.4891074245 - 1735.29702574406i \\
& 22456.4891074245 + 1735.29702574406i \\
& 22456.4891074245 - 1735.29702574406i \\
& 1580.10924379877 \\
& -1580.10924379877 \\
& 1324.23808451944 \\
& -1324.23808451944 \\
x_3 = \ & 430.528578109536, -2783.30427366986 \\
x_2 = \ & 542.260767703823, -711.800947103387.
\end{aligned}
$$

(b) The B. Sturmfels formulation discussed in Sect. 5.3.2 solves the determinant of a 6×6 matrix leading to a univariate polynomial in x_1. The solution of the obtained univariate polynomial equation expressed in [25, Box 3-2] gives identical results as those of Groebner basis, i.e.,

$$
\begin{aligned}
det(A) &= A_8 x_1^8 + A_6 x_1^6 + A_4 x_1^4 + A_2 x_1^2 + A_0 \\
A_0 &= -1.94861995192249^{27}, \quad A_2 = 1.89926218863483^{21} \\
A_4 &= -452439022420067 \\
A_8 &= -0.00172934726897456, \quad A_6 = 1741133.88822977
\end{aligned}
$$

$$
\begin{aligned}
x_1 = \ & -22456.4891075064 + 1735.29702538544i \\
& -22456.4891075064 - 1735.29702538544i \\
& 22456.4891075064 + 1735.29702538544i \\
& 22456.4891075064 - 1735.29702538544i \\
& 1580.10924379877 \\
& -1580.10924379877 \\
& 1324.23808451944 \\
& -1324.23808451944 \\
x_3 = \ & 430.528578109535, -2783.30427366986 \\
x_2 = \ & 542.260767703824, -711.800947103388.
\end{aligned}
$$

The computed distances from F. Macaulay and B. Sturmfels' approaches above tally. The required solutions $\{x_1, x_2, x_3\}$ obtained from

Groebner basis computation and those of multipolynomial resultants are the same {i.e, 1324.2381 m, 542.2608 m, 430.5286 m} respectively. The computed distances are then used to determine the position of $K1$ using ranging techniques discussed in Sect. 9.3.2. The unknown orientation elements are computed from (8.10) on p. 97.

3d-resection to more than Three Known Stations

In the preceding section, only three known stations were required to solve in a closed form the three-dimension resection problem for the position and orientation of the unknown station $K1$. If superfluous observations are available, due to the availability of several known stations, as in the case of the test network Stuttgart Central, closed form three-dimensional resection procedures give way to Gauss-Jacobi combinatorial approach. We illustrate this by means of Example 11.3.

Example 11.3 (Three-dimensional resection given more than three known stations). From the test network Stuttgart Central in Fig. 7.1 of Sect. 7.4, the three-dimensional coordinates $\{X, Y, Z\}$ of the unknown station $K1$ are sought. Using observations in Tables 7.2 and 7.3 on p. 88, the algorithm is applied in four steps as follows:

Step 1 (combinatorial solution):
From Fig. 7.1, 35 minimal combinatorials are formed using (6.26) on p. 69. The systems of nonlinear Grunert's distance equations (11.32) for each of the 35 combinatorials is solved in a closed form to give the distances linking the unknown station $K1$ to the 7 known stations. Use is made of either Groebner basis or polynomial resultants approaches as already discussed in Sect. 11.2.2. Each combinatorial minimal subset results in 3 distances, thus giving rise to a total of 105 (3×35) which are used in the next steps as pseudo-observations.

Step 2 (error propagation to determine the dispersion matrix Σ):
The variance-covariance matrix is computed for each of the combinatorial set $j = 1, \ldots, 35$ using error propagation. Equation (11.32) on p. 177 is applied to obtain the dispersion matrix Σ using (6.31) as discussed in Example 6.4 on p. 72.

Step 3 (rigorous adjustment of the combinatorial solution points):
The 105 combinatorial distances from step 1 are finally adjusted using the linear Gauss-Markov model (6.7) on p. 61. Each of the 105 pseudo-observations is expressed as

$$S_i^j = S_i + \varepsilon_i^j \mid i \in \{1, 2, 3, 4, 5, 6, 7\}, j \in \{1, 2, 3, 4, 5, 6, 7, \ldots, 35\},$$

and placed in the vector of observation $\mathbf{y}$. The coefficients of the unknown distances S_i are placed in the design matrix $\mathbf{A}$. The vector $\boldsymbol{\xi}$ comprises the unknowns S_i. The solutions are obtained via (6.10) and the root-mean-square errors of the estimated parameters through (6.11) on p. 62. The results of the adjusted distances, root-mean-square-errors and the deviations in distances are presented in Table 11.4. These deviations are obtained by subtracting the combinatorial derived distance S_i from its ideal value S in Table 7.2 on p. 88. The adjusted distances in Table 11.4 were: $K1$-Haussmanstr.(S_1) , $K1$-Eduardpfeiffer (S_2), $K1$-Lindenmuseum (S_3) , $K1$-Liederhalle (S_4), $K1$-Dach LVM (S_5) , $K1$-Dach FH (S_6) and $K1$-Haussmanstr.(S_7).

Table 11.4. Gauss-Jacobi combinatorial derived distances

Distance	Value (m)	Root mean square (m)	Deviation $\Delta(m)$
S_1	1324.2337	0.0006	0.0042
S_2	542.2598	0.0006	0.0011
S_3	364.9832	0.0006	-0.0035
S_4	430.5350	0.0008	-0.0063
S_5	400.5904	0.0007	-0.0067
S_6	269.2346	0.0010	-0.0037
S_7	566.8608	0.0005	0.0027

Step 4 (determination of position by ranging method):
The derived distances in Table 11.4 are then used as in Example 9.6 on p. 144 to determine the position of $K1$.

Example 11.4 (Comparison between exact and overdetermined 3d-resection solutions).
In this example, we are interested in comparing the solutions of the position of station $K1$ obtained from;

- closed form procedures of either Groebner basis or polynomial resultants,
- closed form solution of Gauss-Jacobi combinatorial for the overdetermined 3d-resection to the 7-stations.

To achieve this, 11 sets of experiments were carried out. For each experiment, the position of $K1$ was determined using Groebner basis and the obtained values subtracted from the known position in Table 7.1 on p. 88. The experiments were then repeated for the Gauss-Jacobi combinatorial approach. In Table 11.5, the deviation of the positions computed using Gauss-Jacobi combinatorial approach from the real values, for the 11 sets of experiments are presented. In Figs. 11.4, 11.5 and 11.6, the plot of the deviations of the X, Y, Z coordinates respectively are presented.

Table 11.5. Deviation of position of station $K1$ from the real value in Table 7.1

Set No.	$\Delta X(m)$	$\Delta Y(m)$	$\Delta Z(m)$
1	-0.0026	0.0013	-0.0001
2	-0.0034	-0.0001	0.0009
3	0.0016	0.0005	0.0028
4	0.0076	0.0007	0.0016
5	0.0027	0.0020	0.0005
6	-0.0011	0.0004	0.0020
7	0.0027	-0.0000	0.0005
8	0.0014	0.0012	-0.0016
9	0.0010	0.0006	0.0005
10	-0.0005	-0.0039	0.0007
11	0.0016	0.0001	-0.0001

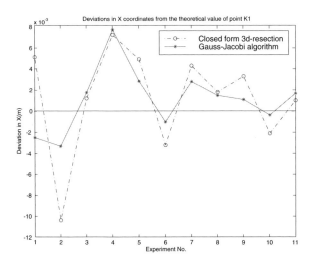

Fig. 11.4. Deviations of $X - Coordinates$ of station $K1$

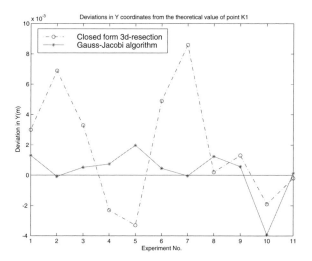

Fig. 11.5. Deviations of $Y - Coordinates$

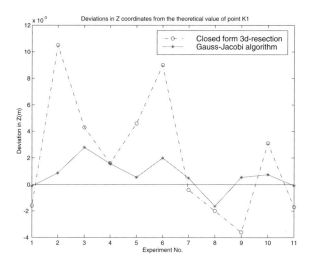

Fig. 11.6. Deviations of $Z - Coordinates$

From the plots of Figs. 11.4, 11.5 and 11.4, it is clearly seen that closed form solutions with more than three known stations yield better results. For less accurate results such as that of locating a station in cadastral and engineering surveys, Groebner basis and polynomial resultants are useful. For more accurate results, resecting to more than three known stations would be desirable. In this case, one could apply the Gauss-Jacobi combinatorial algorithm.

11.3 Photogrammetric Resection

Similar to the case of the scanner resection in Fig. 11.1 on p. 166, photogrammetric resection concerns itself with the determination of the position and orientation of the aerial camera during photography (see e.g., Fig. 11.7). At least three scene objects are required to achieve three-dimensional resection. The coordinates of the perspective center of the camera and the orientation which comprises the elements of exterior orientation are solved by three-dimensional photogrammetric resection. Once the coordinates and the orientation of the camera have been established, they are used in the intersection step (see Sect. 12.3) to compute coordinates of the pass points. Besides, they also find use in transformation procedures where coordinates in the photo plane are transformed to ground system and vice versa. The three-dimensional photogrammetric resection is formulated as follows: Given image coordinates of at least three points $\{p_i, p_j\}$, respectively, $i \neq j$ (e.g., in Fig. 11.7), determine the position $\{X_0, Y_0, Z_0\}$ and orientation $\{\omega, \phi, \kappa\}$ of the perspective center p. In practice, using stereoplotters etc., the bundle of rays projected from the perspective center to the ground are normally translated and rotated until there exist a match between points on the photograph and their corresponding points on the ground. The mathematical relationship between the points $\{\xi_i, \eta_i, f\}$ on the photographs and their corresponding ground points $\{X_i, Y_i, Z_i\}$ are related by

$$
\left[\begin{aligned}
\xi_i &= \xi_0 - f \frac{r_{11}(X_i - X_0) + r_{21}(Y_i - Y_0) + r_{31}(Z_i - Z_0)}{r_{13}(X_i - X_0) + r_{23}(Y_i - Y_0) + r_{33}(Z_i - Z_0)} \\
\eta_i &= \eta_0 - f \frac{r_{12}(X_i - X_0) + r_{22}(Y_i - Y_0) + r_{32}(Z_i - Z_0)}{r_{13}(X_i - X_0) + r_{23}(Y_i - Y_0) + r_{33}(Z_i - Z_0)},
\end{aligned} \right. \tag{11.57}
$$

where f is the focal length, $\{\xi_0, \eta_0\}$ the perspective center coordinates on the photo plane and r_{ij} the elements of the rotation matrix (see Sect. 15.2.1 p. 262 for details)

$$
R = \begin{bmatrix} r_{11} & r_{12} & r_{13} \\ r_{21} & r_{22} & r_{23} \\ r_{31} & r_{32} & r_{33} \end{bmatrix} \Big|_{\omega, \phi, \kappa}. \tag{11.58}
$$

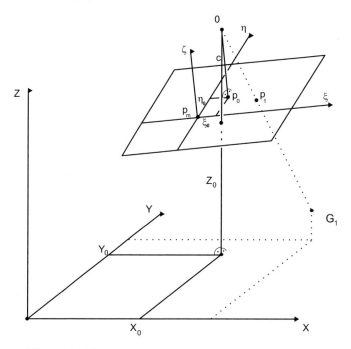

Fig. 11.7. Photogrammetric three-dimensional resection

The solution of (11.57) for the unknown position $\{X_0, Y_0, Z_0\}$ and orientation $\{\omega, \phi, \kappa\}$ is often achieved by;

- first linearizing about approximate values of the unknowns,
- application of least squares approach to the linearized observations,
- iterating to convergence.

In what follows, we present algebraic solutions of (11.57) based on the Grafarend-Shan Möbius and Groebner basis/polynomial resultant approaches.

11.3.1 Grafarend-Shan Möbius Photogrammetric Resection

In this approach of [152], the measured image coordinates $\{x_i, y_i, f\}$ of point p_i and $\{x_j, y_j, f\}$ of point p_j are converted into space angles by

$$\cos \psi_{ij} = \frac{x_i x_j + y_i y_j + f^2}{\sqrt{x_i^2 + y_i^2 + f^2}\sqrt{x_j^2 + y_j^2 + f^2}}. \tag{11.59}$$

Equation (11.59) is the photogrammetric equivalent of (11.30) on p. 177 for geodetic resection. The Grafarend-Shan algorithm operates in five steps as follows:

Step 1: The space angles ψ_{ij} relating angles to image coordinates of at least four known stations are computed from (11.59).

Step 2: The distances $\mathbf{x}_i - \mathbf{x}_j$ from the given cartesian coordinates of points p_i and p_j are computed using

$$s_{ij} = \sqrt{((x_i - x_j)^2 + (y_i - y_j)^2 + (z_i - z_j)^2)|_{\{i \neq j\}}.} \qquad (11.60)$$

Step 3: Using the distances from (11.60) and the space angles from (11.59) in step 1, Grunert's distance equations (11.32) are solved using the Grafarend-Lohse-Schaffrin procedure discussed in Sect. 11.2.2.

Step 4: Once the distances have been obtained in step 3, they are used to compute the perspective center coordinates using the Ansermet's algorithm [8].

Step 5: The orientation is computed by solving (15.3) on p. 262.

11.3.2 Algebraic Photogrammetric Resection

The algebraic algorithms of Groebner basis or polynomial resultants operate in five steps as follows:

Step 1: The space angles ψ_{ij} relating the angles to the image coordinates of at least four known stations are computed from (11.59).

Step 2: The distances $\mathbf{x}_i - \mathbf{x}_j$ from the given cartesian coordinates of points p_i and p_j are computed from (11.60).

Step 3: Using the distances from step 2 and the space angles from step 1, Grunert's distance equations in (11.32) are solved using procedures of Sect. 11.2.2.

Step 4: Once distances have been solved in step 3, they are used to compute the perspective center coordinates using ranging techniques of Chap. 9.

Step 5: The three-dimensional orientation parameters are computed from (8.10) on p. 97.

Example 11.5 (Three-dimensional photogrammetric resection). In this example, we will use data of two photographs adopted from [336]. From these data, we are interested in computing algebraically the perspective center coordinates of the two photographs. The image coordinates of photographs 1010 and 1020 are given in Tables 11.6 and 11.7 respectively. The corresponding ground coordinates are as given in Table 11.8. Table 11.10 gives for control purposes the known coordinates of the projection center adopted from [336]. We use four image coordinates (No. 1,2,3,5) to compute algebraically the perspective center coordinates. From these image coordinates, combinatorials are formed using (6.26) on p. 69 and are as given in Table 11.9. For each combination, distances are solved using reduced Groebner basis (4.38) on p. 44. Once the distances have been computed for each combination, the perspective center coordinates are computed using the ranging techniques of Chap. 9. The mean values are then obtained. The results are summarized in Table 11.9. A comparison between the Groebner basis derived results and those of Table 11.10 is presented in Table 11.11. Instead of the mean value which does not take weights into consideration, the Gauss-Jacobi combinatorial techniques that we have studied can be used to obtain refined solutions. For photo 1020, the first combination 1-2-3 gave complex values for the distances S_2 and S_3. For computation of the perspective center in this Example, the real part was adopted.

Table 11.6. Image coordinates in Photo 1010: $f = 153000.000[\mu m]$

Point	No.	$x(\mu m)$	$y(\mu m)$
100201	1	18996.171	-64147.679
100301	2	113471.749	-73694.266
200201	3	16504.609	16331.646
200301	4	128830.826	21085.172
300201	5	13716.588	106386.802
300301	6	120577.473	128214.823

Table 11.7. Image coordinates in Photo 1020: $f = 153000.000[\mu m]$

Point	No.	$x(\mu m)$	$y(\mu m)$
100201	1	-74705.936	-71895.580
100301	2	5436.953	-78524.687
200201	3	-87764.035	7895.436
200301	4	3212.790	10311.144
300201	5	-84849.923	94110.338
300301	6	802.388	106585.613

Table 11.8. Ground coordinates

Point	X(m)	Y(m)	Z(m)
100201	-460.000	-920.000	-153.000
100301	460.000	-920.000	0.000
200201	-460.000	0.000	0.000
200301	460.000	0.000	153.000
300201	-460.000	920.000	-153.000
300301	460.000	920.000	0.000

Table 11.9. Algebraic computed perspective center coordinates

Photo 1010						
Combination	$S_1(m)$	$S_2(m)$	$S_3(m)$	$X(m)$	$Y(m)$	$Z(m)$
1-2-3	1918.043	2008.407	1530.000	-459.999	0.000	1530.000
1-2-5	1918.043	2008.407	1918.043	-459.999	0.000	1530.000
1-3-5	1918.043	1530.000	1918.043	-459.292	0.000	1530.000
2-3-5	2008.407	1530.000	1918.043	-459.999	0.000	1530.000
mean				-459.822	0.000	1530.000
Photo 1020						
Combination	$S_1(m)$	$S_2(m)$	$S_3(m)$	$X(m)$	$Y(m)$	$Z(m)$
1-2-3	2127.273	1785.301	1785.301i	460.001	0.001	1529.999
1-2-5	2127.273	1785.301	2127.273	460.001	0.000	1530.000
1-3-5	2127.273	1785.301	2127.272	460.036	0.002	1529.978
2-3-5	1785.301	1785.301	2127.273	459.999	0.000	1530.000
mean				460.009	0.000	1529.994

Table 11.10. Ground coordinates of the projection centers

Photo	X(m)	Y(m)	Z(m)
1010	-460.000	0.000	1530.000
1020	460.000	0.000	1530.000

Table 11.11. Deviation of the computed mean from the real value

Photo	$\Delta X(m)$	$\Delta Y(m)$	$\Delta Z(m)$
1010	-0.178	0.000	0.000
1020	-0.009	0.000	0.006

11.4 Concluding Remarks

As evident in the chapter, the problem of resection is still vital in geodesy and geoinformatics. The algebraic algorithms of Groebner basis, polynomial resultants and the Gauss-Jacobi combinatorial are useful where exact solutions are required. They may be used to control the numeric methods used in practice. Instead of the forward and backward steps, straight forward algebraic approaches presented save on computational time. Further references are [8, 54, 124, 125, 131, 140, 187, 197, 317].

Positioning by Intersection Methods

12.1 Intersection Problem and its Importance

The similarity between resection methods presented in the previous chapter and intersection methods discussed herein is their application of angular observations. The distinction between the two however, is that for resection, the unknown station is *occupied* while for intersection, the unknown station is *observed*. Resection uses measuring devices (e.g., theodolite, total station, camera etc.) which occupy the unknown station. Angular (direction) observations are then measured to three or more known stations as we saw in the preceding chapter. Intersection approach on the contrary measures angular (direction) observations to the unknown station; with the measuring device occupying *each* of the three or more known stations. It has the advantage of being able to position an unknown station which can not be physically occupied. Such cases are encountered for instance during engineering constructions or cadastral surveying. During civil engineering construction for example, it may occur that a station can not be occupied because of swampiness or risk of sinking ground. In such a case, intersection approach can be used. The method is also widely applicable in photogrammetry. In aero-triangulation process, simultaneous resection and intersection are carried out where common rays from two or more overlapping photographs intersect at a common ground point (see e.g., Fig. 9.1).

The applicability of the method has further been enhanced by the Global Positioning System (GPS), which the authors also refer to as GPS: Global Problem Solver. With the entry of GPS system, classical geodetic and photogrammetric positioning techniques have reached a new horizon. Geodetic and photogrammetric directional observations (machine vision, total stations) have to be analyzed in a three-dimensional Euclidean space. The challenge has forced positioning techniques such as resection and intersection to operate three-dimensionally. As already pointed out in Chap. 11, closed form solutions of the three-dimensional resection problem exist in a great number. On the contrary, closed form solutions of three-dimensional intersection problem are very rare. For instance [151, 152] solved the two $P4P$ or the combined three-dimensional resection-intersection problem in terms of *Möbius barycentric coordinates* in a closed form. One reason for the rare existence of the closed form solutions of the three-dimensional intersection problem is the nonlinearity of directional observation equations, partially caused by the *external orientation parameters*. One target of this chapter, therefore, is to address the problem of orientation parameters.

The key to overcome the problem of nonlinearity caused by orientation parameters is taken from the *Baarda Doctrine*. Baarda [38, 42] proposed to use *dimensionless quantities* in geodetic and photogrammetric networks: Angles in a three-dimensional *Weitzenböck space*, shortly called *space angles* as well as *distance ratios* are the *dimensionless* structure elements which are *equivalent* under the action of the seven parameter *conformal group*, also called similarity transformation.

12.2 Geodetic Intersection

12.2.1 Planar Intersection

The planar intersection problem is formulated as follows: Given directions or angular measurements from two known stations P_1 and P_2 to an unknown station P_0, determine the position $\{X_0, Y_0\}$. The solution to the problem depends on whether angles or directions are used as discussed in the next section.

Conventional Solution

Closed form solution of planar intersection in terms of angles has a long tradition. Let us consult Fig. 12.1 on p. 206 where we introduce the angles ψ_{12} and ψ_{21} in the planar triangle $\Delta : P_0P_1P_2$, with P_0, P_1, P_2 being the nodes. The Cartesian coordinates $\{X_1, Y_1\}$ and $\{X_2, Y_2\}$ of the points P_1 and P_2 are given while $\{X_0, Y_0\}$ of the point P_0 are unknown. The angles $\psi_{12} = \alpha$ and $\psi_{21} = \beta$ are derived from direction observations by differencing horizontal directions. $\psi_{12} = T_{10} - T_{12}$ or $\psi_{21} = T_{20} - T_{21}$ are examples of observed horizontal directions T_{10} and T_{12} from P_1 to P_0 and P_1 to P_2 or T_{21} and T_{20} from P_2 to P_1 and P_2 to P_0 respectively. By means of taking differences we map direction observations to angles and eliminate orientation unknowns. The solution of the two-dimensional intersection problem in terms of angles, a classical procedure in analytical surveying, is given by (12.1) and (12.2) as

$$X_0 = s_{12} \frac{\cos\alpha\,\sin\beta}{\sin(\alpha+\beta)} \tag{12.1}$$

$$Y_0 = s_{12} \frac{\sin\alpha\,\sin\beta}{\sin(\alpha+\beta)}. \tag{12.2}$$

Note: The Euclidean distance between the nodal points is given by

$$s_{12} = \sqrt{(X_2 - X_1)^2 + (Y_2 - Y_1)^2}.$$

In deriving (12.1) and (12.2), use was made of angular observations. In case directions are adopted, measured values from known stations P_1 and P_2 to unknown station P_0 are designated T_{10} and T_{20} respectively. If the theodolite horizontal circle reading from point P_1 to P_0 is set to zero, then the measured angle α is equal to the directional measurement T_{12} from point P_1 to P_2. Likewise if the direction from P_2 to P_1 is set to zero, the measured angle β is equal to the directional measurement T_{20} from point P_2 to P_0. In this way, we make use of both the angles and directions thus introducing two more unknowns, i.e., the unknown orientation σ_1 and σ_2 in addition to the unknown coordinates $\{X_0, Y_0\}$ of point P_0. This leads to four observation equations in four unknowns, written as:

$$\left[\begin{aligned} \tan(T_{12} + \sigma_1) &= \left\{ \frac{Y_2 - Y_1}{X_2 - X_1} \right\} \\[2mm] \tan(T_{10} + \sigma_1) &= \left\{ \frac{Y_0 - Y_1}{X_0 - X_1} \right\} \\[2mm] \tan(T_{21} + \sigma_2) &= \left\{ \frac{Y_1 - Y_2}{X_1 - X_2} \right\} \\[2mm] \tan(T_{20} + \sigma_2) &= \left\{ \frac{Y_0 - Y_2}{X_0 - X_2} \right\}, \end{aligned} \right. \tag{12.3}$$

where $\{X_1, Y_1, X_2, Y_2\}$ are coordinates of the two known stations $\{P_1, P_2\}$, while $\{T_{12}, T_{10}, T_{21}, T_{20}\}$ are the measured horizontal directions and $\{X_0, Y_0, \sigma_1, \sigma_2\}$ are the desired position and orientation of the unknown station P_0. In (12.3), the first and the third expressions contain the orientation elements σ_1 and σ_2 as the only unknowns. They are solved by obtaining the inverse of the tangents as

$$
\left[
\begin{array}{l}
\sigma_1 = tan^{-1} \left\{ \dfrac{Y_2 - Y_1}{X_2 - X_1} \right\} - T_{12} \\[2mm]
\sigma_2 = tan^{-1} \left\{ \dfrac{Y_1 - Y_2}{X_1 - X_2} \right\} - T_{21}.
\end{array}
\right.
\tag{12.4}
$$

Once the unknown orientation elements have been solved in (12.4), they are substituted in the second and fourth expressions of (12.3) to form simultaneous equation whose solution give the values $\{X_0, Y_0\}$. Next, let us see how (12.3) can be solved using reduced Groebner basis (4.38) on p. 44.

Reduced Groebner Basis Solution

The left-hand-sides of (12.3) are expanded using additions theorem

$$
tan(\alpha + \beta) = \frac{tan\,\alpha + tan\,\beta}{1 - tan\,\alpha\,tan\,\beta},
\tag{12.5}
$$

to give:

$$
\left[
\begin{array}{l}
\dfrac{tan\,T_{12} + tan\,\sigma_1}{1 - tan\,T_{12}tan\,\sigma_1} = \left\{ \dfrac{Y_2 - Y_1}{X_2 - X_1} \right\} \\[4mm]
\dfrac{tan\,T_{10} + tan\,\sigma_1}{1 - tan\,T_{10}\,tan\,\sigma_1} = \left\{ \dfrac{Y_0 - Y_1}{X_0 - X_1} \right\} \\[4mm]
\dfrac{tan\,T_{21} + tan\,\sigma_2}{1 - tan\,T_{21}\,tan\,\sigma_2} = \left\{ \dfrac{Y_1 - Y_2}{X_1 - X_2} \right\} \\[4mm]
\dfrac{tan\,T_{20} + tan\,\sigma_2}{1 - tan\,T_{20}\,tan\,\sigma_2} = \left\{ \dfrac{Y_0 - Y_2}{X_0 - X_2} \right\}.
\end{array}
\right.
\tag{12.6}
$$

Expanding (12.6) and re-arranging gives trigonometric algebraic expressions

$$\begin{bmatrix} (X_2 - X_1 + Y_2 tanT_{12} - Y_1 tanT_{12})tan\,\sigma_1 + X_2 tanT_{12} - X_1 tanT_{12} + Y_1 - Y_2 = 0 \\ \\ X_0 tanT_{10} + X_0 tan\,\sigma_1 + Y_0 tanT_{10} tan\,\sigma_1 - Y_0 - X_1 tan\,\sigma_1 - Y_1 tanT_{10} tan\sigma_1 \\ -X_1 tanT_{10} + Y_1 = 0 \\ \\ (X_1 - X_2 + Y_1 tanT_{21} - Y_2 tanT_{21})tan\,\sigma_2 + X_1 tanT_{21} - X_2 tanT_{21} + Y_2 - Y_1 = 0 \\ \\ X_0 tanT_{20} + X_0 tan\,\sigma_2 + Y_0 tanT_{20} tan\,\sigma_2 - Y_0 - X_2 tan\,\sigma_2 - Y_2 tanT_{20} tan\,\sigma_2 \\ -X_2 tanT_{20} + Y_2 = 0. \end{bmatrix}$$

$$(12.7)$$

Denoting

$$\begin{bmatrix} a_1 = tanT_{12},\ a_2 = tanT_{21} \\ b = tanT_{10} \\ c = tanT_{20} \\ d_1 = tan\sigma_1,\ d_2 = tan\sigma_2, \end{bmatrix}$$

$$(12.8)$$

and substituting in (12.7) leads to four algebraic equations which are arranged in the lexicographic order $\{X_0 > Y_0 > d_2 > d_1\}$ as

$$\begin{bmatrix} f_1 = d_1 X_2 - d_1 X_1 + d_1 Y_2 a_1 - d_1 Y_1 a_1 + X_2 a_1 - X_1 a_1 + Y_1 - Y_2 = 0 \\ f_2 = X_0 b + X_0 d_1 - Y_0 + Y_0 b d_1 - X_1 d_1 - Y_1 b d_1 - X_1 b + Y_1 = 0 \\ f_3 = d_2 X_1 - d_2 X_2 + d_2 Y_1 a_2 - d_2 Y_2 a_2 + X_1 a_2 - X_2 a_2 + Y_2 - Y_1 = 0 \\ f_4 = X_0 c + X_0 d_2 - Y_0 + Y_0 c d_2 - X_2 d_2 - Y_2 c d_2 - X_2 c + Y_2 = 0. \end{bmatrix}$$

$$(12.9)$$

Using reduced Groebner basis (4.38) on p. 44, (12.9) is solved as

$$\begin{bmatrix} Groebner\,Basis\,[\{f_1, f_2, f_3, f_4\}, \{X_0, Y_0, d_2, d_1\}, \{X_0, Y_0, d_2\}] \\ Groebner\,Basis\,[\{f_1, f_2, f_3, f_4\}, \{X_0, Y_0, d_1, d_2\}, \{X_0, Y_0, d_1\}] \\ Groebner\,Basis\,[\{f_1, f_2, f_3, f_4\}, \{d_2, d_1, Y_0, X_0\}, \{Y_0, d_2, d_1\}] \\ Groebner\,Basis\,[\{f_1, f_2, f_3, f_4\}, \{d_2, d_1, X_0, Y_0\}, \{X_0, d_2, d_1\}]. \end{bmatrix}$$

$$(12.10)$$

The first and second expressions of (12.10) give linear equations relating the tangents d_1 and d_2 of the unknown orientations σ_1 and σ_2 and the coordinates $\{X_1, Y_1, X_2, Y_2\}$ of the known stations P_1 and P_2. The third and fourth expressions give linear equations relating the coordinates X_0 and Y_0 of unknown station P_0, coordinates $\{X_1, Y_1, X_2, Y_2\}$ of known stations P_1 and P_2, and the orientation terms d_1 and d_2. The computed reduced Groebner basis (linear functions) are

$$
\left[
\begin{aligned}
d_1 &= \frac{(-a_1 X_1 + a_1 X_2 + Y_1 - Y_2)}{(X_1 - X_2 + a_1 Y_1 - a_1 Y_2)} \\[2mm]
d_2 &= \frac{(-a_2 X_1 + a_2 X_2 + Y_1 - Y_2)}{(X_1 - X_2 + a_2 Y_1 - a_2 Y_2)} \\[2mm]
X_0 &= \frac{\left\{\begin{array}{l} -(Y_1 - Y_2 - d_1 X_1 + d_2 X_2 - bX_1 + cX_2 - bd_1 Y_1 + bcY_1 - cd_2 Y_1 \\ +bd_1 Y_2 - bcY_2 + cd_2 Y_2 - cd_1 Y_2 + cd_1 Y_1 + a_2 cd_1 d_2 Y_2 + a_2 cd_1 X_2 - \\ a_2 cd_1 X_1 + cd_1 d_2 X_2 - bd_1 d_2 X_2 + bcd_2 X_2 + a_2 bcX_2 - a_2 cd_1 d_2 Y_1 - \\ bcd_1 X_2 - a_2 bcX_1 - bcd_1 d_2 Y_2 + bcd_1 d_2 Y_1 + a_2 bcd_2 Y_2 - a_2 bcd_2 Y_1) \end{array}\right\}}{d_1 + bcd_1 - d_2 - bcd_2 + bd_1 d_2 + b - c - cd_1 d_2} \\[2mm]
Y_0 &= \frac{\left\{\begin{array}{l} -(a_2 bX_1 - a_2 bd_2 Y_2 + cd_1 X_2 - a_2 d_1 X_2 + bcX_2 - a_2 bX_2 - cd_1 X_1 + \\ a_2 d_1 X_1 - bcd_1 Y_1 - bcX_1 - a_2 d_1 d_2 Y_2 - bY_1 - bd_1 d_2 Y_1 + bcd_2 Y_2 + \\ cd_1 d_2 Y_2 + a_2 d_1 d_2 Y_1 + d_2 Y_1 + cY_1 + a_2 bd_2 Y_1 - d_1 Y_1) \end{array}\right\}}{d_1 + bcd_1 - d_2 - bcd_2 + bd_1 d_2 + b - c - cd_1 d_2}
\end{aligned}
\right.
$$

$$(12.11)$$

Example 12.1 (Planar intersection problem). Consider the example of [195, p. 292]. In this example, planar Cartesian coordinates of two known stations $F := P_1$ and $E := P_2$ are given as

$$
\{X_1 = 2490.50\,m,\ Y_1 = 2480.79\,m\}_{P_1}
$$
$$
\{X_2 = 780.67\,m,\ Y_2 = 7394.05\,m\}_{P_2}.
$$

The adjusted angles from points $F := P_1$ and $E := P_2$ to the unknown station $G := P_0 \in \mathbb{E}^2$ are $117°\ 11'\ 20.7''$ and $27°\ 35'\ 47.9''$ respectively. Using these angles and Fig. 12.1 on p. 206 one writes the directions as: $T_{10} = 0°\ 00'\ 00.0''$, $T_{12} = 117°\ 11'\ 20.7''$, $T_{21} = 0°\ 00'\ 00.0''$ and $T_{20} = 27°\ 35'\ 47.9''$. These directions are used in (12.8) to compute the constants $\{a_1, a_2, b, c\}$ which are then inserted in the first two expressions of (12.11) to give the values of d_1 and d_2, which are used in the fourth expression of (12.8). This leads to the two unknown orientation parameters σ_1 and σ_2 as; $351°\ 59'\ 56.3''$ and $289°\ 11'\ 17.0''$ respectively. The planar coordinates $\{X_0, Y_0\}\,p_0$ of the unknown station $G := P_0 \in \mathbb{E}^2$ are then computed from the third and fourth expressions of (12.11) as; $\{X_0 = 6629.0952\,m,\ Y_0 = 1899.0728\,m\}_{P_1}$, which compare well with those of [195, p. 292].

Example 12.2 (Planar intersection problem). Let us consider another example of [195, p. 292] where the planar Cartesian coordinates of two known stations $E := P_1$ and $D := P_2$ are given as

$$
\{X_1 = 780.67\,m,\ Y_1 = 7394.05\,m\}_{P_1}
$$
$$
\{X_2 = 5044.25\,m,\ Y_2 = 7752.70\,m\}_{P_2}.
$$

The adjusted angles from points $E := P_1$ and $D := P_2$ to the unknown station $G := P_0 \in \mathbb{E}^2$ are $48^0\ 01'\ 25.3''$ and $100^0\ 20'\ 27.8''$
respectively. Using these angles and Fig. 12.1 as in the previous
example, one writes the directions as: $T_{10} = 0°\ 00'\ 00.0''$, $T_{12} = 48°\ 01'\ 25.3''$, $T_{21} = 0°\ 00'\ 00.0''$ and $T_{20} = 100°\ 20'\ 27.8''$. These
directions are used in (12.8) to compute $\{a_1, a_2, b, c\}$, which are in-
serted in the first two expressions of (12.11) to give the values of
d_1 and d_2. These values of d_1 and d_2 are inserted in the fourth expression of (12.8) to give the two unknown orientation parameters σ_1
and σ_2 as $316°\ 47'\ 04.8''$ and $04°\ 48'\ 30.1''$ respectively. The planar
coordinates $\{X_0, Y_0\}\, p_0$ of the unknown station $G := P_0 \in \mathbb{E}^2$ are
then computed from the third and fourth expressions of (12.11) as;
$\{X_0 = 6629.1007\, m,\ Y_0 = 1899.0635\, m\}_{P_1}$, which compare well with
those of [195, p. 292].

12.2.2 Three-dimensional Intersection

Closed Form Solution

In the case of three-dimensional intersection problem, the triple of three
points $\{P_1, P_2, P_3\}$ in Fig. 12.1 are given by their three-dimensional
Cartesian coordinates $\{X_1, Y_1, Z_1\}$, $\{X_2, Y_2, Z_2\}$, $\{X_3, Y_3, Z_3\}$, but the
coordinates $\{X_0, Y_0, Z_0\}$ of point P_0 are unknown. The dimensionless
quantities $\{\psi_{12}, \psi_{23}, \psi_{31}\}$ are space angles; $\psi_{12} = \angle P_0 P_1 P_2$, $\psi_{23} = \angle P_0 P_2 P_3$, $\psi_{31} = \angle P_1 P_3 P_0$. This problem is formulated as follows; Given
horizontal directions T_i and vertical directions B_i measured from three
known stations to an unknown station, determine the position of the
unknown station P_0. These directional measurements are transformed
into space angles $\{\psi_{12}, \psi_{23}, \psi_{31}\}$ using (11.30) on p. 177 (see e.g., Fig.
12.1). Equation (11.30) is the analytic version of a map of directions to
space coordinates. Indeed, the map eliminates the external orientation
parameters. The space angles are then used to obtain the unknown distances $\{x_1 = S_i, x_2 = S_2, x_3 = S_3\}$. These distances relate the unknown
station $P_0 \in \mathbb{E}^3$ to three known stations $P_i \in \mathbb{E}^3 \mid i = \{1, 2, 3\}$ in the
first step. The nonlinear system of equations relating the unknown dis-
tances $\{x_1 = S_i, x_2 = S_2, x_3 = S_3\}$ to the space angles $\{\psi_{12}, \psi_{23}, \psi_{31}\}$
are given as

$$\begin{bmatrix} x_2^2 = x_1^2 + S_{12}^2 - 2S_{12}\cos(\psi_{12})x_1 \\ x_3^2 = x_2^2 + S_{23}^2 - 2S_{23}\cos(\psi_{23})x_2 \\ x_1^2 = x_3^2 + S_{31}^2 - 2S_{31}\cos(\psi_{31})x_3. \end{bmatrix} \tag{12.12}$$

In the *second step*, the computed distances from *step* 1 are used in the three-dimensional ranging techniques of Chap. 9 to solve for the unknown position $P_0 \in \mathbb{E}^3$.

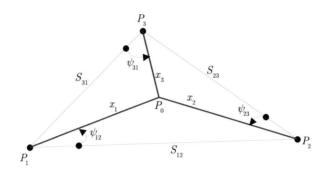

Fig. 12.1. 3d-intersection

Conventional Solution

Equation (12.12) is solved by first adding (12.12)i, (12.12)ii and (12.12)iii to eliminate the squared terms. The resulting expression

$$S_{12}^2 + S_{23}^2 + S_{31}^2 - 2x_1 S_{12} \cos(\psi_{12}) - 2x_2 S_{23} \cos(\psi_{23}) - 2x_3 S_{31} \cos(\psi_{31}) = 0 \tag{12.13}$$

is linear in x_1, x_2 and x_3. The variable x_1 in (12.13) is then expressed in terms of x_2 and x_3 as

$$x_1 = \frac{S_{12}^2 + S_{23}^2 + S_{31}^2 - 2x_2 S_{23} \cos(\psi_{23}) - 2x_3 S_{31} \cos(\psi_{31})}{2S_{12} \cos(\psi_{12})}, \tag{12.14}$$

and substituted in (12.12)i to give an expression in x_2 and x_3 only. The resulting expression in x_2 and x_3 is solved simultaneously with (12.12)ii to give values of x_2 and x_3. On the other hand, if (12.13) is now written such that x_3 is expressed in terms of x_2 and x_1 and substituted in (12.12)iii, an expression in x_2 and x_1 will be given which together with (12.12)i can be solved for the values of x_2 and x_1.

The setback with this approach is that one variable, in this case x_2, is determined twice with different values being given; which clearly is undesirable. A direct solution to the problem based on algebraic approaches of either Groebner basis or polynomial resultants alleviates the problem.

Reduced Groebner Basis Solution

Reduced Groebner basis (4.38) on p. 44 is performed in two steps as follows:

- Step 1 (derivation of distances):
  Equation (12.12) is re-written algebraically as

$$\begin{bmatrix} f_1 := x_1^2 + b_1 x_1 - x_2^2 + a_0 = 0 \\ f_2 := x_2^2 + b_2 x_2 - x_3^2 + b_0 = 0 \\ f_3 := x_3^2 + b_3 x_3 - x_1^2 + c_0 = 0, \end{bmatrix} \qquad (12.15)$$

with $b_1 = -2S_{12}\cos(\psi_{12}), b_2 = -2S_{23}\cos(\psi_{23}), b_3 = -2S_{31}\cos(\psi_{31})$ and $a_0 = S_{12}^2, b_0 = S_{23}^2, c_0 = S_{31}^2$. The reduced Groebner basis of (12.15) is then computed as

$$\begin{bmatrix} GroebnerBasis[\{f_1, f_2, f_3\}, \{x_1, x_2, x_3\}, \{x_2, x_3\}] \\ GroebnerBasis[\{f_1, f_2, f_3\}, \{x_1, x_2, x_3\}, \{x_1, x_3\}] \\ GroebnerBasis[\{f_1, f_2, f_3\}, \{x_1, x_2, x_3\}, \{x_2, x_3\}], \end{bmatrix} \qquad (12.16)$$

which leads to three quartic polynomials for determining the unknown distances $\{x_1 = S_i, x_2 = S_2, x_3 = S_3\}$;

$$\begin{bmatrix} x_1 := d_4 x_1^4 + d_3 x_1^3 + d_2 x_1^2 + d_1 x_1 + d_0 = 0 \\ x_2 := e_4 x_2^4 + e_3 x_2^3 + e_2 x_2^2 + e_1 x_2 + e_0 = 0 \\ x_3 := f_4 x_3^4 + f_3 x_3^3 + f_2 x_3^2 + f_1 x_3 + f_0 = 0. \end{bmatrix} \qquad (12.17)$$

The coefficients of (12.17) are as given in [30, Appendix].

- Step 2 (position determination):
  In this step, the computed distances from (12.17) are used to determine the unknown position $P_0 \in \mathbb{E}^3$ as discussed in Sect. 9.3.2.

Sturmfels' Resultants Solution

Algorithm presented in Sect. 5.3.2 proceeds in two steps as follows:

- Step 1 (derivation of distances):
  Following (5.12) on p. 53, (12.15) is homogenized using the variable
  x_4 and re-written for the solutions of x_1, x_2 and x_3 in (12.18), (12.19)
  and (12.20) respectively as
  - Solving for x_1 by treating it as a constant (polynomial of degree
    zero)

$$\left[\begin{array}{l} g_1 := (x_1^2 + b_1 x_1 + a_0)x_4^2 - x_2^2 = 0 \\ g_2 := x_2^2 + b_2 x_2 x_4 - x_3^2 + b_0 x_4^2 = 0 \\ g_3 := x_3^2 + b_3 x_3 x_4 + (c_0 - x_1^2)x_4^2 = 0. \end{array}\right. \qquad (12.18)$$

  - Solving for x_2 by treating it as a constant (polynomial of degree
    zero)

$$\left[\begin{array}{l} h_1 := x_1^2 + b_1 x_1 x_4 + (a_0 - x_2^2)x_4^2 = 0 \\ h_2 := (x_2^2 + b_2 x_2 + b_0)x_4^2 - x_3^2 = 0 \\ h_3 := x_3^2 + b_3 x_3 x_4 - x_1^2 + c_0 x_4^2 = 0. \end{array}\right. \qquad (12.19)$$

  - Solving for x_3 by treating it as a constant (polynomial of degree
    zero)

$$\left[\begin{array}{l} k_1 := x_1^2 + b_1 x_1 x_4 - x_2^2 + a_0 x_4^2 = 0 \\ k_2 := x_2^2 + b_2 x_2 x_4 + (b_0 - x_3^2)x_4^2 = 0 \\ k_3 := (x_3^2 + b_3 x_3 + c_0)x_4^2 - x_1^2 = 0. \end{array}\right. \qquad (12.20)$$

From (12.18), (12.19) and (12.20), expressing $a_1 = (x_1^2 + b_1 x_1 + a_0)$
and $a_2 = (c_0 - x_1^2)$ in (12.18), $a_3 = (x_2^2 + b_2 x_2 + b_0)$ and $c_2 = (a_0 - x_2^2)$
in (12.19), and finally $c_3 = (b_0 - x_3^2)$ and $c_1 = (x_3^2 + b_3 x_3 + c_0)$ in
(12.20), one forms the Jacobian determinant matrices with (5.13)
on p. 53 respectively as

$$J_{x1} = \det \begin{bmatrix} \dfrac{\partial g_1}{\partial x_2} & \dfrac{\partial g_1}{\partial x_3} & \dfrac{\partial g_1}{\partial x_4} \\[2mm] \dfrac{\partial g_2}{\partial x_2} & \dfrac{\partial g_2}{\partial x_3} & \dfrac{\partial g_2}{\partial x_4} \\[2mm] \dfrac{\partial g_3}{\partial x_2} & \dfrac{\partial g_3}{\partial x_3} & \dfrac{\partial g_3}{\partial x_4} \end{bmatrix} = \det \begin{bmatrix} -2x_2 & 0 & 2a_1 x_4 \\[2mm] 2x_2 + b_2 x_4 & -2x_3 & b_2 x_2 + 2b_0 x_4 \\[2mm] 0 & 2x_3 + b_3 x_4 & b_3 x_3 + 2a_2 x_4 \end{bmatrix},$$

$$(12.21)$$

$$
J_{x2} = \det \begin{bmatrix} \dfrac{\partial h_1}{\partial x_1} & \dfrac{\partial h_1}{\partial x_3} & \dfrac{\partial h_1}{\partial x_4} \\[2mm] \dfrac{\partial h_2}{\partial x_1} & \dfrac{\partial h_2}{\partial x_3} & \dfrac{\partial h_2}{\partial x_4} \\[2mm] \dfrac{\partial h_3}{\partial x_1} & \dfrac{\partial h_3}{\partial x_3} & \dfrac{\partial h_3}{\partial x_4} \end{bmatrix} = \det \begin{bmatrix} 2x_1 + b_1 x_4 & 0 & b_1 x_1 + 2c_2 x_4 \\[2mm] 0 & -2x_3 & 2a_3 x_4 \\[2mm] -2x_1 & 2x_3 + b_3 x_4 & b_3 x_3 + 2c_0 x_4 \end{bmatrix},
$$

$$(12.22)$$

and

$$
J_{x3} = \det \begin{bmatrix} \dfrac{\partial k_1}{\partial x_1} & \dfrac{\partial k_1}{\partial x_2} & \dfrac{\partial k_1}{\partial x_4} \\[2mm] \dfrac{\partial k_2}{\partial x_1} & \dfrac{\partial k_2}{\partial x_2} & \dfrac{\partial k_2}{\partial x_4} \\[2mm] \dfrac{\partial k_3}{\partial x_1} & \dfrac{\partial k_3}{\partial x_2} & \dfrac{\partial k_3}{\partial x_4} \end{bmatrix} = \det \begin{bmatrix} 2x_1 + b_1 x_4 & -2x_2 & b_1 x_1 + 2a_0 x_4 \\[2mm] 0 & 2x_2 + b_2 x_4 & b_2 x_2 + 2c_3 x_4 \\[2mm] -2x_1 & 0 & 2c_1 x_4 \end{bmatrix}.
$$

$$(12.23)$$

The resulting determinants are cubic polynomials:

$$
J_{x1} = 4x_2 b_3 x_3^2 + 8x_2 x_3 a_2 x_4 + 4b_2 x_3^2 x_3 + 2b_2 x_2^2 b_3 x_4 + 8x_2 b_0 x_4 x_3 + 4x_2 b_0 x_4^2 b_3 + 8a_1 x_4 x_2 x_3 + 4a_1 x_4^2 x_2 b_3 + 4a_1 x_4^2 b_2 x_3 + 2a_1 x_4^3 b_2 b_3.
$$

$$
J_{x2} = -4x_1 b_3 x_3^2 - 8x_1 x_3 c_0 x_4 - 8x_1 a_3 x_4 x_3 - 4x_1 a_3 x_4^2 b_3 - 2b_1 x_4 b_3 x_3^2 - 4b_1 x_4^2 x_3 c_0 - 4b_1 x_4^2 a_3 x_3 - 2b_1 x_4^3 a_3 b_3 - 4x_1^2 x_3 b_1 - 8x_1 x_3 c_2 x_4.
$$

$$
J_{x3} = 8c_1 x_4 x_1 x_2 + 4c_1 x_4^2 x_1 b_2 + 4c_1 x_4^2 b_1 x_2 + 2c_1 x_4^3 b_1 b_2 + 4x_1 b_2 x_2^2 + 8x_1 x_2 c_3 x_4 + 4b_1 x_1^2 x_2 + 2b_1 x_1^2 b_2 x_4 + 8x_1 a_0 x_4 x_2 + 4x_1 a_0 x_4^2 b_2.
$$

Making use of (5.14) and (5.15) on p. 54 lead to

$$
\begin{bmatrix} x_1 := d_4 x_1^4 + d_3 x_1^3 + d_2 x_1^2 + d_1 x_1 + d_0 = 0 \\[1mm] x_2 := e_4 x_2^4 + e_3 x_2^3 + e_2 x_2^2 + e_1 x_2 + e_0 = 0 \\[1mm] x_3 := f_4 x_3^4 + f_3 x_3^3 + f_2 x_3^2 + f_1 x_3 + f_0 = 0. \end{bmatrix}
\qquad (12.24)
$$

The coefficients of (12.24) are given in [31, Appendix].

- Step 2 (position determination):
  In this step, the computed distances from (12.24) are used to determine the unknown position $P_0 \in \mathbb{E}^3$ as discussed in Sect. 9.3.2.

Example 12.3 (3d-intersection from three known stations). Using the computed quartic polynomials (12.17) or (12.24), the distances $S_i = x_i \in \mathbb{R},^+ i = \{1, 2, 3\} \in \mathbb{Z}_+^3$ between an unknown station $K1 \in \mathbb{E}^3$ and known stations $P_i \in \mathbb{E}^3$ for the test network Stuttgart Central in

Fig. 7.1 on p. 86 are determined. Points P_1, P_2, P_3 of the tetrahedron $\{P_0 P_1 P_2 P_3\}$ in Fig. 12.1 correspond to the chosen known GPS stations Schlossplatz, Liederhalle, and Eduardpfeiffer (see Fig. 7.1). The distance from $K1$ to Schlossplatz. is designated $S_1 = x_1 \in \mathbb{R},^+$ $K1$ *to Liederhalle* $S_2 = x_2 \in \mathbb{R},^+$ while that of $K1$ *to Eduardpfeiffer* is designated $S_3 = x_3 \in \mathbb{R}.^+$ The distances between the known stations $\{S_{12}, S_{23}, S_{31}\} \in \mathbb{R}^+$ are computed from their respective GPS coordinates in Table 7.1 on p. 88. Using the horizontal directions T_i and vertical directions B_i from Table 7.3 on p. 88, space angles $\{\psi_{12}, \psi_{23}, \psi_{31}\}$ are computed using (11.30) on p. 177 and presented in Table 12.1. From (12.17), we see that $S_1 = x_1$, $S_2 = x_2$ and $S_3 = x_3$ each has four roots. The solutions are real as depicted in Figs. 12.2, 12.3 and 12.4. The desired distances are selected with the help of prior information (e.g., from Fig. 7.1) as $S_1 = 566.8635$, $S_2 = 430.5286$, and $S_3 = 542.2609$. These values compare well with their real values in Fig. 7.1. Once the distances have been established, they are used to determine the coordinates of the unknown station $K1$ in step 2 via ranging techniques. In this example, the computed Cartesian coordinates of $K1$ are $X = 4157066.1116\,m$, $Y = 671429.6655\,m$ and $Z = 4774879.3704\,m$; which tallies with the GPS coordinates in Table 7.1.

Table 12.1. Space angles

Observation from	Space angle (gon)
$K1$-Schlossplatz-Liederhalle ψ_{12}	35.84592
$K1$-Liederhalle-Eduardpfeiffer ψ_{23}	49.66335
$K1$-Eduardpfeiffer-Schlossplatz ψ_{31}	14.19472

Intersection to more than Three Known Stations

The formulation of the *overdetermined* three-dimension intersection problem is as follows; given space angles from *more than three known stations*, i.e., $P_1, P_2, P_3, ..., P_n$, determine the unknown position $P_0 \in \mathbb{E}^3$. In this case, the observations will comprise horizontal directions T_i and vertical directions B_i from P_1 to P_0, P_2 to P_0, P_3 to P_0,...,P_n to P_0, with the unknowns being $\{X, Y, Z\}$.

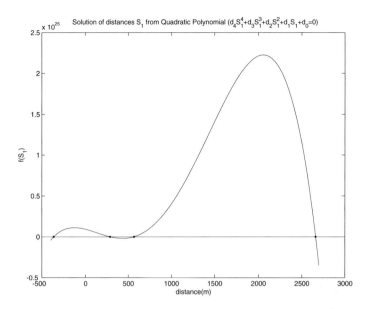

Fig. 12.2. Solution for distance S_1

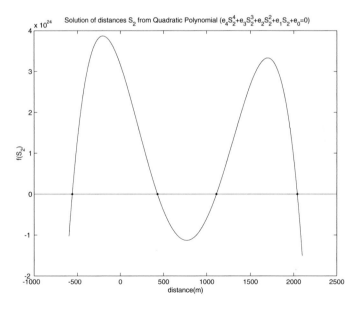

Fig. 12.3. Solution for distance S_2

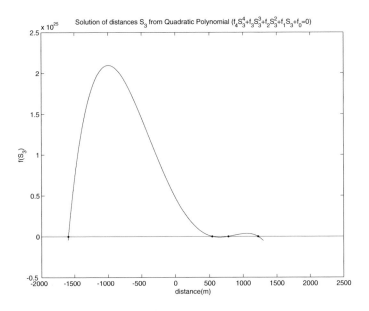

Fig. 12.4. Solution for distance S_3

Example 12.4 (Three-dimensional intersection from more than three known stations). For the test network Stuttgart Central in Fig. 7.1, the three-dimensional coordinates $\{X, Y, Z\}$ of the unknown station $K1$ are desired. Using all the observation data of Table 7.2 on p. 88, one proceeds to compute the position of $K1$ in four steps as follows:

Step 1 (combinatorial solution):
  From Fig. (7.1) on p. 86, and using (6.26) on p. 69, 35 minimal combinatorials are formed whose nonlinear systems of equations (12.12) are solved for the distances $\{S_i | i = 1, 2, 3\}$ to the unknown station $K1$ in closed form using either (12.17) or (12.24). Each combinatorial minimal subset results in 3 distances thus giving rise to a total of (3×35) 105 distances which we consider in the subsequent steps as pseudo-observations.

Step 2 (error propagation to determine the dispersion matrix $\boldsymbol{\Sigma}$):
  The variance-covariance matrix is computed for each of the combinatorial set $j = 1, \ldots, 35$ using error propagation. Equation (12.12) is used to obtain the dispersion matrix $\boldsymbol{\Sigma}$ in (6.31) as discussed in Example 6.4 on p. 72.

Step 3 (rigorous adjustment of the combinatorial solution points in a polyhedron):

Once the 105 combinatorial solution points in a polyhedron have been obtained in step 1, they are finally adjusted using the linear Gauss-Markov model (6.7) on p. 61 with the dispersion matrix Σ obtained via the nonlinear error propagation law in step 2.

Step 4 (position determination by ranging):

The position is then determined from values of steps 1 to 3 as in Example 9.6 on p. 144.

Using the data of Table 7.2, space angles for the network are computed and used to determine the position of the unknown station $K1$. Figure 12.5 presents the deviation of the computed scatter of the distance Haussmanstr.-$K1$ around its adjusted value. The plot of deviations of the adjusted distances from those derived from GPS coordinates are presented in Fig. 12.6. The numbers in the X-axis of Fig. 12.6 represent distances as follows; Haussmanstr.-$K1$ (1), Schlossplatz-$K1$ (2), Dach FH-$K1$ (3), Dach LVM-$K1$ (4), Liederhalle-$K1$ (5), Lindenmuseum-$K1$ (6) and Eduardpfeiffer-$K1$ (7).

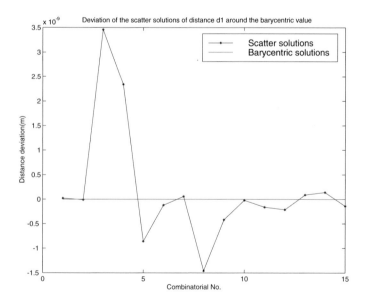

Fig. 12.5. Deviation of the scatter solutions of the distance Haussmanstr.-$K1$ from the adjusted value

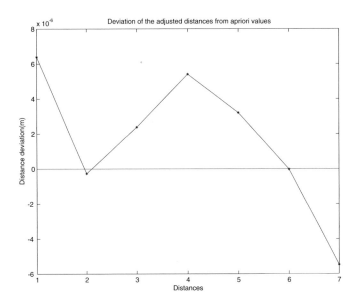

Fig. 12.6. Deviation of the 7-adjusted distances from real measured values

12.3 Photogrammetric Intersection

In Chap. 11, the exterior elements of orientation were determined as discussed in Sect. 11.3. Using these exterior elements, we demonstrate in this section how they are applied to determine algebraically the ground coordinates of unknown station. The problem of photogrammetric intersection is formulated as follows: Given the position and orientation of two or more photographs, let the conjugate image rays from the photographs intersect at a common ground point (e.g., Fig. 12.7). Determine the ground coordinates of the unknown station P. Let us examine two possible ways of solving this problem algebraically.

Grafarend-Shan Möbius Approach

Let us assume that the Cartesian coordinates $\{X_l, Y_l, Z_l\}$ and $\{X_r, Y_r, Z_r\}$, respectively, for the left perspective center P_l and the right perspective center P_r in Fig. 12.7 have been obtained from the photogrammetric resection approach in Sect. 11.3. The perspective center equations are

$$
\begin{bmatrix} X - X_l \\ Y - Y_l \\ Z - Z_l \end{bmatrix} = s_l \mathbf{R}_l \begin{bmatrix} x_l \\ y_l \\ -f_l \end{bmatrix}, \tag{12.25}
$$

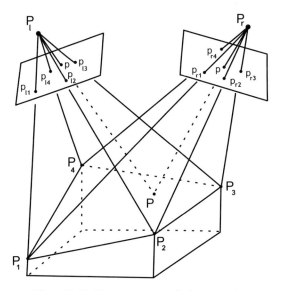

Fig. 12.7. Photogrammetric intersection

and

$$
\begin{bmatrix} X - X_r \\ Y - Y_r \\ Z - Z_r \end{bmatrix} = s_r \mathbf{R}_r \begin{bmatrix} x_r \\ y_r \\ -f_r \end{bmatrix}.
\tag{12.26}
$$

In (12.25) and (12.26), f_l and f_r are the left and right focal lengths and $\mathbf{R}$ the rotation matrix. The distance ratios s_l and s_r are given respectively by

$$
s_l := \frac{\|\overrightarrow{PP_l}\|}{\|\overrightarrow{pp_l}\|} = \frac{\sqrt{(X - X_l)^2 + (Y - Y_l)^2 + (Z - Z_l)^2}}{\sqrt{x_l^2 + y_l^2 + z_l^2}},
\tag{12.27}
$$

and

$$
s_r := \frac{\|\overrightarrow{PP_r}\|}{\|\overrightarrow{pp_r}\|} = \frac{\sqrt{(X - X_r)^2 + (Y - Y_r)^2 + (Z - Z_r)^2}}{\sqrt{x_r^2 + y_r^2 + z_r^2}}.
\tag{12.28}
$$

Equations (12.25) and (12.26) could be expanded into the 7-parameter similarity transformation equation

$$
\begin{bmatrix} X \\ Y \\ Z \end{bmatrix} = s_l \mathbf{R}_l \begin{bmatrix} x_l \\ y_l \\ -f_l \end{bmatrix} + \begin{bmatrix} X_l \\ Y_l \\ Z_l \end{bmatrix}
$$
$$
= s_r \mathbf{R}_r \begin{bmatrix} x_r \\ y_r \\ -f_r \end{bmatrix} + \begin{bmatrix} X_r \\ Y_r \\ Z_r \end{bmatrix}.
\tag{12.29}
$$

Grafarend and Shan [152] propose a *three-step* solution approach based on Möbius coordinates as follows:

- In the *first step*, the ratio of distances $\{s_l, s_r\}$ between the perspective centers and the unknown intersected point are determined from a linear system of equations. The area elements of the left and right images are employed to form the linear system of equations.
- In the *second step*, the computed distance ratios are used to compute the Möbius coordinates.
- These coordinates are converted to the three-dimensional cartesian coordinates $\{X, Y, Z\}$ in the *third step*.

Commutative Algebraic Approaches

Whereas the Grafarend-Shan approach discussed in Sect. 12.3 solves the intersection of rays from two photographs, the algebraic approaches solves the intersection of rays from three photographs. Consider the case in Fig. 9.1, the unknown station is intersected from three photographs. The distances $\{S_1, S_2, S_3\}$ to the unknown stations are determined using either using (12.17) or (12.24). The coordinates of the unknown stations are determined from procedures of Sect. 9.3.2.

12.4 Concluding Remarks

The techniques presented in this chapter could provide direct approaches for obtaining positions from direction (angular) measurements to stations that can not be physically occupied. Intersection techniques discussed could be useful in structural deformation monitoring in industries. For surfaces or structures that are harmful when physical contact is made, intersection techniques come in handy. the methods can also be used for quick station search during cadastral and engineering surveying operations. These methods can be augmented with resection and ranging techniques to offer a wide range of possibilities. Further reference are [14, 27, 37, 132, 142, 152].

13

GPS Meteorology in Environmental Monitoring

13.1 Satellite Environmental Monitoring

In 1997, the Kyoto protocol to the United Nation's framework convention on climate change spelt out measures that were to be taken to reduce the greenhouse gas emission that has contributed to *global warming*. Global warming is just but one of the many challenges facing our environment today. The rapid increase in *desertification* on one hand and *flooding* on the other hand are environmental issues that are increasingly becoming of concern. For instance, the torrential rains that caused havoc and destroyed properties in USA in 1993 is estimated to have totalled to $15 billion, 50 people died and thousands of people were evacuated, some for months [216]. Today, the threat from torrential rains and flooding still remains real as was seen in 1997 El'nino rains that swept roads and bridges in Kenya, the 2000 Mozambique flood disaster, 2002 Germany flood disaster or the Hurricane Isabel in the US coast[1]. The melting of polar ice thus raising the sea level is creating fear of submerssion of beaches and cities surrounded by the oceans and those already below sea level. In-order to be able to predict and model these occurrences so as to minimize damages such as those indicated by [216], atmospheric studies have to be undertaken with the aim of improving on mechanism for providing *reliable, accurate* and *timely* data. These data are useful in Numerical Weather Prediction (NWP) models for weather forecasting and climatic models for monitoring climatic changes. Besides, *accurate* and *reliable information* on weather is essential for other applications such as agriculture, flight navigation, etc.

---

[1]BBC 19th Sept. 2003 online report: http://news.bbc.co.uk/

Data for NWP and climatic models are normally collected using balloon filled radiosondes, satellites (polar and geostationary) and other sources e.g., flight data from aeroplanes. Whereas [232, p. 94] points out that about 9500 surface based stations and 7000 merchant ships exist that send up weather balloons, [335] noted that most of these data cover the northern hemisphere, with the southern hemisphere (mainly Africa and South America) lacking adequate data due to financial constraints. Lack of radiosonde data is also noted in the oceanic areas hence leading to shortage of adequate data for NWP and climatic models. These models require precise and accurate data for estimating initial starting values in-order to give accurate and reliable weather forecast, and to be of use for climate monitoring. The shortage of radiosonde data is complemented with the polar and geostationary satellite data. Polar satellites include for instance the US owned National Ocean and Atmospheric Administration NOAA-14 and NOAA-15, while the geostationary satellites include US based Geostationary Operational Environmental Satellite (GEOS) and Europe owned METEOrological SATellite (METEOSAT).

Polar and geostationary satellites (e.g., NOAA, GOES and METEOSAT) used for temperature and water vapour profile measurements have their own limitations however. In high altitude winter conditions for instance, use of passive Infra Red (IR) is difficult due to *very cold temperatures, common near surface thermal inversion*, and *high percentage of ice cloud* that play a role in limiting the IR sounding [248]. In volcanic areas, low flying remote sensing satellites are also affected by the presence of dust and aerosol. Large-scale volcanic eruption normally injects large amount of aerosols into the lower stratosphere and thus limiting the IR observation of the stratosphere and lower regions. In-order therefore to enhance global weather and climatic prediction, current systems have to be complemented by a system that will provide global coverage and whose signals will be able to penetrate clouds and dust to remote sense the atmosphere. Such system, already proposed as early as 1965 by Fischbach [108], and which is currently an active area of research, is the new field of **GPS-Meteorology**. It involves the use of GPS satellites to obtain atmospheric profiles of *temperature, pressure* and *water vapour/humidity*.

As we saw in Chap. 7, Global Positioning System (GPS) satellites were primarily designed to be used by the US military. their main task was to obtain the position of any point on Earth from space. The signals emitted by GPS satellites traverse the ionosphere and neutral atmosphere to be received by ground based GPS receivers. One of the major obstacles to positioning or navigating with GPS is the signal delay caused by atmospheric refraction. Over the years, research efforts have been dedicated to modelling atmospheric refraction in-order to improve on positioning accuracy. In the last decade however, [248] suggested that this negative effect of the atmosphere on GPS signals could be inverted to remote sense the atmosphere using space borne techniques. Melbourne [248], proposed that Low Earth Orbiting Satellites LEO be fitted with GPS receivers and be used to track the signals of rising or setting GPS satellites (occulting satellites). The signal delay could then be measured and used to infer on the atmospheric profiles of temperature, pressure, water vapour and geopotential heights.

This new technology of GPS atmospheric remote sensing has the advantages of;

(a) being *global*,
(b) *stable* owing to the stable GPS oscillators and
(c) having radio frequencies that can *penetrate* clouds and dusts.

The new technology therefore plays a major role in complementing the existing techniques, e.g., radiosondes. Atmospheric profiles from GPS remote sensing have been tested in NWP models and preliminary results so far are promising [180]. Indeed, [205] have already demonstrated using the data of the pilot project GPS/MET that the accuracy of global and regional analysis of weather prediction can significantly be improved. Also motivating are the results of [300] who showed that high accuracy of measurements and vertical resolution around the tropopause would be relevant to monitor climatic changes in the next decades. Several atmospheric sounding missions have been launched, e.g., the CHAllenging Minisatellite Payload mission (CHAMP), Gravity Recovery And Climate Experiment (GRACE) and SAC-C. Constellation Observing System for Meteorology, Ionosphere and Climate (COSMIC) mission that will provide up to 3000 occultation data daily is proposed to be launched by University Corporation of Atmospheric Research UCAR in 2005 [9]. EQUatorial Atmosphere Research Satellite (EQUARS) mission that will provide equatorial coverage is also proposed to be launched in 2005 [316]. Currently, studies are being undertaken at Jet Propulsion Laboratory (JPL) on possibilities of having future atmospheric sounding missions that will have satellites of the sizes of a laptop with GPS receivers of the sizes of a credit card [351].

Plans are also underway to have the European owned EUropean organization for the exploitation of METeorological SATellites (EUMETSAT) and American NOAA owned National Polar Orbiting Environmental Satellite System (NPOESS) installed with GPS occultation receivers GRAS (GNSS Receiver for Atmospheric Sounding) and GPSOS (GPS Occultation Sensor) [335]. The planned satellite missions together with the proposed GALILEO satellites scheduled to be operational by 2010 [335] and the Russian GLONASS promises a brighter future for environmental monitoring. Indeed, that these atmospheric sounding missions promise to provide daily global coverage of thousands of remotely sensed data which will be vital for weather, climatic and atmospheric sciences studies will be a revolution in the history of environmental studies.

Space borne GPS meteorology which we discuss in detail in Sect. 13.2.1 is just but one part of this new technique. The other component is the ground based GPS meteorology which will be discussed in detail in Sect. 13.2.2. Collection of articles on this new technique has been presented for instance in [10]. In ground based GPS meteorology, a dense GPS network is used to measure precisely GPS path delays caused by the ionosphere and the neutral troposphere traversed by the GPS signals. These path delays are then converted into Total Electronic Contents (TEC) and Integrated Precipitate Water Vapour IPWV. Conversion to IPWV requires prior information of surface pressure or estimates along the GPS ray path. These create a continuous, accurate, all weather, real time lower and upper atmospheric data with a variety of opportunities for atmospheric research [330].

Clearly, GPS meteorology promises to be a real boost to atmospheric studies with expected improvements on weather forecasting and climatic change monitoring, which directly impact on our day to day lives. In [18], the possible use of IPWV for flood prediction is proposed, while [44] have outlined the potential of water vapour for meteorological forecasting. For environmental monitoring, GPS meteorology will further play the following roles:

1. Precisely derive vertical temperature and pressure profiles: These will be useful in the following ways [248]:
   (a) By combining them with other observations of ozone densities and dynamic models, our understanding of conditions which lead to the formation of polar stratosphere clouds will be improved. We will also be able to understand how particles in which het-

erogeneous chemical reactions lead to ozone loss are believed to occur.

(b) The precise measured temperature will enable the monitoring of global warming and the effect of greenhouse gases. This is made possible as the change in surface temperatures caused by an increase in the greenhouse gas densities is generally predicted to be the largest and therefore most apparent at high latitudes. Precise temperature can be used to map the structure of the stratosphere, particularly in the polar region where temperature is believed to be an important factor in the minimum levels of ozone observed in spring.

(c) Accurate high vertical resolution temperature reconstruction in the upper troposphere will increase our understanding on the conditions which cirrus clouds form. The cirrus clouds will generate for instance a positive feed back effect if global warming displaces a given cloud layer to a higher and colder region. The colder cloud will then emit less radiation forcing the troposphere to warm in-order to compensate for the decrease.

(d) Accurate temperature retrievals from GPS meteorological measurements combined with high horizontal resolution temperatures derived from the nadir-viewing microwave radiometers will provide a powerful data set for climate studies of the Earth's lower atmosphere. This can be achieved by using the derived profiles to monitor trends in the upper troposphere and lower stratosphere where the GPS meteorological techniques yield its most accurate results.

(e) The measured pressure is expected to contribute to the monitoring of global warming. This is because pressure versus geometrical height is potentially an interesting diagnostic of troposphere's climatic change since the height of any pressure surface is a function of the integrated temperature below.

(f) The temperature in the upper troposphere/tropopause influences the amount of energy radiated to space. Accurate measurements of temperature in this region over a long period of time will provide data for global warming and climatologic studies.

2. Derive water vapour: Precise analysis of the water vapour will contribute to the data required by hydrologists to enhance the prediction of local torrential rain that normally cause damage and havoc

(see e.g., [18]). Besides, the knowledge of water vapour density in the lower troposphere will be useful in;

- providing data that will be directly assimilated into meteorological models to enhance predictability and forecasting of weather,
- applicable for creation of distribution of water vapour via tomographic techniques (e.g., [111]),
- applied to correct the wet delay component in both Synthetic Aperture Radar (SAR) and GPS positioning thus benefiting applications requiring precise positioning such as deformation monitoring,
- beneficial to low altitude aircraft navigation, since limitation in the mitigation of tropospheric delay is a major source of positioning error,
- global warming monitoring by determining the latent heat suspended in the atmosphere where water vapour comprise one of the greenhouse gases,
- the radiative forcing due to vapour and cloud inferred from humidity,
- improved inputs for weather forecasting, climate and hydrology. Water vapour will be essential for short term (0-24hrs) forecasting of precipitation. Currently, lack of atmospheric water vapour is the major source of error in short term weather forecasting [175].

3. Contribute towards climatic studies: By comparing the observed temperatures against the predicted model values, a method for detecting and characterizing stratospheric climatic variations as well as a means for evaluating the performance of model behaviour at stratospheric altitudes will be developed and the existing ones tested.

4. Enhance geodynamic studies: The study of the gravitation effects of the atmospheric pressure, water vapour and other phenomenons will contribute towards the determination of high-resolution local geoid, which is vital for monitoring crustal deformation. The transient drift that occurs per week in estimate of crustal deformation from GPS measurement will be corrected.

5. Enhance disaster mitigation measures: Its information will contribute to the much-needed information required to improve forecasting of catastrophic weather around the world.

6. With abundance of GPS remote sensing data, accuracy better than 1 -2K in temperature given by GPS meteorological missions (e.g., CHAMP, GRACE etc.) will be realized.

In-order to fully realize the potential of the GPS atmospheric remote sensing listed above, estimated profiles have to be of high quality. Already, comparative results with the existing models such as European Centre for Medium Weather Forecast (ECMWF) and National Centre for Environmental Prediction (NCEP) are promising as seen from the works of [281, 335] with respect to GPS/MET and CHAMP missions, respectively.

13.2 GPS Remote Sensing

13.2.1 Space Borne GPS Meteorology

Radio occultation with GPS takes place when a transmitting GPS satellite, setting or rising behind the Earth's limb, is viewed by a LEO satellite as illustrated in Fig. 13.1[2]. GPS satellites send navigation signals, which passes through successively deeper layer of the Earth's atmosphere and are received by LEO satellites. These signals are bent and retarded causing a delay in the arrival at the Leo (see Fig.13.1[2]). Figure 13.3 indicates the occultation geometry where the signal is sent from GPS to the LEO satellite passing through dispersive layers of the ionosphere and atmosphere remote sensing them. As the signal is bent, the total bending angle α, an impact parameter a and a tangent radius r_t define the ray passing through the atmosphere. *Refraction angle* is accurately measured and related to atmospheric parameters of temperature, pressure and water vapour via the refractive index. Use is made of radio waves where the LEO receiver measures, at the required sampling rate, the dual band carrier phase, the C/A and P-code group delay and the signal strength made by the flight receiver [248]. The data is then processed to remove errors arising from short time oscillator and instabilities in; satellites and receivers. This is achieved by using at least one ground station and one satellite that is not being occulted. Once the observations have been corrected for possible sources of errors, the resulting *Doppler shift* is used to determine the refraction angle α (see Fig. 13.3).

---

[2]source: http://geodaf.mt.asi.it/html/GPSAtmo/space.html

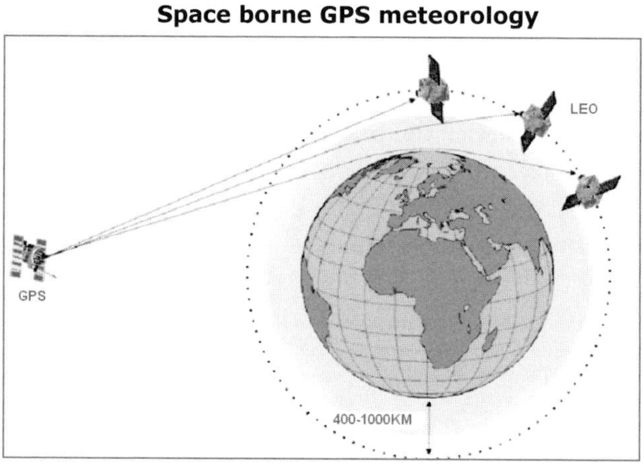

Space borne GPS meteorology

Fig. 13.1. GPS Radio occultation

The variation of α with a during an occultation depends primarily on the vertical profile of atmospheric refractive index, which is determined globally by *Fermat's principle* of least time and locally by *Snell's law*

$$n sin\phi = constant, \qquad (13.1)$$

where ϕ denotes the angle between the gradient of refraction and the ray path. Doppler shift is determined by projecting spacecraft velocities onto the ray paths at the transmitter and receiver, so that atmospheric bending contributes to its measured value. Data from several GPS transmitters and post-processing ground stations are used to establish the precise positions and velocities of the GPS transmitters and LEO satellites. These derived positions and velocities are used to calculate the Doppler shift expected in the absence of atmospheric bending (i.e., were the signal to travel in vacuo). By subtracting the *expected* shift from the measured shift, one obtains the excess Doppler shift. Assuming local symmetry and with Snell's law, the excess Doppler shift together with satellites' *positions* and *velocities* are used to compute the values of the bending angles α with respect to the impact parameters a. In Sect. 13.3, we will present an algebraic approach for computing bending angles and impact parameters. Once computed, these bending (refraction) angles are related to the refractive index by

$$\alpha(a) = 2a \int_{r=r_0}^{r=\infty} \frac{1}{\sqrt{n^2 r^2 - a^2}} \frac{dln(n)}{dr} dr, \qquad (13.2)$$

which is inverted using Abel's transformation to give the desired refractive index

$$n(r_0) = \exp\left[\frac{1}{\pi} \int_{a=a_0}^{a=\infty} \frac{\alpha(a)}{\sqrt{a^2 - a_0^2}} da\right]. \qquad (13.3)$$

Rather than the refractive index in (13.3), refractivity is used as

$$N = (n-1)10^6 = 77.6\frac{P}{T} + 3.73 \times 10^5 \frac{P_w}{T^2} - 40.3 \times 10^6 \frac{n_e}{f^2} + 1.4w. \quad (13.4)$$

In (13.4), P denotes the atmospheric pressure in {mbar}, T the atmospheric temperature in K, P_w the water vapour in {mbar}, n_e the electron number density per cubic meter {number of electron/m^3}, f the transmitter frequency in Hz and w the liquid water content in g/m^3. Three main contributors to refractivity are:

- The *dry neutral atmosphere* (called the dry component, i.e., the first component on the right-hand-side of (13.4)).
- Water vapour (also called the wet or moist components, i.e., the second component on the right-hand-side of (13.4))
- The free electrons in the ionosphere (i.e., the third component on the right-hand-side of (13.4)).

If the atmospheric temperature T and pressure P are provided from external source, e.g., from models and synoptic meteorological data over tropical oceanic regions, then the vertical water vapour density may be recovered from satellite remote sensing data [248]. The refraction effects on the signals in the ionosphere must be corrected using signals at two frequencies at which these effects are substantially different.

13.2.2 Ground based GPS meteorology

Whereas GPS receivers are onboard low flying (LEO) satellites (e.g., CHAMP, GRACE etc.) in space borne GPS remote sensing, they are fixed on ground stations in the case of ground-based GPS meteorology. These receivers track the transmitted signals which have traversed the atmosphere as indicated in Fig.13.2[3]. As the signals travel through

---

[3]Source: http://apollo.lsc.vsc.edu/classes/remote/lecture_notes/gps/theory/theoryhtml.htm

the atmosphere from the satellites to the ground based receivers, they are delayed by the troposphere. The tropospheric delay comprise the hydrostatic and the wet parts as seen in (13.4). The contribution of hydrostatic part which can be modeled and eliminated very accurately using surface pressure data or three-dimensional numerical models is about 90% [87, 103]. The wet delay however is highly variable with little correlation to surface meteorological measurements. Assuming that the wet delay can be accurately derived from GPS data, and that reliable surface temperature data are available, the wet delay can be converted into the estimation of the total atmospheric water vapour P_w present along the GPS ray path as already suggested by [59]. This atmospheric water vapour P_w is termed precipitable water in GPS meteorology.

The precipitable water as opposed to the vertical profile is estimated with a correction made for the fact that the radio beams normally are slanted from the zenith. The phase delay along the zenith direction is called the "zenith delay" and is related to the atmospheric refractivity by:

$$zenith\ Delay = 10^6 \int_{antenna}^{\infty} N(z)\mathrm{d}z, \qquad (13.5)$$

where the integral is in the zenith direction. Substituting (13.4) in (13.5) leads to the calculation of the zenith wet delay which is related to the total amount of water vapour along the zenith direction. The zenith delay is considered to be constant over a certain time interval. It is the average of the individual slant ray path delays that are projected to the zenith using the mapping functions (e.g., [257]) which are dependent on the receiver to satellite elevation angle, altitude and time of the year.

The significant application of GPS satellites in ground based GPS meteorology is the determination of the slant water. If one could condense all the water vapour along the ray path of a GPS signal (i.e., from the GPS satellite to the ground receiver), the column of the liquid water after condensation is termed slant water. By converting the GPS derived tropospheric delay during data processing, slant water is obtained. By using several receivers to track several satellites (e.g., Fig.13.2[3]), a three-dimensional distribution of water vapour and its time variation can be quantified. In Japan, there exist (by 2004) more than 1200 GPS receivers within the framework of GPS Earth Observing NETwork (GEONET) with a spatial resolution of 12-25km dedicated to GPS meteorology (see e.g., [10, 315]). These dense network of GPS receivers are capable of delivering information on water vapour that are useful as already stated in Sect. 13.1.

Ground based GPS Remote sensing

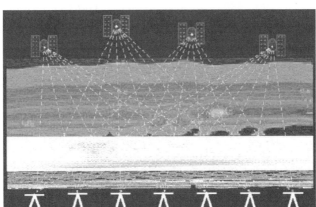

Fig. 13.2. Water vapour from ground based GPS receivers

13.3 Refraction (Bending) Angles

In space borne GPS meteorology, the measured quantities are normally the excess path delay of the signal. It is obtained by measuring the excess phase of the signal owing to atmospheric refraction during the traveling period. The determination of the refraction angle α from the measured excess phase therefore marks the beginning of the computational process to retrieve the atmospheric profiles of temperature, pressure, water vapour and geopotential heights. The unknown refraction angle α is related to the measured excess phase by a system of two nonlinear trigonometric equations;

1. an equation relating the doppler shift at the Low Earth Orbiting (LEO) satellite (e.g., CHAMP, GRACE etc.) expressed as the difference in the projected velocities of the two moving satellites on the ray path tangent on one hand, and the doppler shift expressed as the sum of the atmosphere free propagation term and a term due to atmosphere on the other hand,

2. an equation that makes use of Snell's law in a spherically layered medium [299, p. 59].

Equations formed from (1) and (2) are nonlinear e.g., (13.6) and have been solved using iterative numerical methods such as Newton's (see e.g., [170, 211, 299, 335]. In-order to solve the trigonometric nonlinear system of equations (13.6), Newton's approach assumes the refractive angles to be small enough such that the relationship between the doppler shift and the bending angles formed from (1) and (2) are linear. The linearity assumption of the relationship between the doppler shift and refraction angles introduces some small *nonlinearity errors*. Vorob'ev and Krasil'nikova [325] have pointed out that neglecting the nonlinearity in (13.6) causes an error of 2% when the beam perigee is close to the Earth's ground and decrease with the altitude of the perigee. The extent of these errors in the dry part of the atmosphere, i.e., the upper troposphere and lower stratosphere, particularly the height 5-30 km, whose bending angle data are directly used to compute the atmospheric profiles or directly assimilated in Numerical Weather Prediction Models (NWPM) (e.g., [180]) is however not precisely stated. The effects of nonlinearity error on the impact parameters to which the refraction bending angles are related is also not known.

In an attempt to circumvent the nonlinearity of (13.6), [325] expand it into series of V/c, where V is the velocity of the artificial satellite and c the velocity of light in vacuum. This corrects for relativistic effects and introduce the concept of perturbation. The angle between the relative position vectors of the two satellites and the tangent velocity vector at GPS is expressed in quadratic terms of the corresponding angle at LEO (also expanded to the second order). The refraction angle is then obtained by making use of its infinitesimal values that are less than 10^{-2}. Though the approach attempts to provide an analytic (direct) solution to nonlinear system of equations for bending angles, it is still nevertheless "quasi-nonlinear" and as such does not offer a complete, exact solution to the problem. The fact that there existed no direct (exact) solution to the nonlinear system of bending angle's equations of space borne GPS meteorology had already been pointed out by [335].

Motivated by Wickert's [335] observation, we will demonstrate in the next sections how the algebraic techniques of Sylvester resultant and reduced Groebner basis offer direct solution to bending angles' nonlinear system of equations (13.6).

13.3.1 Transformation of Trigonometric Equations to Algebraic

The system of nonlinear trigonometric equations for determining the refraction angles comprises of two equations given as

$$\begin{bmatrix} v_L cos(\beta_L - \phi_L) - v_G cos(\phi_G + \beta_G) = \dfrac{dL_i}{dt} + v_L cos(\beta_L - \psi_L) - v_G cos(\psi_G + \beta_G) \\ r_G sin\phi_G = r_L sin\phi_L, \end{bmatrix}$$

$$(13.6)$$

where v_L, v_G are the projected LEO and GPS satellite velocities in the occultation plane, r_L, r_G the radius of tangent points at LEO and GPS respectively, and $\dfrac{dL_i}{dt}$, the doppler shift. The angles in (13.6) are as shown in Fig. 13.3.

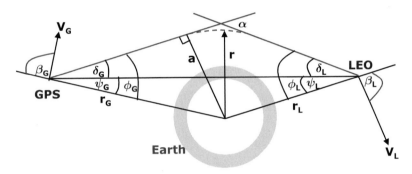

Fig. 13.3. Geometry of space borne GPS meteorology

Let us denote

$$\begin{bmatrix} x = sin\phi_G, y = sin\phi_L, a_1 = v_L cos\beta_L, a_2 = v_L sin\beta_L \\ a_3 = -v_G cos\beta_G, a_4 = v_G sin\beta_G, a_5 = r_G, a_6 = -r_L, \end{bmatrix}$$

$$(13.7)$$

where the signs of the velocities change depending on the directions of the satellites. Using;

- Theorem 3.1 on p. 20,
- the *trigonometric addition formulae*,
- and (13.7),

(13.6) simplifies to

$$\begin{bmatrix} a_1 cos\phi_L + a_2 y + a_3 cos\phi_G + a_4 x = a \\ a_5 x + a_6 y = 0. \end{bmatrix}$$

$$(13.8)$$

In (13.8), the right-hand-side of the first expression of (13.6) has been substituted with a. In-order to eliminate the trigonometric terms $cos\phi_L$ and $cos\phi_G$ appearing in (13.8), they are taken to the right-hand-side and the resulting expression squared as

$$(a_2y + a_4x - a)^2 = (-a_1cos\phi_L - a_3cos\phi_G)^2. \tag{13.9}$$

The squared trigonometric values $cos^2\phi_G$ and $cos^2\phi_L$ from (13.9) are then replaced by variables $\{x, y\}$ from (13.7). This is done following the application of trigonometric Pythagorean theorem of a unit circle $\{cos^2\phi_G + sin^2\phi_G = 1\}$ and $\{cos^2\phi_L + sin^2\phi_L = 1\}$ which convert the cosine terms into sines. The resulting expression has only trigonometric product $\{2a_1a_3cos\phi_Lcos\phi_G\}$ on the right-hand-side. Squaring both sides of the resulting expression and replacing the squared trigonometric values $cos^2\phi_G$ and $cos^2\phi_L$, with $\{x, y\}$ from (13.7) completes the conversion of (13.6) into algebraic

$$\left[\begin{array}{l} d_1x^4 + d_2x^3 + d_3x^3y + d_4x^2 + d_5x^2y^2 + d_6x^2y + d_7x + d_8xy^3 + d_9xy^2 + d_{00} = 0 \\ \qquad\qquad a_5x + a_6y = 0, \end{array} \right. \tag{13.10}$$

where $d_{00} = d_{10}xy + d_{11}y^4 + d_{12}y^3 + d_{13}y^2 + d_{14}y + d_{15}$. The coefficients $d_1, ..., d_{15}$ are:

$$
\begin{aligned}
d_1 &= b_4^2 & d_9 &= 2b_1b_5 + 2b_2b_3 \\
d_2 &= 2b_4b_5 & d_{10} &= 2b_3b_6 + 2b_5b_2 \\
d_3 &= 2b_4b_3 & d_{11} &= b_1^2 \\
d_4 &= 2b_6b_4 + b_5^2 + b_7^2 & d_{12} &= 2b_1b_2 \\
d_5 &= 2b_1b_4 + b_3^2 - b_7^2 & d_{13} &= 2b_1b_6 + b_2^2 + b_7^2 \\
d_6 &= 2b_3b_5 + 2b_2b_4 & d_{14} &= 2b_2b_6 \\
d_7 &= 2b_6b_5 & d_{15} &= b_6^2 - b_7^2, \\
d_8 &= 2b_1b_3 &
\end{aligned}
$$

with

$$
\begin{aligned}
b_1 &= a_1^2 + a_2^2 \\
b_2 &= -2aa_2 \\
b_3 &= 2a_2a_4 \\
b_4 &= (a_3^2 + a_4^2) \\
b_5 &= -2aa_4 \\
b_6 &= a^2 - a_1^2 - a_3^2 \\
b_7 &= 2a_1a_3.
\end{aligned}
$$

The algebraic equation (13.10) indicates that the solution of the non-linear bending angle equation (13.6) is given by the intersection of a *quartic polynomial* (e.g., p. 27) and a *straight line* (see e.g., Fig. 13.4).

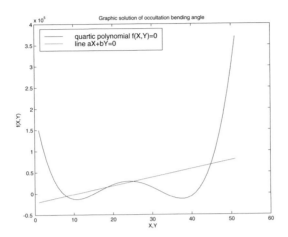

Fig. 13.4. Algebraic curves for the solution of nonlinear system of bending angle equations

13.3.2 Algebraic Determination of Bending Angles

Application of Groebner Basis

Denoting the nonlinear system of algebraic (polynomial) equations (13.10) by

$$\left[\begin{array}{l} f_1 := d_1 x^4 + d_2 x^3 + d_3 x^3 y + d_4 x^2 + d_5 x^2 y^2 + d_6 x^2 y + d_7 x + d_8 x y^3 + d_9 x y^2 + d_{00} \\ f_2 := a_5 x + a_6 y, \end{array}\right.$$

(13.11)

reduced Groebner basis (4.38) on p. 44 is computed for x and y as

$$\left[\begin{array}{l} Groebner\,Basis[\{f_1, f_2\}, \{x, y\}, \{y\}] \\ Groebner\,Basis[\{f_1, f_2\}, \{x, y\}, \{x\}]. \end{array}\right.$$

(13.12)

The terms $\{f_1, f_2\}$ in (13.12) indicate the polynomials in (13.11), $\{x, y\}$ the variables with lexicographic ordering x comes before y, and $\{y\}$, $\{x\}$ the variables to be eliminated. The first expression of (13.12), i.e., $Groebner\,Basis[\{f_1, f_2\}, \{x, y\}, \{y\}]$ gives a quartic polynomial in x (the first expression of 13.13), while the second expression gives a quartic polynomial in y (the second expression of 13.13). The results of (13.12) are:

$$\left[\begin{array}{l} h_4 x^4 + h_3 x^3 + h_2 x^2 + h_1 x + h_0 = 0 \\ g_4 y^4 + g_3 y^3 + g_2 y^2 + g_1 y + g_0 = 0, \end{array}\right. \tag{13.13}$$

with the coefficients as

$$\begin{aligned} h_4 &= (a_6^4 d_1 + a_5^4 d_{11} - a_5 a_6^3 d_3 + a_5^2 a_6^2 d_5 - a_5^3 a_6 d_8) \\ h_3 &= (-a_5^3 a_6 d_{12} + a_6^4 d_2 - a_5 a_6^3 d_6 + a_5^2 a_6^2 d_9) \\ h_2 &= (-a_5 a_6^3 d_{10} + a_5^2 a_6^2 d_{13} + a_6^4 d_4) \\ h_1 &= (-a_5 a_6^3 d_{14} + a_6^4 d_7) \\ h_0 &= a_6^4 d_1 5, \end{aligned}$$

and

$$\begin{aligned} g_4 &= (a_6^4 d_1 + a_5^4 d_{11} - a_5 a_6^3 d_3 + a_5^2 a_6^2 d_5 - a_5^3 a_6 d_8) \\ g_3 &= (a_5^4 d_{12} - a_5 a_6^3 d_2 + a_5^2 a_6^2 d_6 - a_5^3 a_6 d_9) \\ g_2 &= (-a_5^3 a_6 d_{10} + a_5^4 d_{13} + a_5^2 a_6^2 d_4) \\ g_1 &= (a_5^4 d_{14} - a_5^3 a_6 d_7) \\ g_0 &= a_5^4 d_{15}. \end{aligned}$$

Four solutions are obtained from (13.13) for both x and y using Matlab's roots command (e.g., 4.40 on p. 44) as $x = roots([h_4\ h_3\ h_2\ h_1\ h_0])$ and $y = roots([g_4\ g_3\ g_2\ g_1\ g_0])$. From (13.7) and the roots of (13.13), the required solutions can now be obtained from

$$\left[\begin{array}{l} \phi_G = sin^{-1} x, \\ \phi_L = sin^{-1} y. \end{array}\right. \tag{13.14}$$

The desired bending angle α in Fig. 13.3 is then obtained by first computing δ_G and δ_L as

$$\left[\begin{array}{l} \delta_G = \phi_G - \psi_G \\ \delta_L = \phi_L - \psi_L, \end{array}\right. \tag{13.15}$$

leading to

$$\left[\begin{array}{l} \alpha = \delta_G + \delta_L \\ p = \frac{1}{2}(r_L sin\phi_L + r_G sin\phi_G), \end{array}\right. \tag{13.16}$$

where $\alpha(p)$ is the bending angle and p the impact parameter.

Sylvester Resultants Solution

The quartic polynomials (13.13) can also be obtained using Sylvester resultants technique as follows:

- Step 1: From the nonlinear system of equations (13.10), hide y by treating it as a constant (i.e., polynomial of degree zero). Using (5.1) and (5.2) on p. 48, one computes the resultant of a 5×5 matrix

$$Res\,(f_1,\ f_2, y) = det \begin{bmatrix} a_5 & a_6y & 0 & 0 & 0 \\ 0 & a_5 & a_6y & 0 & 0 \\ 0 & 0 & a_5 & a_6y & 0 \\ 0 & 0 & 0 & a_5 & a_6y \\ d_1 & d_2 + d_3y & b_{53} & b_{54} & b_{55} \end{bmatrix}, \qquad (13.17)$$

with $b_{53} = d_4 + d_5y^2 + d_6y$, $b_{54} = d_7 + d_8y^3 + d_9y^2 + d_{10}y$ and $b_{55} = d_{00}$. The solution of (13.17) leads to the first expression of (13.13).

- Step 2: Now hide x by treating it as a constant (i.e., polynomial of degree zero) and compute the resultant of a 5×5 matrix as

$$Res\,(f_1,\ f_2, x) = det \begin{bmatrix} a_6 & a_5x & 0 & 0 & 0 \\ 0 & a_6 & a_5x & 0 & 0 \\ 0 & 0 & a_6 & a_5x & 0 \\ 0 & 0 & 0 & a_6 & a_5x \\ d_{11} & d_{12} + d_8x & c_{53} & c_{54} & c_{55} \end{bmatrix}, \qquad (13.18)$$

with $c_{53} = d_{13} + d_5x^2 + d_9x$, $c_{54} = d_{14} + d_3x^3 + d_6x^2 + d_{10}x$ and $c_{55} = d_{15} + d_1x^4 + d_2x^3 + d_4x^2 + d_7x$. The solution of (13.18) leads to the second expression of (13.13) from which the bending angles and the impact parameters can be solved as already discussed.

In summary, the algebraic solution of refraction angles in space borne GPS meteorology proceeds in five steps as follows:

Step 1 (coefficients computation):
  Using (13.7), compute the coefficients $\{h_4\ h_3\ h_2\ h_1\ h_0\}$ and $\{g_4\ g_3\ g_2\ g_1\ g_0\}$ of the quartic polynomials in (13.13).
Step 2 (solution of variables $\{x, y\}$):
  Using the coefficients $\{h_i, g_i\}|i = 1, 2, 3, 4$ computed from step 1, obtain the roots of the univariate polynomials in (13.13) for $\{x, y\}$.
Step 3 (determine the angles $\{\phi_G, \phi_L\}$):
  With the admissible values of $\{x, y\}$ from step 2, compute the angles $\{\phi_G, \phi_L\}$ using (13.14).

Step 4 (obtain the angles $\{\delta_G, \delta_L\}$):

Using the values of $\{\phi_G, \phi_L\}$ from step 3, compute the angles $\{\delta_G, \delta_L\}$ using (13.15).

Step 5 (determine the angle α and the impact parameter p):

Finally, the bending angle α and the impact parameter p are computed using the values of $\{\phi_G, \phi_L\}$ from step 4 in (13.16).

13.4 Algebraic Analysis of some CHAMP Data

Let us now apply the algebraic algorithm outlined in steps 1 to 5 to assess the effect of neglecting nonlinearity (i.e., nonlinearity error) in using Newton's iterative approach, which assumes (13.6) to be linear. In-order carry out the analysis, bending angles from CHAMP satellite level 2 data for two satellite occultations were computed and compared to those obtained from iterative approach in [299]. The occultations were chosen at different times of the day and years. Occultation number 133 of 3rd May 2002 occurred past mid-day at 13:48:36. For this period of the day, the solar radiation is maximum and so is the ionospheric noise. In contrast, occultation number 3 of 14th May 2001 occurred past mid-night at 00:39:58.00. For this period, the solar radiation is minimum and the effect of ionospheric noise is also minimum.

The excess phase length data are smoothed using polyfit function of Matlab software and the resulting doppler shift values for L1 and L2 used together with (13.7), (13.13) and (13.14) to obtain the angles ϕ_G and ϕ_L. These angles were then used in (13.16) to compute the refraction angle α and the impact parameter p (also denoted in this analysis as a). Let us examine in detail the computation of occultation number 133 of 3rd May 2002 which occurred during the maximum solar radiation period. The results of occultation number 3 of 14th May 2001 will thereafter be briefly presented. For occultation number 133 of 3rd May 2002, which occurred from the time 13:48:36 to 13:49:51.98, the bending angles were computed using both algebraic and the classical Newton's (e.g., [299]) algorithms. Since the algebraic procedure leads to four solutions of (13.13), a criteria for choosing the admissible solution had to be developed. This was done by using the bending angles from the classical Newton's approach as prior information. Time $t = 24.66\,sec$ was randomly chosen and its solutions from both algebraic and classical Newton's methods for the L1 signal compared. Figures (13.5) and (13.7) show the plot of the four solutions for x and

y computed from (13.13) respectively. These solutions are converted into angular values $\{\delta_G, \delta_L\}$ using (13.14) and (13.15) respectively and plotted in Figs. 13.6 and 13.8. From the values of Figs. 13.6 and 13.8, the smallest values (encircled) were found to be close to those of the classical Newton's solution. The algebraic algorithm was then set to select the smallest value amongst the four solutions. Though Newton's approach converged after three iterations, a fixed value of 20 was set for this analysis. The threshold was set such that the difference between two consecutive solutions were smaller than 1×10^{-6}.

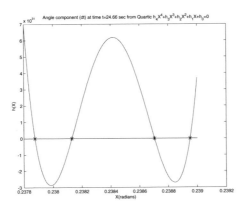

Fig. 13.5. x values for computing the bending angle component δ_G for L1 at $t = 24.66sec$

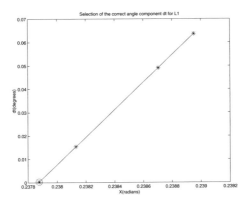

Fig. 13.6. Selection of the admissible x value for computing the component δ_G for L1 at $t = 24.66sec$

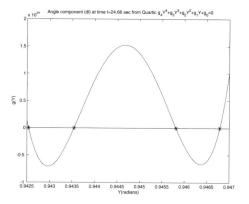

Fig. 13.7. y values for computing the bending angle component δ_L for L1 at $t = 24.66sec$

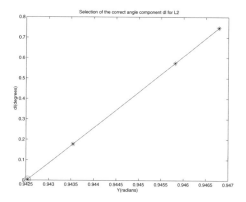

Fig. 13.8. Selection of the admissible y value for computing the component δ_L for L1 at $t = 24.66sec$

For the entire occultation, the bending angles $\{\alpha = \delta_G + \delta_L\}$ for both L1 and L2 signals were computed using algebraic algorithm and are plotted in Fig. 13.9. A magnifications of Fig. 13.9 above the height 30 km is plotted in Figs. 13.10 to show the effect of the residual iono-spheric errors on bending angles.

Since bending angle's data above 40 km are augmented with model values and those below 5 km are highly influenced by the presence of water vapour (see e.g., Figs. 13.9 and 13.10), we will restricted our analysis to the region between 5-40 km. Data in this region are normally

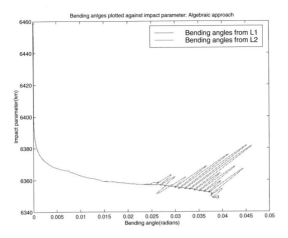

Fig. 13.9. Bending angles for L1 and L2 from algebraic algorithm

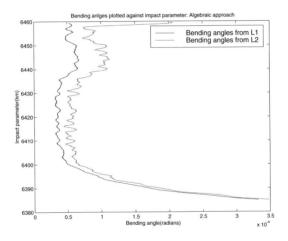

Fig. 13.10. Magnification of the bending angles in Fig. 13.9 above 6380 km

used directly to derive the atmospheric profiles required for Numerical Weather Prediction (NWP) models . In-order to assess the effect of non-linearity assumptions, we subtract the results of the classical Newton's approach from those of algebraic approach. This is performed for both the bending angles α and the impact parameter p. The computations were carried out separately for both L1 and L2 signals. In-order to compare the results, the computed differences are plotted in Figs. 13.11, 13.12, 13.13 and 13.14. In these Figures, the vertical axes are fixed while the horizontal axes indicate the range of the computed differences. In Figures 13.11 and 13.12, the computed differences in bending angles due to nonlinearity assumption for L1 are in the range $\pm 6 \times 10^{-5}(degrees)$ with the maximum absolute value of $5.14 \times 10^{-5}(degree)$. For L2, they are in the range $\pm 5 \times 10^{-5}(degrees)$, with the maximum absolute value of $4.85 \times 10^{-5}(degree)$. The effects of nonlinearity error on the impact parameters for L1 are in the range $\pm 1.5\,m$ with the maximum absolute value of $1.444\,m$, while those of L2 are in the range $\pm 2\,m$ with the maximum absolute value of $1.534\,m$. The large differences in the impact parameters are due to the large distances of the GPS satellites $(r_G > 20,000\,km)$. They are used in the second equation of (13.16) to compute the impact parameters to which the bending angles are related. Any small difference in the computed bending angles due to nonlinearity therefore contributes significantly to the large differences in the impact parameters. For this particular occultation therefore, the bending angles of L1 and L2 signals could probably be related to impact parameters that are off by up to $\pm 2\,m$.

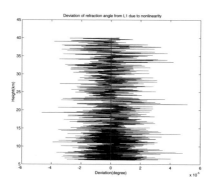

Fig. 13.11. Differences in bending angles from L1 due to nonlinearity for occultation 133 of 3rd May 2002.

In-order to assess the overall effect of nonlinearity on the bending angles, both bending angles from algebraic and Newton's procedures have to be related to the same impact parameters. In this analysis, the bending angles of L2 from algebraic approach and those of L1 and L2 from Newton's approach are all matched through interpolation to the impact parameters P1 of L1 from algebraic approach. The resulting total bending angles from both algebraic and iterative procedures are then obtained by the linear correction method of [325] as

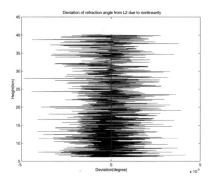

Fig. 13.12. Differences in bending angles from L2 due to nonlinearity for occultation 133 of 3rd May 2002.

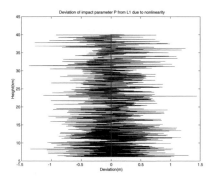

Fig. 13.13. Differences in impact parameters from L1 due to nonlinearity for occultation 133 of 3rd May 2002.

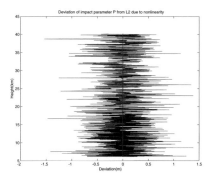

Fig. 13.14. Differences in impact parameters from L2 due to nonlinearity for occultation 133 of 3rd May 2002.

$$\alpha(a) = \frac{f_1^2 \alpha_1(a) - f_2^2 \alpha_2(a)}{f_1^2 - f_2^2}, \tag{13.19}$$

where f_1, f_2 are the frequencies of L1 and L2 signals respectively and, $\alpha_1(a)$ and $\alpha_2(a)$ the bending angles from L1 and L2 signals respectively. The resulting bending angles $\alpha(a)_i$ from the Newton's approach and $\alpha(a)_a$ from algebraic approach using (13.19) are plotted in Fig. 13.15. The deviation $\bigtriangledown \alpha = \alpha(a)_a - \alpha(a)_i$ obtained are plotted in Fig. 13.16 which indicates the nonlinearity error to increase with decreasing atmospheric height. From 40km to 15km, the deviation is within $\pm 2 \times 10^{-4} (degrees)$ but increases to $\pm 7 \times 10^{-4} (degrees)$ for the region below 15km with the maximum absolute deviation of $0.00069°$ for this particular occultation. This maximum absolute error is less than 1%. Vorob'ev and Krasil'nikova [325] pointed out that the error due to nonlinearity increases downwards to a maximum of about 2% when the beam perigee is close to the Earth's ground. The large difference in computed bending angles with decrease in height is expected as the region below 5km is affected by the presence of water vapour, and as seen from Fig. 13.9, the bending angles due to L2 are highly nonlinear.

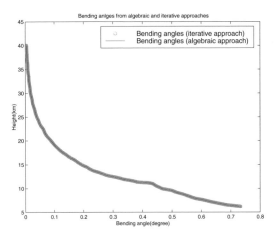

Fig. 13.15. Bending angles from iterative and algebraic algorithms matched to the same impact parameters for occultation 133 of 3rd May 2002.

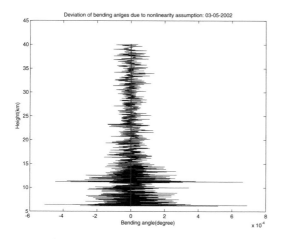

Fig. 13.16. Differences of computed bending angles due to nonlinearity for occultation 133 of 3rd May 2002.

The algebraic approach was next used to compute the bending angles of occultation number 3 of 14th May 2001 which occurred past mid-night at 00:39:58.00. For this period, as stated earlier, the solar radiation is minimum and the effect of ionospheric noise is also minimum. The results from this occultation show the differences in bending angles from the algebraic and Newton's methods to be smaller (see Fig. 13.17) compared to those of solar maximum period. The maximum absolute difference value for bending angles was $0.00001°$. For the computed impact parameters, the differences were in the range $±5\,cm$ for L1 signal (Fig. 13.18) and $±6\,cm$ for L2 (Fig. 13.19). The maximum absolute values were $4\,cm$ and $5\,cm$ respectively. In comparison to the results of occultation 133 of 3rd May 2002, the results of occultation 3 of 14th May 2001 indicate the effect of ionospheric noise during low solar radiation period to be less. The ionospheric noise could therefore increase the errors due to nonlinearity. In [33], further analysis of nonlinear bending angles have shown that there could exist other factors that influence the nonlinearity error other than the ionospheric noise.

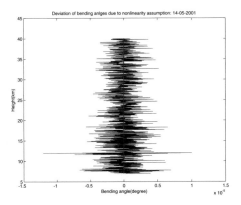

Fig. 13.17. Differences in bending angles due to nonlinearity for occultation number 3 of 14th May 2001.

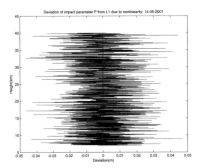

Fig. 13.18. Differences in impact parameters from L1 due to nonlinearity for occultation number 3 of 14th May 2001

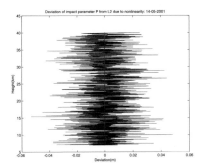

Fig. 13.19. Differences in impact parameters from L2 due to nonlinearity for occultation number 3 of 14th May 2001

13.5 Concluding Remarks

The new concept of GPS meteorology and its application to environmental monitoring is still new and an active area of research. The data that has been collected so far have unearthed several atmospheric properties that were hitherto difficult to fathom. The new technique clearly promises to contribute significantly and enormously to environmental and atmospheric studies. When the life span of the various missions (e.g., CHAMP, GRACE) will have reached, thousands of data will have been collected which will help unravel some of the hidden and complicated atmospheric and environmental phenomenon. Satellite missions such as EQUARS will contribute valuable equatorial data that have long been elusive due to poor radiosonde coverage. From the analysis of water vapour trapped in the atmosphere and the tropopause temperature, global warming studies will be enhanced a great deal.

We have also successfully presented an independent algebraic algorithm for solving the system of nonlinear bending angles for space borne GPS meteorology and shown that nonlinearity correction should be taken into account if the accuracy of the desired profiles are to be achieved to 1%. In particular, it has been highlighted how the nonlinearity errors in bending angles contribute to errors in the impact parameters to which the bending angles are related. Occultation number 133 of 3rd May 2002 which occurred past noon and occultation number 3 of 14th May 2001 which occurred past mid-night indicated the significance of ionospheric noise on nonlinearity error. When ionospheric noise is minimum, e.g., during mid-night, the computed differences in bending angles between the two procedures are almost negligible. During maximum solar radiation in the afternoons with increased ionospheric noise, the computed differences in bending angles between algebraic and classical Newton's methods increases.

The proposed algebraic method could therefore be used to control the results of the classical Newton's method especially when the ionospheric noise is suspected to be great, e.g., for occultations that occur during maximum solar radiation periods. The hurdle that must be overcome however is to concretely identify the criteria for selecting the admissible solution amongst the four algebraic solutions. In this analysis, the smallest values amongst the four algebraic solutions turned out to be the admissible in comparison with values of the classical Newton's approach. Whether this applies in general is still subject to investigation. In terms of computing time, the algebraic approach would probably have an advantage over the classical Newton's iterative procedure in cases where thousands of occultations are to be processed. For single occultations however, the classical Newton' s approach generally converges after few iterations and as such, the advantage of the algebraic approach in light of modern computers may not be so significant. For further literature on GPS meteorology, we refer to [10, 276].

Algebraic Diagnosis of Outliers

14.1 Outliers in Observation Samples

In Chap. 6, we introduced parameter estimation from observational data sample and defined the models applicable to linear and nonlinear cases. In-order for the estimates to be meaningful however;

(a) proper field observations must be carried out to minimize chances of gross errors,
(b) the observed data sample must be processed to minimize or eliminate the effects of systematic errors,
(c) appropriate estimation procedures that account for random errors have to be applied.

Despite the care taken during observation period and the improved models used to correct systematic errors, observations may still be contaminated with outliers or gross errors. Outliers are those observations that are inconsistent with the rest of the observation sample. They often degrade the quality of the estimated parameters and render them unreliable for any meaningful inferences (deductions). Outliers find their way into observational data sample through:

- Miscopying during data entry, i.e., a wrong value can be entered during data input into the computer or other processing machines.
- Misreading during observation period, e.g., number 6 can erroneously be read as 9.
- Instrumental errors (e.g., problems with centering, vertical and horizontal circles, unstable tripod stands etc.)
- Rounding and truncation errors (e.g., during data processing)

- Poor models applied to correct systematic errors and estimate parameters (e.g., a linear model may be assumed where a nonlinear model could be suitable). This error is also common during data smoothing where a linear fit is used where actually a cubic fit could have been most suitable etc.
- Key punch errors during data input etc.

A special problem faced by users while dealing with outliers is the basis on which to discard observations from a set of data on the grounds that they are contaminated with outliers.

The least squares method used to estimate parameters assume the observational errors to be independent and normally distributed. In the presence of gross errors in the observational data sample, these assumptions are violated and hence render the estimators, such as least squares, ineffective. Earlier attempts to circumvent the problem of outlier involved procedures that would first detect and isolate the outliers before adjusting the remaining data sample. Such procedures were both statistical as seen in the works of [39, 40, 41, 252, 342, 343, 344], and non statistical e.g., [183]. Other outlier detection procedures have been presented by [4, 7].

The detection and isolation approach to the outlier problem comes with its own shortcoming. On one hand, there exists the danger of *false deletion* and *false retention* of the assumed outliers. On the other hand, there exists the problem that the detection techniques are based on the residuals computed initially using the least squares method which has the tendency of masking the outliers by pulling their residuals closer to the regression fit. This makes the detection of outliers difficult. These setbacks had been recognized by the father of robust statistics P. J. Huber [190], [191, p. 5] and also [172, pp. 30–31] who suggested that the best option to deal with the outlier problem was to use *robust estimation procedures*. Such procedures would proceed safely despite the presence of outliers, isolate them and give admissible estimates that could have been achieved in the absence of outliers (i.e., if underlying distribution was normal). Following the fundamental paper by P. J. Huber in 1964 [189] and [191], several robust estimation procedures have been put forward that revolve around the robust M-estimators, L-estimators and R-estimators. In geodesy and geoinformatics, use of robust estimation techniques to estimate parameters has been presented e.g., in [5, 16, 17, 88, 164, 200, 202, 203, 254, 284, 344, 347, 348, 350] among others.

In this chapter, we present a non-statistical algebraic approach to outlier diagnosis that uses the Gauss-Jacobi combinatorial algorithm presented in Chap. 6. The combinatorial solutions are analyzed and those containing falsified observations identified. In-order to test the capability of the algorithm to diagnose outliers, we inject outliers of different magnitudes and signs on planar ranging and GPS pseudo-ranging problems. The algebraic approach is then employed to diagnose the outlying observations.

For GPS pseudo-range observations, the case of multipath effect is considered. Multipath is the error that occurs when the GPS signal is reflected (mostly by reflecting surfaces in built up areas) towards GPS receivers, rather than travelling directly to the receiver. This error still remains a menace which hinders full exploitation of the GPS system. Whereas other GPS observational errors such as ionospheric and at-mospheric refractions can be modelled, the error due to multipath still poses some difficulties in being contained thus necessitating a search for procedures that can deal with it. In proposing procedures that can deal with the error due to multipath, [338] have suggested the use of robust estimation approach that is based on iteratively weighted least squares (e.g., a generalization of the Danish method to heterogeneous and correlated observations). Awange [15] proposed the use of algebraic deterministic approach to diagnose outliers of type multipath.

14.2 Algebraic Diagnosis of Outliers

Let us illustrate by means of a simple linear example how the algebraic algorithm diagnoses outliers.

Example 14.1 (Outlier diagnosis using Gauss-Jacobi combinatorial algorithm). Consider a case where three linear equations have been given for the purpose of solving the two unknowns(x, y) in Fig. 14.1. Three possible combinations, each containing two equations necessary for solving the two unknowns, can be formed as shown in the box labelled "combination". Each of these systems of two linear equations is either solved by substitution, graphically or matrix inversion to give three pairs of solutions $\{x_{1,2}, y_{1,2}\}, \{x_{2,3}, y_{2,3}\}$ and $\{x_{1,3}, y_{1,3}\}$. The final step involves adjusting these pseudo-observations $\{x_{1,2}, y_{1,2}, x_{2,3}, y_{2,3}, x_{1,3}, y_{1,3}\}$ as indicated in the box labelled "adjustment of the combi-natorial subsets solutions". The weight matrix Σ or the weight el-ements $\{\pi_{1,2}, \pi_{2,3}, \pi_{1,3}\}$ are obtained via nonlinear error/variance-covariance propagation. Assuming now that observation y_1 is contaminated by gross error ∂y, then the first two combinatorial sets of the system of equations containing observation y_1 will have their results $\{x_{1,2}, y_{1,2}\}, \{x_{1,3}, y_{1,3}\}$ changed to $\{x_{1,2}^*, y_{1,2}^*\}, \{x_{1,3}^*, y_{1,3}^*\}$ respectively because of the change of observation y_1 to $(y_1 + \partial y)$. The third combination set $\{x_{2,3}, y_{2,3}\}$ without observation $(y_1 + \partial y)$ remains unchanged. If one computes the *combinatorial positional norms*

$$\begin{bmatrix} p_1 = \sqrt{(x_{1,2}^{*2} + y_{1,2}^{*2})} \\ p_2 = \sqrt{(x_{1,3}^{*2} + y_{1,3}^{*2})} \\ p_3 = \sqrt{(x_{2,3}^{2} + y_{2,3}^{2})}, \end{bmatrix}$$

and subtract them from the norms of the adjusted positional values, median or a priori values (say from maps), one can analyze the deviations to obtain the falsified observation y_1 which is common in the first two sets. It will be noticed that the deviation of the first two sets containing the contaminated value is larger than the uncontaminated set. The *median* is here used as opposed to the *mean* as it is less sensitive to extreme scores.

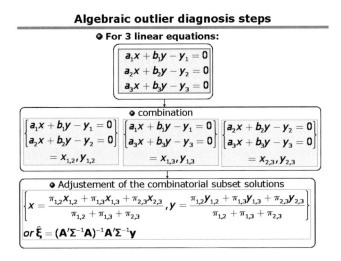

Fig. 14.1. Algebraic outlier diagnosis steps

The program operates in the following steps:

Step 1: Given an overdetermined system with n observations in m unknowns, k combinations are formed using (6.26) on p. 69.

Step 2: Each of the minimal combination is solved in closed form using either Groebner basis or polynomial resultant algebraic technique of Chaps. 4 or 5. From the combinatorial solutions, compute the positional norm.

Step 3: Perform the nonlinear error/variance-covariance propagation to obtain the weight matrix of the pseudo-observations resulting from step 2.

Step 4: Using these pseudo-observations and the weight matrix from step 3, perform an adjustment using linear Gauss-Markov model (6.10) on p. 62.

Step 5: Compute the adjusted barycentric coordinate values together with its positional norm and the median positional norm from step 2. Subtract these positional norms from those of the combinatorial solutions to diagnose outliers from the deviations.

14.2.1 Outlier Diagnosis in Planar Ranging

In Sect. 9.3 of Chap. 9, we discussed the planar ranging problem and presented the solution to the overdetermined case. We demonstrated by means of Example 9.4 on p. 130 how the position of unknown station could be obtained from distance measurements to more than two stations. In this section, we use the same example to demonstrate how the algebraic combinatorial algorithm can be used to diagnose outliers. From observational data of the overdetermined planar ranging problem of [195] given in Table 9.9 on p. 131, the position of the unknown station is determined. The algorithm is then applied to diagnose outlying observations. Let us consider three cases as follows; first, the algorithm is subjected to outlier free observations and used to compute the positional norms. Next, an outlier of 0.95 m is injected to the distance observation to station 2 and the algorithm applied to diagnose that particular observation. Finally, the distance observed to station 4 is considered to have been miss-booked with 6 typed as 9, thus leading to an error of 3 m.

Example 14.2 (Outlier free observations). From the values of Table 9.9 and using (6.26) on p. 69, 6 combinations, each consisting of two observation equations are formed. The aim is to obtain the unknown position from the nonlinear ranging observations equations. From the computed positions in step 2, the positional norms are given by

$$P_i = \sqrt{X_i^2 + Y_i^2}\,|_{i=1,\dots,6},\qquad(14.1)$$

where $(X_i, Y_i)\,|_{i=1,\dots,6}$ are the two-dimensional geocentric coordinates of the unknown station computed from each combinatorial pair. Table 14.1 indicates the combinations, their computed positional norms, and deviations from the norm of the adjusted value (48941.769 m) from step 4. The results are for the case of outlier free observations. These deviations are plotted against combinatorial numbers in Fig. 14.2.

Table 14.1. Combinatorial positional norms and their deviations from that of the adjusted value (outlier free observations)

Combination	Positional norm (m)	Deviation from adjusted value (m)
1-2	48941.776	0.007
1-3	48941.760	-0.009
1-4	48941.767	-0.002
2-3	48941.769	0.000
2-4	48941.831	0.062
3-4	48941.764	-0.005

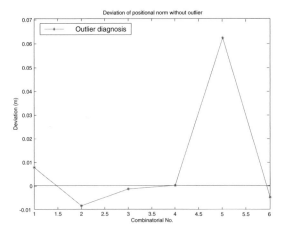

Fig. 14.2. Deviations of the combinatorial positional norms from that of the adjusted value (outlier free observation)

Example 14.3 (Outlier of 0.950 m in observation to station 2). Let us now consider a case where the distance observation to station 2 in Table 9.9 has an outlier of 0.950 m. For this case, the distance is falsified such that the observed value is recorded as 1530.432 m instead of the correct value appearing in Table 9.9. The computed deviations of the combinatorial positional norms from the adjusted value (48941.456 m) and the median value of (48941.549 m) are presented in Table 14.2 and plotted in Fig. 14.3.

Given that there exists an outlier in the distance observation to station 2, which appears in combinations (1-2), (2-3) and (2-5), i.e., the *first, fourth,* and *fifth* combinations respectively, one expects the deviations in positional norms in these combinations to be larger than those combinations without station 2. Values of Table 14.2 columns *three* and *four* indicate that whereas combinations (1-3), (1-4) and (3-4), i.e., the *second, third,* and *sixth* combinations respectively have deviations in the same range (c.a. 0.3 m and 0.2 m in columns three and four respectively), the other combinations with outliers are clearly seen to have varying deviations. Figure 14.3 clearly indicates the *first, fourth,* and *fifth* combinations respectively with outlying observation to have larger deviations as compared to the rest. This is attributed to observation to station 2 containing gross error.

Table 14.2. Combinatorial positional norms and their deviations from that of the adjusted value and median (error of 0.950 m in observation to station 2)

Combination	Positional norm-(m)	Deviation from adjusted value (m)	Deviation from median value (m)
1-2	48940.964	-0.492	-0.585
1-3	48941.760	0.304	0.211
1-4	48941.767	0.311	0.218
2-3	48941.338	-0.118	-0.211
2-4	48936.350	-5.106	-5.199
3-4	48941.764	0.308	0.215

Example 14.4 (Outlier of 3 m in observation to station 4). Next, we consider a case where observation to station 4 in Table 9.9 has an outlier of 3 m, which erroneously resulted from miss-booking of the number 6 as 9. This falsified the distance such that the recorded value was 1209.524 m. The computed deviations of the combinatorial positional norms from the norm of the adjusted value (48942.620 m) and

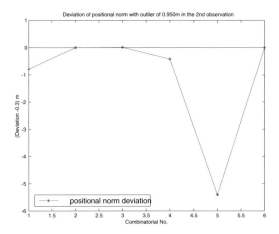

Fig. 14.3. Deviations of the combinatorial positional norms from that of the adjusted value (error of 0.950 m in observation to station 2)

the median value (48941.772 m) are given as in Table 14.3 and plotted in Fig. 14.4. Given that there exists an outlier in the distance observation to station 4, which appears in combinations (1-4), (2-4) and (3-4) i.e., the *third*, *fifth*, and *sixth* combinations, Table 14.3 columns three and four together with Fig. 14.4 clearly indicates the deviations from these combinations to be larger than those of the combinations without observation 4, thus attributing it to observation to station 4 containing gross error.

Table 14.3. Combinatorial positional norms and deviations from the norms of adjusted value and median (error of 3m in observation to station 4)

Combination	Positional norm-(m)	Deviation from adjusted value (m)	Deviation from median value (m)
1-2	48941.776	-0.844	0.004
1-3	48941.760	-0.860	-0.012
1-4	48944.388	1.768	2.616
2-3	48941.769	-0.851	-0.003
2-4	48927.912	-14.708	-13.860
3-4	48943.061	0.441	1.289

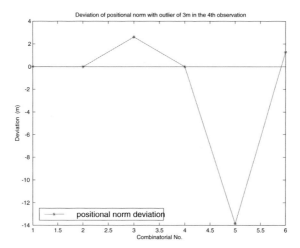

Fig. 14.4. Deviations of the combinatorial positional norms from that of the median value (error of 3m in observation to station 4)

14.2.2 Diagnosis of Multipath Error in GPS Positioning

For GPS pseudo-ranging, consider that a satellite signal meant to travel straight to the receiver was reflected by a surface as shown in Fig. 14.5. The measured pseudo-range reaching the receiver ends up being longer than the actual pseudo-range, had the signal travelled directly. In-order to demonstrate how the algorithm can be used to detect outlier

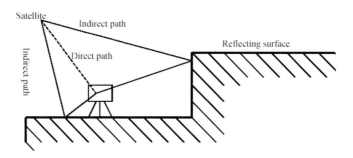

Fig. 14.5. Multipath effect

of type multipath, let us make us of Example 9.2 on p. 116. Using the six satellites, 15 combinations are formed whose positional norms are computed using

$$P_i = \sqrt{X_i^2 + Y_i^2 + Z_i^2}\,|_{i=1,\ldots,15},\qquad(14.2)$$

where $(X_i, Y_i, Z_i)\,|_{i=1,\ldots,15}$ are the three-dimensional geocentric coordinates of the unknown station computed from each combinatorial set. The computed positional norm are then used to diagnose outliers. Three cases are presented as follows: In case A, outlier free observations are considered while for cases B and C, outliers of 500 m and 200 m are injected in pseudo-range measurements from satellites 23 and 9 respectively.

Example 14.5 (Case A: Multipath free pseudo-ranges). From the values of Table 9.2 and using (6.26) on p. 69, 15 combinations, each consisting of four satellites, are formed with the aim of solving for the unknown position. For each combination, the position of the receiver is computed as discussed in Example 9.2 on p. 116. Table 14.4 indicates the combinations, the computed combinatorial positional norms from (14.2) and the deviations from the norms of the adjusted value of 6369. 582 m and the median from step 4, for outlier free case. The combinatorial algorithm diagnoses the poor geometry of the 10th combination. Figure 14.6 indicates the plotted deviations versus combinatorials.

Table 14.4. Positional norms and deviations from the norms of adjusted value and median (multipath free)

Combination Number	Combination	Positional norm (km)	Deviation from the norm of the adjusted value (m)	the norm of Deviation from the median (m)
1	23-9-5-1	6369.544	-39.227	-17.458
2	23-9-5-21	6369.433	-149.605	-127.837
3	23-9-5-17	6369.540	-43.255	-21.487
4	23-9-1-21	6369.768	185.342	207.110
5	23-9-1-17	6369.538	-44.603	-22.835
6	23-9-21-17	6369.561	-21.768	0.000
7	23-5-1-21	6369.630	47.449	69.217
8	23-5-1-17	6369.542	-41.229	-19.461
9	23-5-21-17	6369.507	-76.004	-54.235
10	23-1-21-17	6373.678	4094.748	4116.516
11	9-5-1-21	6369.724	140.976	162.744
12	9-5-1-17	6369.522	-60.746	-38.978
13	9-5-21-17	6369.648	64.830	86.598
14	9-1-21-17	6369.712	128.522	150.2908
15	5-1-21-17	6369.749	166.096	187.865

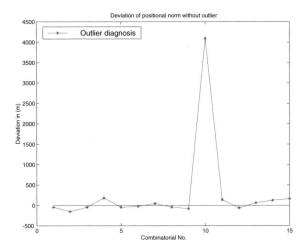

Fig. 14.6. Deviations of the combinatorial positional norms from that of the adjusted value (outlier free observation).

Example 14.6 (Case B: Multipath error of 500 m in pseudo-range measurements from satellite 23). Let us assume that satellite number 23 had its pseudo-range longer by 500 m owing to multipath effect. Once the positions have been computed for the various combinations in Table 9.2, the positional norms for the 15 combinatorials are then computed via (14.2). The computed deviations of the positional norms from the norm of the adjusted value 6368.785 m, norm of the median value of 6368.638 m and a priori norm from case A are presented in Table 14.5. The deviations from a priori norm in case A are plotted in Fig. 14.7.

Given that there exists an outlier in the pseudo-range measurements from satellite 23, which appears in combinations 1 to 10, one expects the deviation in positional norms in these combinations to contain higher fluctuations than the combinations without satellite 23. Values of Table 9.2 columns four, five and six indicate that whereas combinations 11, 12, 13, 14 and 15 without satellite number 23 have values with less fluctuation of positional norms, the variation of the first 10 combinations containing satellite 23 were having larger fluctuations. The case is better illustrated by Fig. 14.7, where prior information is available on the desired position (e.g., from the norm of outlier free observations in Example 14.5). In such case, it becomes clearer which combinations are contaminated. From the figure, the first 10 combinations have larger deviations as opposed to the last 5, thus diagnosing satellite 23 as the

outlying satellite. In practice, such prior information can be obtained
from existing maps.

Table 14.5. Positional norms and deviations from the norms of adjusted value,
median norm and the norm of case A (Multipath error of 500 m in satellite 23)

Comb. No.	Combination	Positional norm (km)	Deviation from norm of adjusted value (m)	norm of Deviation from the median (m)	Deviation from a priori norm (m)
1	23-9-5-1	6368.126	-658.763	-512.013	-1456.925
2	23-9-5-21	6367.147	-1637.513	-1490.763	-2435.675
3	23-9-5-17	6368.387	-397.366	-250.616	-1195.528
4	23-9-1-21	6370.117	1332.597	1479.347	534.435
5	23-9-1-17	6368.475	-309.906	-163.155	-1108.068
6	23-9-21-17	6368.638	-146.750	0.000	-944.912
7	23-5-1-21	6368.895	110.069	256.820	-688.093
8	23-5-1-17	6368.256	-528.806	-382.055	-1326.967
9	23-5-21-17	6368.006	-779.197	-632.447	-1577.359
10	23-1-21-17	6368.068	-716.569	-569.818	-1514.730
11	9-5-1-21	6369.724	939.136	1085.888	140.976
12	9-5-1-17	6369.522	737.416	884.166	-60.746
13	9-5-21-17	6369.648	862.991	1009.742	64.829
14	9-1-21-17	6369.712	926.684	1073.434	128.522
15	5-1-21-17	6369.749	964.258	1111.008	166.096

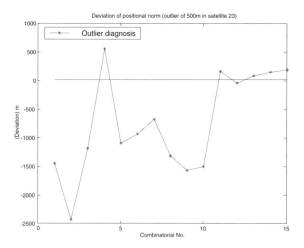

Fig. 14.7. Deviations of combinatorial positional norms from the norm of the a
priori value in case A (error of 500m in pseudo-range of satellite 23)

Example 14.7 (Case C: Multipath error of 200 m in satellite 9). Let us now suppose that satellite number 9 appearing in the last 5 combinations in case B (i.e., combinations 11, 12, 13, 14 and 15) has outlier of 200 m. The positional norms for the 15 combinatorials are then computed from (14.2). The computed deviations of the positional norms from the norm of the adjusted value 6369.781 m, norm of the median value 6369.804 m and a priori norm from case A are presented in Table 14.6 and plotted in Fig. 14.8. Satellite 9 appears in combinations 1, 2, 3, 4, 5, 6, 11, 12, 13, 14 and 15, with larger deviations in positional norms as depicted in Table 14.6 columns four, five and six and plotted in Fig. 14.8. Whereas combinations 7, 8, 9 and 10 without satellite number 9 have values with less deviations of positional norms, the deviations of the first 6 combinations and those of combinations 11-15 containing satellite 9 were larger. With prior information as shown in Fig. 14.8, satellite 9 can then be isolated to be the satellite with outlier. The value of combinatorial 10 is due to poor geometry as opposed to outlier since this particular combination does not contain satellite 9. This can be confirmed by inspecting the coefficients of the quadratic equations used to solve the unknown pseudo-range equations as already discussed in Example 9.2 on p. 116.

Table 14.6. Positional norms and deviations from the norms of adjusted value, median norm and the norm of case A (Multipath error of 200m in satellite 9)

Comb. No.	Combination	Positional norm (km)	Deviation from norm of adjusted value (m)	norm of Deviation from the median (m)	Deviation from a priori norm (m)
1	23-9-5-1	6369.386	-394.656	-417.522	-196.748
2	23-9-5-21	6369.075	-705.644	-728.511	-507.737
3	23-9-5-17	6369.699	-81.796	-104.6639	116.111
4	23-9-1-21	6370.019	238.062	215.195	435.969
5	23-9-1-17	6369.804	22.866	0.000	220.774
6	23-9-21-17	6369.825	44.237	21.371	242.145
7	23-5-1-21	6369.630	-150.459	-173.325	47.449
8	23-5-1-17	6369.542	-239.137	-262.003	-41.229
9	23-5-21-17	6369.507	-273.911	-296.778	-76.004
10	23-1-21-17	6373.678	3896.840	3873.974	4094.748
11	9-5-1-21	6369.894	113.076	90.209	310.983
12	9-5-1-17	6371.057	1276.444	1253.578	1474.352
13	9-5-21-17	6370.333	552.230	529.364	750.138
14	9-1-21-17	6369.966	184.890	162.024	382.798
15	5-1-21-17	6369.749	-31.811	-54.678	166.096

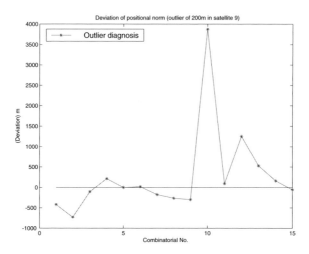

Fig. 14.8. Deviations of combinatorial positional norms from the norm of the a priori value (error of 200m in pseudo-range of satellite 9)

The diagnosed outliers in planar ranging observations as well as the pseudo-ranges of satellites 23 and 9 could therefore either;

- be eliminated from the various combinations and the remaining observations used to estimate the desired parameters or,
- the effect of the outlier could be managed using robust estimation techniques such as those discussed in [16, 17, 338], but with the knowledge of the contaminated observations.

14.3 Concluding Remarks

The success of the algebraic Gauss-Jacobi combinatorial algorithm to diagnose outliers in the cases considered is attributed to its computing engine. The capability of the powerful algebraic tools of Groebner basis and polynomial resultants to solve in a close form the nonlinear systems of equations is the key to the success of the algorithm. With prior information from e.g., existing maps, the method can further be enhanced. For the 7-parameter datum transformation problem discussed in the next chapter, Procrustes algorithm II could be used as the computing engine instead of Groebner basis or polynomial resultants. The algebraic approach presented could be developed to further enhance the statistical approaches for detecting outliers.

Transformation Problem: Procrustes Algorithm II

15.1 7-Parameter Datum Transformation and its Importance

The 7-parameter datum transformation $\mathbb{C}_7(3)$ problem comprises the determination of seven parameters required to transform coordinates from one system to another. Transformation of coordinates is a computational procedure that maps one set of coordinates in a given system onto another. This is achieved by translating the given system so as to cater for its origin with respect to the final system, and rotating the system about its own axes so as to orient it to the final system. In addition to translation and rotation, scaling is performed in-order to match the corresponding baseline lengths in the two systems. The *three translation parameters, three rotation parameters* and the *scale element* comprise the *7 parameters of the datum transformation $\mathbb{C}_7(3)$ problem*. They are required to transform a set of three-dimensional coordinates from one system A onto another system B. In the 7-parameter datum transformation problem $\mathbb{C}_7(3)$, one understands $\mathbb{C}_7(3)$ to be the notion of the *seven parameter conformal group in* $\mathbb{R}^3$, leaving "*space angles*" and "*distance ratios*" *equivariant* (invariant). A mathematical introduction to conformal field theory is given by [114, 290], while a systematic approach of geodetic datum transformation including geometrical and physical terms is has been presented by [157]. For a given network, it suffices to compute the transformation parameters using three or more coordinates in both systems. These parameters are then later used for subsequent conversions.

In geodesy and geoinformatics, the 7-parameter datum transformation problem has gained significance following the advent of Global

Navigation Satellites System (GNSS), and particularly GPS. Since satellite positioning operates on a global reference frame (see e.g., Chap. 7), there often exists the need to transform coordinates from local systems onto GPS's World Geodetic System 84 (WGS-84). Specifically coordinates can be transformed;

- from map systems to digitizing tables (e.g., in Geographical Information System GIS),
- from photo systems (e.g., photo coordinates) to ground systems (e.g., WGS-84),
- from local (national) systems to global reference systems (e.g., WGS-84) as in (12.29) on p. 215,
- from regional (e.g., European Reference Frame EUREF system) to global reference systems (e.g., WGS-84),
- from local (national) systems to regional reference systems, and
- from one local system onto another local system. In some countries, there exist different systems depending on political boundaries.

This problem, also known as 7-parameter similarity transformation, has its 7 unknown transformation parameters related to the known coordinates in the two systems by nonlinear equations. These equations are often solved using numerical methods which as already pointed out in the preceding chapters, rely on linearization, approximate starting values and iterations. In this chapter, we solve the problem algebraically using;

(a) Procrustes II
(b) Groebner basis, and
(c) Gauss-Jacobi combinatorial algorithms.

Before we present the usage of these algebraic algorithms, let us see how the 7-parameter datum transformation problem is formulated.

Formulation of the Problem

Consider a case where coordinates have been given in two systems A and B. For clarity purposes, let us assume the two coordinate systems to be e.g., photo image coordinates in system A and ground coordinates in system B (see e.g., Fig. 12.7 on p. 215). The ground coordinates $\{X_i, Y_i, Z_i | i, \ldots, n\}$ of the objects are obtained from say GPS measurements. Given the photo coordinates $\{a_i = x_i, b_i = y_i, c_i = -f | i, \ldots, n\}$

and their equivalent ground coordinates $\{X_i, Y_i, Z_i | i, \ldots, n\}$, the 7-parameter datum transformation problem concerns itself with the determination of;

(1) scale parameter $x_1 \in \mathbb{R}$,
(2) three translation parameters $\mathbf{x}_2 \in \mathbb{R}^3$, and
(3) rotation matrix $\mathbf{X}_3 \in \mathbb{R}^{3 \times 3}$ comprising three rotation elements.

Once the listed unknowns have been determined, coordinates can subsequently be transformed from one system onto another. The nonlinear equations relating these unknowns and coordinates from both systems are given by (cf., equation 12.29 on p. 215)

$$\begin{bmatrix} a_i \\ b_i \\ c_i \end{bmatrix} = x_1 \mathbf{X}_3 \begin{bmatrix} X_i \\ Y_i \\ Z_i \end{bmatrix} + \mathbf{x}_2 \mid i = 1, 2, 3, \ldots, n, \qquad (15.1)$$

subject to

$$\boxed{\mathbf{X}_3' \mathbf{X}_3 = \mathbf{I}_3}. \qquad (15.2)$$

In (15.1), $\{a_i, b_i, c_i\}$ and $\{X_i, Y_i, Z_i\}$ are coordinates of the same points, e.g., in both photo and ground coordinate systems respectively. The determination of the unknowns $x_1 \in \mathbb{R}$, $\mathbf{x}_2 \in \mathbb{R}^3$, $\mathbf{X}_3 \in \mathbb{R}^{3 \times 3}$ require a minimum of *three points* in both systems whose coordinates are known. Owing to the nonlinearity of (15.1), the solutions have always been obtained using least squares approach iteratively. With this approach, (15.1) is first linearized and some initial approximate starting values of the unknown parameters used. The procedure then iterates, each time improving on the solutions of the preceding iteration step. This is done until a convergence criteria is achieved.

Where the rotation angles are small e.g., in photogrammetry, the starting values of zeros are normally used. In other fields such as geodesy, the rotation angles are unfortunately not small enough to be initialized by zeros thereby making the solution process somewhat difficult and cumbersome. Bad choice of initial starting values often leads to many iterations for the convergence criteria to be achieved. In some cases, where the initial starting values are far from those of the unknown parameters, iteration processes may fail to converge. With these uncertainties of initial starting values, cumbersomeness of linearization and iterations, procedures that would offer exact solution to the 7-parameter datum transformation problem would be desirable. In answer to this challenge, we propose algebraic approaches whose advantages over the approximate numerical methods have already been mentioned.

Apart from the computational difficulties associated with numerical procedures, the 7-parameter datum transformation problem poses another challenge to the existing algorithms. This is; the incorporation of the variance-covariance (weight) matrices of the two systems involved. Communications between [201, 220, 277] on the subject, following the work of [219], provides an insight to this problem. In practice, users have been forced to rely on iterative procedures and linearized least squares solution which are incapable of incorporating the variance-covariance matrices of both systems in play. We will attempt to address this challenge algebraically in this chapter.

15.2 Algebraic (Analytic) Determination of Transformation Parameters

15.2.1 Groebner Basis Transformation

By making use of the *skew-symmetric* matrix $\mathbf{S}$, the rotation matrix $\mathbf{X}_3 \in \mathbb{R}^{3 \times 3}$ in (15.1) is expressed as

$$\mathbf{X}_3 = (\mathbf{I}_3 - \mathbf{S})^{-1}(\mathbf{I}_3 + \mathbf{S}), \tag{15.3}$$

where $\mathbf{I}_3$ is the identity matrix and the *skew-symmetric* matrix $\mathbf{S}$ given by

$$\mathbf{S} = \begin{bmatrix} 0 & -c & b \\ c & 0 & -a \\ -b & a & 0 \end{bmatrix}. \tag{15.4}$$

The rotation matrix $\mathbf{X}_3 \in \mathbb{R}^{3 \times 3}$ is parameterized using *Euler* or *Cardan* angles. With Cardan angles, we have:

Solution 15.1 (Parametrization of the rotation matrix by Cardan angles).

$$\mathbf{X}_3 = \mathbf{R}_1(\alpha)\mathbf{R}_2(\beta)\mathbf{R}_3(\gamma) \tag{15.5}$$

with

$$\mathbf{R}_1 = \begin{bmatrix} 1 & 0 & 0 \\ 0 & cos\alpha & sin\alpha \\ 0 & -sin\alpha & cos\alpha \end{bmatrix}, \mathbf{R}_2 = \begin{bmatrix} cos\beta & 0 & -sin\beta \\ 0 & 1 & 0 \\ sin\beta & 0 & cos\beta \end{bmatrix}, \mathbf{R}_3 = \begin{bmatrix} cos\gamma & sin\gamma & 0 \\ -sin\gamma & cos\gamma & 0 \\ 0 & 0 & 1 \end{bmatrix},$$

leading to

$$\mathbf{R}_1(\alpha)\mathbf{R}_2(\beta)\mathbf{R}_3(\gamma) =$$
$$\begin{bmatrix} cos\beta cos\gamma & cos\beta sin\gamma & -sin\beta \\ sin\alpha sin\beta cos\gamma - cos\alpha sin\gamma & sin\alpha sin\beta sin\gamma + cos\alpha cos\gamma & sin\alpha cos\beta \\ cos\alpha sin\beta cos\gamma + sin\alpha sin\gamma & cos\alpha sin\beta sin\gamma - sin\alpha cos\gamma & cos\alpha cos\beta \end{bmatrix}.$$
$$(15.6)$$

The Cardan angles are then obtained from the rotation matrix $\mathbf{X}_3 \in \mathbb{R}^{3\times3}$ through:

$$\begin{bmatrix} \alpha = tan\left\{\dfrac{r_{23}}{r_{33}}\right\} \\[3mm] \gamma = tan\left\{\dfrac{r_{12}}{r_{11}}\right\} \\[3mm] \beta = tan\left\{\dfrac{-r_{31}}{\sqrt{r_{11}^2 + r_{12}^2}}\right\} = tan\left\{\dfrac{-r_{31}}{\sqrt{r_{23}^2 + r_{33}^2}}\right\}. \end{bmatrix} \qquad (15.7)$$

For parametrization using Euler angles, we have:

Solution 15.2 (Parametrization of the rotation matrix by Euler angles $\Lambda_\Gamma, \Phi_\Gamma, \Sigma_\Gamma$).

$$\mathbf{R}_E(\Lambda_\Gamma, \Phi_\Gamma, \Sigma_\Gamma) := \mathbf{R}_3(\Sigma_\Gamma)\mathbf{R}_2(\frac{\pi}{2} - \Phi_\Gamma)\mathbf{R}_3(\Lambda_\Gamma) \qquad (15.8)$$

$$\mathbf{R}_1 := \begin{bmatrix} 1 & 0 & 0 \\ 0 & \cos 1 & \sin 1 \\ 0 & -\sin 1 & \cos 1 \end{bmatrix}, \quad \mathbf{R}_2 := \begin{bmatrix} \cos 2 & 0 & -\sin 2 \\ 0 & 1 & 0 \\ \sin 2 & 0 & \cos 2 \end{bmatrix}, \quad \mathbf{R}_3 := \begin{bmatrix} \cos 3 & \sin 3 & 0 \\ -\sin 3 & \cos 3 & 0 \\ 0 & 0 & 1 \end{bmatrix}$$
$$(15.9)$$

$$\mathbf{R}_3(\Lambda_\Gamma) = \begin{bmatrix} \cos \Lambda_\Gamma & \sin \Lambda_\Gamma & 0 \\ -\sin \Lambda_\Gamma & \cos \Lambda_\Gamma & 0 \\ 0 & 0 & 1 \end{bmatrix}, \quad \mathbf{R}_2(\frac{\pi}{2} - \Phi_\Gamma) = \begin{bmatrix} \sin \Phi_\Gamma & 0 & -\cos \Phi_\Gamma \\ 0 & 1 & 0 \\ \cos \Phi_\Gamma & 0 & \sin \Phi_\Gamma \end{bmatrix} \quad (15.10)$$

$$\mathbf{R}_2(\frac{\pi}{2} - \Phi_\Gamma)\mathbf{R}_3(\Lambda_\Gamma) = \begin{bmatrix} \sin \Phi_\Gamma \cos \Lambda_\Gamma & \sin \Phi_\Gamma \sin \Lambda_\Gamma & -\cos \Phi_\Gamma \\ -\sin \Lambda_\Gamma & \cos \Lambda_\Gamma & 0 \\ \cos \Phi_\Gamma \cos \Lambda_\Gamma & \cos \Phi_\Gamma \sin \Lambda_\Gamma & \sin \Phi_\Gamma \end{bmatrix}$$
$$(15.11)$$

$$\mathbf{R} := \mathbf{R}_3(\Sigma_\Gamma)\mathbf{R}_2(\tfrac{\pi}{2} - \Phi_\Gamma)\mathbf{R}_3(\Lambda_\Gamma) =$$

$$
\begin{bmatrix}
\begin{matrix} \cos \Sigma_\Gamma \sin \Phi_\Gamma \cos \Lambda_\Gamma \\ - \sin \Sigma_\Gamma \sin \Lambda_\Gamma \end{matrix} &
\begin{matrix} \cos \Sigma_\Gamma \sin \Phi_\Gamma \sin \Lambda_\Gamma \\ + \sin \Sigma_\Gamma \cos \Lambda_\Gamma \end{matrix} &
- \cos \Sigma_\Gamma \cos \Phi_\Gamma \\[2mm]
\begin{matrix} - \sin \Sigma_\Gamma \sin \Phi_\Gamma \cos \Lambda_\Gamma \\ - \cos \Sigma_\Gamma \sin \Lambda_\Gamma \end{matrix} &
\begin{matrix} - \sin \Sigma_\Gamma \sin \Phi_\Gamma \sin \Lambda_\Gamma \\ + \cos \Sigma_\Gamma \cos \Lambda_\Gamma \end{matrix} &
\sin \Sigma_\Gamma \cos \Phi_\Gamma \\[2mm]
\cos \Phi_\Gamma \cos \Lambda_\Gamma &
\cos \Phi_\Gamma \sin \Lambda_\Gamma &
\sin \Phi_\Gamma
\end{bmatrix}
\tag{15.12}
$$

$$0 \leq \Lambda_\Gamma < 2\pi, -\tfrac{\pi}{2} < \Phi_\Gamma < +\tfrac{\pi}{2}, 0 \leq \Sigma_\Gamma < 2\pi$$

The inverse map of

$$\mathbf{R} = [r_{kl}], k,l \in \{1, 2, 3\},$$

to

$$(\Lambda_\Gamma, \Phi_\Gamma, \Sigma_\Gamma)$$

is given by Lemma 15.1.

Lemma 15.1 (Inverse map $\mathbf{R} \mapsto \Lambda_\Gamma, \Phi_\Gamma, \Sigma_\Gamma$). *Let the direct Euler map of the rotation matrix be given by (15.12), namely*

$$\mathbf{R} := \mathbf{R}_3(\Sigma_\Gamma)\mathbf{R}_2\left(\frac{\pi}{2} - \Phi_\Gamma\right)\mathbf{R}_3(\Lambda_\Gamma),$$

$$(\Lambda_\Gamma, \Phi_\Gamma, \Sigma_\Gamma) \in \left\{\mathbf{R}^3 \,\middle|\, 0 \leq \Lambda_\Gamma < 2\pi, -\frac{\pi}{2} < \Phi_\Gamma < +\frac{\pi}{2}, 0 \leq \Sigma_\Gamma < 2\pi\right\}.$$

The inverse Euler map is parameterized by

$$
\begin{bmatrix}
\tan \Lambda_\Gamma = \dfrac{r_{32}}{r_{31}} \Rightarrow \Lambda_\Gamma = \arctan \dfrac{r_{32}}{r_{31}} \\[4mm]
\tan \Phi_\Gamma = \dfrac{r_{33}}{\sqrt{r_{31}^2 + r_{32}^2}} \Rightarrow \Phi_\Gamma = \arctan \dfrac{r_{33}}{\sqrt{r_{31}^2 + r_{32}^2}} \\[4mm]
\tan \Sigma_\Gamma = -\dfrac{r_{23}}{r_{13}} \Rightarrow \Sigma_\Gamma = \arctan -\dfrac{r_{23}}{r_{13}}.
\end{bmatrix}
\tag{15.13}
$$

The properties of the rotation matrix $\mathbf{X}_3 \in \mathbb{R}^{3\times3}$ expressed as in (15.3) have been examined by [352] and shown to fulfill (15.2). Only a minimum of three corresponding points in both systems are required for the transformation parameters to be obtained. For these points, (15.1) is now written for $i = 1, 2, 3$ using (15.3) as

$$\begin{bmatrix} 1 & c & -b \\ -c & 1 & a \\ b & -a & 1 \end{bmatrix} \begin{bmatrix} a_i \\ b_i \\ c_i \end{bmatrix} = x_1 \begin{bmatrix} 1 & -c & b \\ c & 1 & -a \\ -b & a & 1 \end{bmatrix} \begin{bmatrix} X_i \\ Y_i \\ Z_i \end{bmatrix} + \begin{bmatrix} X_0 \\ Y_0 \\ Z_0 \end{bmatrix}, \quad (15.14)$$

with $\{X_0, Y_0, Z_0\} \in \mathbf{x}_2$ being the translation parameters. For these three corresponding points in both systems, the observation equations for solving the 7 datum transformation parameters are expressed from (15.14) as:

$$\begin{cases} f_1 := x_1 X_1 - x_1 c Y_1 + x_1 b Z_1 + X_0 - a_1 - c b_1 + b c_1 = 0 \\ f_2 := x_1 c X_1 + x_1 Y_1 - x_1 a Z_1 + Y_0 + c a_1 - b_1 - a c_1 = 0 \\ f_3 := -x_1 b X_1 + x_1 a Y_1 + x_1 Z_1 + Z_0 - b a_1 + a b_1 - c_1 = 0 \\ f_4 := x_1 X_2 - x_1 c Y_2 + x_1 b Z_2 + X_0 - a_2 - c b_2 + b c_2 = 0 \\ f_5 := x_1 c X_2 + x_1 Y_2 - x_1 a Z_2 + Y_0 + c a_2 - b_2 - a c_2 = 0 \\ f_6 := -x_1 b X_2 + x_1 a Y_2 + x_1 Z_2 + Z_0 - b a_2 + a b_2 - c_2 = 0 \\ f_7 := x_1 X_3 - x_1 c Y_3 + x_1 b Z_3 + X_0 - a_3 - c b_3 + b c_3 = 0 \\ f_8 := x_1 c X_3 + x_1 Y_3 - x_1 a Z_3 + Y_0 + c a_3 - b_3 - a c_3 = 0 \\ f_9 := -x_1 b X_3 + x_1 a Y_3 + x_1 Z_3 + Z_0 - b a_3 + a b_3 - c_3 = 0, \end{cases} \quad (15.15)$$

where $\{x_i, y_i, z_i\} := \{a_i, b_i, c_i\} \mid i \in \{1, 2, 3\}$ are coordinates of the three points in one of the systems (e.g., local system) and $\{X_i, Y_i, Z_i\} \mid i \in \{1, 2, 3\}$ are the corresponding coordinates in the other system (e.g., global system). In (15.15), $\{f_1, f_2, f_3\}$ are algebraic expressions formed from the *first point* with coordinates in both systems, $\{f_4, f_5, f_6\}$ from the *second point* and $\{f_7, f_8, f_9\}$ from the *third point*. From (15.15), one requires only seven equations for a closed form solution of the 7 parameter datum transformation problem.

Let us consider the system of nonlinear equations extracted from (15.15) to be formed by $\{f_1, f_2, f_3, f_4, f_5, f_6, f_9\}$. Our target now is to solve algebraically this nonlinear system of equations using Groebner basis approach to provide symbolic solutions. We proceed as follows: First, the translation parameters $\{X_0, Y_0, Z_0\}$ are eliminated by differencing

$$\begin{cases} f_{14} := f_1 - f_4 = x_1 X_{12} - x_1 c Y_{12} + x_1 b Z_{12} - a_{12} - c b_{12} + b c_{12} \\ f_{25} := f_2 - f_5 = x_1 c X_{12} + x_1 Y_{12} - x_1 a Z_{12} + c a_{12} - b_{12} - a c_{12} \\ f_{39} := f_3 - f_9 = -x_1 b X_{13} + x_1 a Y_{13} + x_1 Z_{13} - b a_{13} + a b_{13} - c_{13} \\ f_{69} := f_6 - f_9 = -x_1 b X_{23} + x_1 a Y_{23} + x_1 Z_{23} - b a_{23} + a b_{23} - c_{23}, \end{cases} \quad (15.16)$$

where

$$
\left. \begin{array}{l}
X_{ij} = X_i - X_j,\ Y_{ij} = Y_i - Y_j,\ Z_{ij} = Z_i - Z_j \\[4pt]
a_{ij} = a_i - a_j,\quad b_{ij} = b_i - b_j,\quad c_{ij} = c_i - c_j
\end{array} \right] \;|\; i, j \in \{1, 2, 3\}, i \neq j.
$$

The reduced Groebner basis of (15.16) is then obtained using (4.38) on p. 44 by

$$
Groebner\,Basis[\{f_{14}, f_{25}, f_{37}, f_{67}\},\ \{x_1, a, b, c\},\ \{a, b, c\}].
$$

This gives only the elements of *Groebner basis* in which the variables $\{a, b, c\}$ have been eliminated and only the scale factor x_1 left. The scale parameter is then given by the following quartic polynomial:

$$
a_4 x_1^4 + a_3 x_1^3 + a_2 x_1^2 + a_1 x_1 + a_0 = 0, \tag{15.17}
$$

with the coefficients as in [20, Boxes 2-2] or [34, Appendix A]. Once the admissible value of scale parameter $x_1 \in \mathbb{R}^+$ has been chosen from the four roots in (15.17), the elements of the skew-symmetric matrix **S** in (15.4) can then be obtained via the linear functions in [20, Boxes 2-3] or [34, Appendix B]. Substituting the skew-symmetric matrix **S** in (15.1) gives the rotation matrix $\mathbf{X}_3$, from which the Cardan rotation angles are deduced using (15.7) in Solution 15.1. The translation elements $\mathbf{x}_2$ can then be computed by substituting the scale parameter x_1 and the rotation matrix $\mathbf{X}_3$ in (15.1). Three sets of translation parameters are obtained from which their mean is taken.

Example 15.1 (Computation of transformation parameters using Groebner basis algorithm). Cartesian coordinates of seven stations are given in the local and global system (WGS-84) in Tables 15.1 and 15.2 respectively. Desired are the seven parameters of datum transformation. Using explicit solutions in [20, Boxes 2-2 and 2-3] or [34, Appendices A and B], the 7 transformation parameters are computed and presented in Table 15.3. These parameters are then used to transform the three points involved in the computations from the local reference System in Table 15.1 to the WGS-84 as shown in Table 15.4.

Table 15.1. Coordinates for system A (local system)

Station Name	$X(m)$	$Y(m)$	$Z(m)$
Solitude	4157222.543	664789.307	4774952.099
Buoch Zeil	4149043.336	688836.443	4778632.188
Hohenneuffen	4172803.511	690340.078	4758129.701
Kuehlenberg	4177148.376	642997.635	4760764.800
Ex Mergelaec	4137012.190	671808.029	4791128.215
Ex Hof Asperg	4146292.729	666952.887	4783859.856
Ex Kaisersbach	4138759.902	702670.738	4785552.196

Table 15.2. Coordinates for system B (WGS-84)

Station Name	$X(m)$	$Y(m)$	$Z(m)$
Solitude	4157870.237	664818.678	4775416.524
Buoch Zeil	4149691.049	688865.785	4779096.588
Hohenneuffen	4173451.354	690369.375	4758594.075
Kuehlenberg	4177796.064	643026.700	4761228.899
Ex Mergelaec	4137659.549	671837.337	4791592.531
Ex Hof Asperg	4146940.228	666982.151	4784324.099
Ex Kaisersbach	4139407.506	702700.227	4786016.645

Table 15.3. Groebner basis' 7 transformation parameters

Transformation parameter	Value	unit
Scale $k - 1$	-1.4343	[ppm]
Rotation $\mathbf{X}_1(a)$	0.32575149	["]
Rotation $\mathbf{X}_2(b)$	-0.46037399	["]
Rotation $\mathbf{X}_3(c)$	-0.00810606	["]
Translation ΔX	643.0953	[m]
Translation ΔY	22.6163	[m]
Translation ΔZ	481.6023	[m]

15.2.2 Gauss-Jacobi Combinatorial Transformation

When more than three points in both systems are given and the transformation parameters desired, Gauss-Jacobi combinatorial algorithm is applied. In such a case, the dispersion matrix has to be obtained via nonlinear error propagation law/variance-covariance propagation law. From the algebraic system of equations (15.15), the Jacobi matrices are given (using e.g., (6.28) and (6.29) on p. 70) as

Table 15.4. Transformed Cartesian coordinates of System A (*Table* 15.1) into System B using parameters in Table 15.3

Site	$X(m)$	$Y(m)$	$Z(m)$
System A: Solitude	4157222.5430	664789.3070	4774952.0990
System B	4157870.2370	664818.6780	4775416.5240
Transformed value	4157870.3070	664818.6742	4775416.5240
Residual	- 0.0700	0.0038	0.0000
System A: Buoch Zeil	4149043.3360	688836.4430	4778632.1880
System B	4149691.0490	688865.7850	4779096.5880
Transformed value	4149691.1190	688865.7812	4779096.5880
Residual	- 0.0700	0.0038	0.0000
System A: Hohenneuffen	4172803.5110	690340.0780	4758129.7010
System B	4173451.3540	690369.3750	4758594.0750
Transformed value	4173451.2141	690369.3826	4758594.0750
Residual	0.1399	-0.0076	0.0000

$$
\mathbf{J}_x = \begin{bmatrix}
\dfrac{\partial f_1}{\partial x_1} & \dfrac{\partial f_1}{\partial a} & \dfrac{\partial f_1}{\partial b} & \dfrac{\partial f_1}{\partial c} & \dfrac{\partial f_1}{\partial X_0} & \dfrac{\partial f_1}{\partial Y_0} & \dfrac{\partial f_1}{\partial Z_0} \\[2mm]
\dfrac{\partial f_2}{\partial x_1} & \dfrac{\partial f_2}{\partial a} & \dfrac{\partial f_2}{\partial b} & \dfrac{\partial f_2}{\partial c} & \dfrac{\partial f_2}{\partial X_0} & \dfrac{\partial f_2}{\partial Y_0} & \dfrac{\partial f_2}{\partial Z_0} \\[2mm]
\dfrac{\partial f_3}{\partial x_1} & \dfrac{\partial f_3}{\partial a} & \dfrac{\partial f_3}{\partial b} & \dfrac{\partial f_3}{\partial c} & \dfrac{\partial f_3}{\partial X_0} & \dfrac{\partial f_3}{\partial Y_0} & \dfrac{\partial f_3}{\partial Z_0} \\[2mm]
\dfrac{\partial f_4}{\partial x_1} & \dfrac{\partial f_4}{\partial a} & \dfrac{\partial f_4}{\partial b} & \dfrac{\partial f_4}{\partial c} & \dfrac{\partial f_4}{\partial X_0} & \dfrac{\partial f_4}{\partial Y_0} & \dfrac{\partial f_4}{\partial Z_0} \\[2mm]
\dfrac{\partial f_5}{\partial x_1} & \dfrac{\partial f_5}{\partial a} & \dfrac{\partial f_5}{\partial b} & \dfrac{\partial f_5}{\partial c} & \dfrac{\partial f_5}{\partial X_0} & \dfrac{\partial f_5}{\partial Y_0} & \dfrac{\partial f_5}{\partial Z_0} \\[2mm]
\dfrac{\partial f_6}{\partial x_1} & \dfrac{\partial f_6}{\partial a} & \dfrac{\partial f_6}{\partial b} & \dfrac{\partial f_6}{\partial c} & \dfrac{\partial f_6}{\partial X_0} & \dfrac{\partial f_6}{\partial Y_0} & \dfrac{\partial f_6}{\partial Z_0} \\[2mm]
\dfrac{\partial f_7}{\partial x_1} & \dfrac{\partial f_7}{\partial a} & \dfrac{\partial f_7}{\partial b} & \dfrac{\partial f_7}{\partial c} & \dfrac{\partial f_7}{\partial X_0} & \dfrac{\partial f_7}{\partial Y_0} & \dfrac{\partial f_7}{\partial Z_0}
\end{bmatrix}, \tag{15.18}
$$

and

$$
\mathbf{J}_y =
\begin{bmatrix}
\dfrac{\partial f_1}{\partial a_1} & \dfrac{\partial f_1}{\partial b_1} & \dfrac{\partial f_1}{\partial c_1} & \dfrac{\partial f_1}{\partial a_2} & \dfrac{\partial f_1}{\partial b_2} & \dfrac{\partial f_1}{\partial c_2} & \dfrac{\partial f_1}{\partial a_3} & \cdots\cdots\cdots & \dfrac{\partial f_1}{\partial Z_3} \\[2mm]
\dfrac{\partial f_2}{\partial a_1} & \dfrac{\partial f_2}{\partial b_1} & \dfrac{\partial f_2}{\partial c_1} & \dfrac{\partial f_2}{\partial a_2} & \dfrac{\partial f_2}{\partial b_2} & \dfrac{\partial f_2}{\partial c_2} & \dfrac{\partial f_2}{\partial a_3} & \cdots\cdots\cdots & \dfrac{\partial f_2}{\partial Z_3} \\[2mm]
\dfrac{\partial f_3}{\partial a_1} & \dfrac{\partial f_3}{\partial b_1} & \dfrac{\partial f_3}{\partial c_1} & \dfrac{\partial f_3}{\partial a_2} & \dfrac{\partial f_3}{\partial b_2} & \dfrac{\partial f_3}{\partial c_2} & \dfrac{\partial f_3}{\partial a_3} & \cdots\cdots\cdots & \dfrac{\partial f_3}{\partial Z_3} \\[2mm]
\dfrac{\partial f_4}{\partial a_1} & \dfrac{\partial f_4}{\partial b_1} & \dfrac{\partial f_4}{\partial c_1} & \dfrac{\partial f_4}{\partial a_2} & \dfrac{\partial f_4}{\partial b_2} & \dfrac{\partial f_4}{\partial c_2} & \dfrac{\partial f_4}{\partial a_3} & \cdots\cdots\cdots & \dfrac{\partial f_4}{\partial Z_3} \\[2mm]
\dfrac{\partial f_5}{\partial a_1} & \dfrac{\partial f_5}{\partial b_1} & \dfrac{\partial f_5}{\partial c_1} & \dfrac{\partial f_5}{\partial a_2} & \dfrac{\partial f_5}{\partial b_2} & \dfrac{\partial f_5}{\partial c_2} & \dfrac{\partial f_5}{\partial a_3} & \cdots\cdots\cdots & \dfrac{\partial f_5}{\partial Z_3} \\[2mm]
\dfrac{\partial f_6}{\partial a_1} & \dfrac{\partial f_6}{\partial b_1} & \dfrac{\partial f_6}{\partial c_1} & \dfrac{\partial f_6}{\partial a_2} & \dfrac{\partial f_6}{\partial b_2} & \dfrac{\partial f_6}{\partial c_2} & \dfrac{\partial f_6}{\partial a_3} & \cdots\cdots\cdots & \dfrac{\partial f_6}{\partial Z_3} \\[2mm]
\dfrac{\partial f_7}{\partial a_1} & \dfrac{\partial f_7}{\partial b_1} & \dfrac{\partial f_7}{\partial c_1} & \dfrac{\partial f_7}{\partial a_2} & \dfrac{\partial f_7}{\partial b_2} & \dfrac{\partial f_7}{\partial c_2} & \dfrac{\partial f_7}{\partial a_3} & \cdots\cdots\cdots & \dfrac{\partial f_7}{\partial Z_3}
\end{bmatrix}, \tag{15.19}
$$

where the doted points in $\mathbf{J}_y$ represent the partial derivatives of (15.15) with respect to

$$\{b_3, c_3, X_1, Y_1, Z_1, X_2, Y_2, Z_2, X_3, Y_3\}.$$

From the dispersion $\boldsymbol{\Sigma}_y$ of the vector of observations $\mathbf{y}$ and with (15.18) and (15.19) forming $\mathbf{J} = \mathbf{J}_x^{-1}\mathbf{J}_y$, the dispersion matrix $\boldsymbol{\Sigma}_x$ is then obtained using (6.29). Finally we obtained the dispersion matrix $\boldsymbol{\Sigma}$ from (6.31) on p. 71. The solution is performed stepwise as discussed on p. 212. There exist two possibilities of using the combinatorial algorithm. These are:

(1) Forming combinations of the given coordinates, each minimal set comprising 3 points. Given n number of points in both systems, combinations can be formed from (6.26) on p. 69, each set containing $m = 3$ points. For each combination, the desired transformation parameters are computed using the explicit formulae in [20, Boxes 2-2 and 2-3] or [34, Appendices A and B]. The resulting combinatorial solutions are then adjusted using the special linear Gauss-Markov model.

(2) Alternatively, instead of forming combinations from points alone and solving as in (1) above, combinations are formed both for the points and also from the 9 equations in (15.15). In this case, each minimal combinatorial in points will have three stations from which a further combinatorial in terms of equations are formed. From the 9 equations in (15.15), combinations are formed with a minimum

of seven equations per set. The solution of the seven equations of each combinatorial set delivers equations of the form in [20, Boxes 2-2 and 2-3]. Once the solution is completed for a minimum combinatorial set for three points, the procedure is repeated for other points until all the combinations have been solved. The resulting combinatorial solutions are then adjusted using the special linear Gauss-Markov model as already explained. This approach is labour intensive but may offer improved accuracy as compared to the approach in (1) as all the available information is exploited. We leave it as an exercise for an interested reader.

Example 15.2 (Computation of transformation parameters using Gauss-Jacobi combinatorial algorithm). We repeat Example 15.1 by computing the 7 transformation parameters for the overdetermined case using the combinatorial algorithm. All the 7 points of Tables 15.1 and 15.2 are used, unlike in Example 15.1 where only three points were used (e.g., the minimal case). The computed transformation parameters are presented in Table 15.5. In-order to check the accuracy of these parameters, they are used to transform the Cartesian coordinates from the local reference system in Table 15.1 to WGS-84. Table 15.6 gives the residuals computed by subtracting the transformed values from the actual GPS coordinates of Table 15.2. Table 15.7 gives for comparison purposes the residuals obtained using least squares method. The residuals from both procedures are of the same magnitude. We also compute the residual norm (square root of the sum of squares of residuals) and present them in Table 15.8. The computed norms from the combinatorial solutions are somewhat better than those of the linearized least squares solutions. Figure 15.1 presents the scatter of the computed 36 minimal combinatorial solutions of scale indicated by doted points (•) around the adjusted value indicated by a line (−). Figures 15.2 and 15.3 plots the scatter of the computed 36 minimal combinatorial solutions of translation and rotation parameters indicated by doted points (•) around the adjusted values indicated by star (⋆).

Table 15.5. Gauss-Jacobi combinatorial's 7 transformation parameters

Transformation parameter	Value	Root-mean-square	unit
Scale $k - 1$	4.92377597	0.350619414	[ppm]
Rotation $\mathbf{X}_1(a)$	-0.98105498"	0.040968549	["]
Rotation $\mathbf{X}_2(b)$	0.68869774"	0.047458707	["]
Rotation $\mathbf{X}_3(c)$	0.96671738"	0.044697434	["]
Translation ΔX	639.9785	2.4280	[m]
Translation ΔY	68.1548	3.0123	[m]
Translation ΔZ	423.7320	2.7923	[m]

Table 15.6. Residuals of the transformed Cartesian coordinates of System A (Table 15.1) into System B using parameters in Table 15.5

Site	$X(m)$	$Y(m)$	$Z(m)$
Solitude	0.0739	0.1381	0.1397
Buoch Zeil	0.0328	-0.0301	0.0095
Hohenneuffen	-0.0297	-0.0687	-0.0020
Kuelenberg	0.0246	-0.0347	-0.0793
Ex Mergelaec	-0.1405	0.0228	-0.0148
Ex Hof Asperg	-0.0477	0.0116	-0.0599
Ex Keisersbach	-0.0673	0.0335	-0.0070

Table 15.7. Residuals of the transformed Cartesian coordinates of System A (Table 15.1) into System B using parameters computed by least squares method

Site	$X(m)$	$Y(m)$	$Z(m)$
Solitude	0.0940	0.1351	0.1402
Buoch Zeil	0.0588	-0.0497	0.0137
Hohenneuffen	-0.0399	-0.0879	-0.0081
Kuelenberg	0.0202	-0.0220	-0.0874
Ex Mergelaec	-0.0919	0.0139	-0.0055
Ex Hof Asperg	-0.0118	0.0065	-0.0546
Ex Keisersbach	-0.0294	0.0041	-0.0017

Table 15.8. Computed residual norms

Method	$X(m)$	$Y(m)$	$Z(m)$
Linearized Least Squares Solution	0.1541	0.1708	0.1748
Gauss-Jacobi Combinatorial	0.1859	0.1664	0.1725

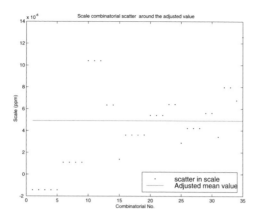

Fig. 15.1. Scatter of the computed 36 minimal combinatorial values of scale around the adjusted value

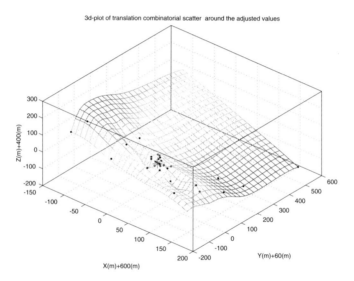

Fig. 15.2. Scatter of the 36 computed translations around the adjusted values

15.2.3 Procrustes Algorithm II

Application to the 7-Parameter Datum Transformation Problem

In Chap. 8, we presented the partial Procrustes algorithm and referred to it as "partial" because it was applied to solve only the rotation component of the datum transformation problem. The analysis of the

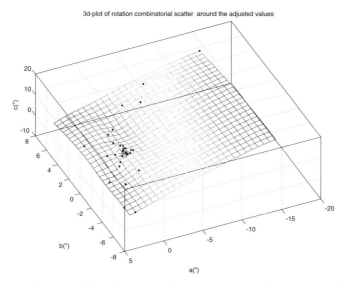

Fig. 15.3. Scatter of the 36 computed rotations around the adjusted values

parameterized conformal group $\mathbb{C}_7(3)$ (7-parameter similarity transformation) as we saw in Sect. 15.1, however, requires the estimation of the *dilatation* unknown (scale factor), unknown vector of *translation* and the *unknown matrix of rotation*. These unknowns are determined from a given matrix data set of Cartesian coordinates as pseudo-observations. In addition to the unknown rotation matrix which was determined in Chap. 8, therefore, one has to determine the scale and translation elements. The partial Procrustes algorithm gives way to the *general Procrustes algorithm* which we call Procrustes algorithm II. The transpose which was indicated by $\{'\}$ in Chap. 8 will be denoted by $\{*\}$ in this section. In Sect. 15.1, we formulated the 7-parameter datum transformation problem such that the solution of (15.1) led to the desired seven parameters. We will now introduce the weight component to (15.1) and solve it using Procrustes algorithm II; an alternative to the procedure that we presented in (15.2.2).

Let us revisit the unknown parameters that we introduced in Sect. 15.1. These were; a scalar-valued unknown $x_1 \in \mathbb{R}$, a vector-valued unknown $\mathbf{x}_2 \in \mathbb{R}^{3\times1}$ (column vector) and a matrix valued unknown $\mathbf{X}_3 \in O^+(3) := \{\mathbf{X}_3 \in \mathbb{R}^{3\times3} \mid \mathbf{X}_3^*\mathbf{X}_3 = \mathbf{I}_3, \mid \mathbf{X}_3 \mid = +1\}$, which in total constitute the 7-dimensional *parameter space*. x_1 represents the *dilatation unknown* (scale factor), $\mathbf{x}_2$ the *translation vector unknown* (3 parameters) and $\mathbf{X}_3$ the *unknown orthonormal matrix* (rotation matrix) which is an element of the special orthogonal group in three dimension. In other words, the $O^+(3)$ differentiable manifold can be coordinated by three parameters. In (8.23) on p. 101, relative position vectors were used to form the two matrices $\mathbf{A}$ and $\mathbf{B}$ in the same dimensional space. If the actual coordinates were used instead, the matrix-valued pseudo-observations $\{\mathbf{Y}_1, \mathbf{Y}_2\}$ become

$$
\begin{bmatrix} x_1 \ x_2 \ \ldots \ x_n \\ y_1 \ y_2 \ \ldots \ y_n \\ z_1 \ z_2 \ \ldots \ z_n \end{bmatrix}^* =: \mathbf{Y}_1, \quad \mathbf{Y}_2 := \begin{bmatrix} X_1 \ X_2 \ \ldots \ X_n \\ Y_1 \ Y_2 \ \ldots \ Y_n \\ Z_1 \ Z_2 \ \ldots \ Z_n \end{bmatrix}^*, \qquad (15.20)
$$

with $\{\mathbf{Y}_1 \text{ and } \mathbf{Y}_2\}$ replacing $\{\mathbf{A} \text{ and } \mathbf{B}\}$. The coordinate matrices of the n points ($n-$dimensional simplex) of a *left three-dimensional Weitzenböck space* as well as a *right three-dimensional Weitzenböck space*, namely $\mathbf{Y}_1 \in \mathbb{R}^{n\times 3}$ and $\mathbf{Y}_2 \in \mathbb{R}^{n\times 3}$ constitute the $6n$ dimensional *observation space*. Left and right matrices $\{\mathbf{Y}_1, \mathbf{Y}_2\}$ are related by means of the *passive 7-parameter conformal group* $\mathbb{C}_7(3)$ in three dimensions (similarity transformation, orthogonal Procrustes transformation) by (cf., 15.1 on p. 261)

$$
\mathbf{Y}_1 \doteq F(x_1, \mathbf{x}_2, \mathbf{X}_3 \mid \mathbf{Y}_2) = \mathbf{Y}_2 \mathbf{X}_3^* x_1 + \mathbf{1}\mathbf{x}_2^*, \mathbf{1} \in \mathbb{R}^{n\times 1}. \qquad (15.21)
$$

The nonlinear matrix-valued equation $F(x_1, \mathbf{x}_2, \mathbf{X}_3 \mid \mathbf{Y}_2) \doteq \mathbf{Y}_1$ is inconsistent since the image $\mathfrak{R}(F) \underset{\neq}{\subset} D(\mathbf{Y}_1)$ of F (range space $\mathfrak{R}(F)$) is constrained in the domain $D(\mathbf{Y}_1)$ of $\mathbf{Y}_1 \in \mathbb{R}^{n\times 3}$ (domain space $D(\mathbf{Y}_1)$). First, as a mapping, F is *"not onto, but into"* or *"not surjective"*. Second, by means of the error matrix $\mathbf{E} \in \mathbb{R}^{n\times 3}$ which accounts for errors in the pseudo-observation matrices $\mathbf{Y}_1$ as well as $\mathbf{Y}_2$, respectively, we are able to make the nonlinear matrix-valued equation $F(x_1, \mathbf{x}_2, \mathbf{X}_3 \mid \mathbf{Y}_2) \doteq \mathbf{Y}_1$ as identity. In this case,

$$
\mathbf{Y}_1 = F(x_1, \mathbf{x}_2, \mathbf{X}_3 \mid \mathbf{Y}_2) + \mathbf{E} = \mathbf{Y}_2 \mathbf{X}_3^* x_1 + \mathbf{1}\mathbf{x}_2^* + \mathbf{E}. \qquad (15.22)
$$

Furthermore, excluding configuration defect which can be detected a priori we shall assume $\aleph(F) = \{0\}$, the kernel of F (null space $\aleph(F)$) to contain only the zero element (empty null space $\aleph(F)$). A simplex of minimal dimension which allows the computation of the seven parameters of the space $\mathbb{X}$ is constituted by $n = 4$ points, namely a *tetrahedron* which is presented in the next examples.

Example 15.3 (Simplex of minimal dimension, $n = 4$ points, tetrahedron).

$$\mathbf{Y}_1 = \begin{bmatrix} x_1 \ x_2 \ x_3 \ x_4 \\ y_1 \ y_2 \ y_3 \ y_4 \\ z_1 \ z_2 \ z_3 \ z_4 \end{bmatrix}^* \in \mathbb{R}^{n \times 3}, \quad \mathbf{Y}_2 = \begin{bmatrix} X_1 \ X_2 \ X_3 \ X_4 \\ Y_1 \ Y_2 \ Y_3 \ Y_4 \\ Z_1 \ Z_2 \ Z_3 \ Z_4 \end{bmatrix}^* \in \mathbb{R}^{n \times 3},$$

$$\begin{bmatrix} x_1 \ y_1 \ z_1 \\ x_2 \ y_2 \ z_2 \\ x_3 \ y_3 \ z_3 \\ x_4 \ y_4 \ z_4 \end{bmatrix} = \begin{bmatrix} X_1 \ Y_1 \ Z_1 \\ X_2 \ Y_2 \ Z_2 \\ X_3 \ Y_3 \ Z_3 \\ X_4 \ Y_4 \ Z_4 \end{bmatrix} \mathbf{X}_3^* x_1 + \mathbf{1} x_2^* + \begin{bmatrix} e_{11} \ e_{12} \ e_{13} \\ e_{21} \ e_{22} \ e_{23} \\ e_{31} \ e_{32} \ e_{33} \\ e_{41} \ e_{42} \ e_{43} \end{bmatrix}.$$

*Example 15.4 (**W**eighted **LE**ast **S**quares' **S**olution W-LESS).* We depart from the set up of the pseudo-observation equations given in Example 15.3 (simplex of minimal dimension, $n = 4$ points, tetrahedron). For a diagonal weight $\mathbf{W} = Diag\,(w_1, \ldots, w_4) \in \mathbb{R}^{4 \times 4}$ we compute the Frobenius error matrix W-semi-norm

$$\| \mathbf{E} \|_{\mathbf{W}}^2 := tr(\mathbf{E}^* \mathbf{W} \mathbf{E}) =$$

$$tr \left\{ \begin{bmatrix} e_{11} \ e_{21} \ e_{31} \ e_{41} \\ e_{12} \ e_{22} \ e_{32} \ e_{42} \\ e_{13} \ e_{23} \ e_{33} \ e_{43} \end{bmatrix} \begin{bmatrix} w_1 \ 0 \ 0 \ 0 \\ 0 \ w_2 \ 0 \ 0 \\ 0 \ 0 \ w_3 \ 0 \\ 0 \ 0 \ 0 \ w_4 \end{bmatrix} \begin{bmatrix} e_{11} \ e_{12} \ e_{13} \\ e_{21} \ e_{22} \ e_{23} \\ e_{31} \ e_{32} \ e_{33} \\ e_{41} \ e_{42} \ e_{43} \end{bmatrix} \right\}$$

$$= tr \left\{ \begin{bmatrix} e_{11}w_1 \ e_{21}w_2 \ e_{31}w_3 \ e_{41}w_4 \\ e_{12}w_1 \ e_{22}w_2 \ e_{32}w_3 \ e_{42}w_4 \\ e_{13}w_1 \ e_{23}w_2 \ e_{33}w_3 \ e_{43}w_4 \end{bmatrix} \begin{bmatrix} e_{11} \ e_{12} \ e_{13} \\ e_{21} \ e_{22} \ e_{23} \\ e_{31} \ e_{32} \ e_{33} \\ e_{41} \ e_{42} \ e_{43} \end{bmatrix} \right\}$$

$$= \begin{bmatrix} w_1 e_{11}^2 + w_2 e_{21}^2 + w_3 e_{31}^2 + w_4 e_{41}^2 \\ +w_1 e_{12}^2 + w_2 e_{22}^2 + w_3 e_{32}^2 + w_4 e_{42}^2 \\ +w_1 e_{13}^2 + w_2 e_{23}^2 + w_3 e_{33}^2 + w_4 e_{43}^2 \end{bmatrix}.$$

Obviously the coordinate errors (e_{11}, e_{12}, e_{13}) have the same weight w_1, $(e_{21}, e_{22}, e_{23}) \rightarrow w_2$, $(e_{31}, e_{32}, e_{33}) \rightarrow w_3$ and finally $(e_{41}, e_{42}, e_{43}) \rightarrow w_4$. We may also say that the *error weight is pointwise isotropic*, namely weight e_{11} =weight e_{12}=weight e_{13}=weight w_1 etc. But the error weight is *not homogeneous* since w_1 =weight $e_{11} \neq$weight $e_{21} = w_2$. Of course, an ideal *homogeneous and isotropic weight distribution* is guaranteed by the criterion $w_1 = w_2 = w_3 = w_4 = w$.

By means of Solution 15.3 we have summarized the parameter space $(x_1, \mathbf{x}_2, \mathbf{X}_3) \in \mathbb{R} \times \mathbb{R}^3 \times \mathbb{R}^{3 \times 3}$. In contrast, Solution 15.4 reviews the pseudo-observation space $(\mathbf{Y}_1, \mathbf{Y}_2) \in \mathbb{R}^{n \times 3} \times \mathbb{R}^{n \times 3}$ equipped with the Frobenius matrix $\mathbf{W}$–semi-norm.

Solution 15.3 (The parameter space $\mathbb{X}$).

$x_1 \in \mathbb{R}$ dilatation parameter (scale factor)

$\mathbf{x}_2 \in \mathbb{R}^{3 \times 1}$ column vector of translation parameters

$\mathbf{X}_3 \in O^+(3) := \{\mathbf{X}_3 \in \mathbb{R}^{3 \times 3} \mid \mathbf{X}_3^* \mathbf{X}_3 = \mathbf{I}_3, \mid \mathbf{X}_3 \mid = +1\}$
                  orthonormal matrix,
                  rotation matrix of three
                  parameters

Solution 15.4 (The observation space $\mathbb{Y}$).

$$\mathbf{Y}_1 = \begin{bmatrix} x_1 \ x_2 \ \dots \ x_n \\ y_1 \ y_2 \ \dots \ y_n \\ z_1 \ z_2 \ \dots \ z_n \end{bmatrix}^* \in \mathbb{R}^{n \times 3}, \qquad \mathbf{Y}_2 = \begin{bmatrix} X_1 \ X_2 \ \dots \ X_n \\ Y_1 \ Y_2 \ \dots \ Y_n \\ Z_1 \ Z_2 \ \dots \ Z_n \end{bmatrix}^* \in \mathbb{R}^{n \times 3}$$

left three-dimensional coordinate matrix of an $n-$ dimensional simplex
     right three-dimensional coordinate matrix of an $n-$ dimensional simplex

The immediate problem that one is faced with is how to solve the inconsistent matrix-valued nonlinear equation (15.22). Essentially, this is the same problem that we introduced in (15.1) on p. 261. The difference between (15.1) and (15.22) is the incorporation of the error matrix $\mathbf{E}$ in the latter. This takes into consideration the stochasticity of the systems $\mathbf{Y}_1$ and $\mathbf{Y}_2$. In what follows W-LESS (i.e., the **W**eighted **LE**ast **S**quares' **Solution**) is defined and materialized by the *Procrustes algorithm II* presented by means of:

- Corollary 15.1 (partial W-LESS for the unknown vector $\mathbf{x}_{2l}$).
- Corollary 15.2 (partial W-LESS for the unknown scalar x_{1l}).
- Corollary 15.3 (partial W-LESS for the unknown matrix $\mathbf{X}_{3l}$).

The partial optimization results are collected in Theorem 15.1 (W-LESS of $\mathbf{Y}_1 = \mathbf{Y}_2 \mathbf{X}_3^* x_1 + \mathbf{1} \mathbf{x}_2^* + \mathbf{E}$) and Corollary 15.4 (I-LESS of $\mathbf{Y}_1 = \mathbf{Y}_2 \mathbf{X}_3^* x_1 + \mathbf{1} \mathbf{x}_2^* + \mathbf{E}$). Solution 15.5 summarizes the general Procrustes algorithm II.

Definition 15.1 (W-LESS). *The parameter array $\{x_{1l}, \mathbf{x}_{2l}, \mathbf{X}_{3l}\}$ is called W-LESS (least squares solution with respect to Frobenius matrix $\mathbf{W}$–semi-norm) of the inconsistent nonlinear matrix-valued system of equations*

$$\mathbf{Y}_2\mathbf{X}_3^*x_1 + \mathbf{1}\mathbf{x}_2^* + \mathbf{E} = \mathbf{Y}_1, \tag{15.23}$$

subject to

$$\mathbf{X}_3^*\mathbf{X}_3 = \mathbf{I}_3, \mid \mathbf{X}_3 \mid = +1, \tag{15.24}$$

if for the parameter array in comparison to all other parameter arrays $\{x_{1l}, \mathbf{x}_{2l}, \mathbf{X}_{3l}\}$, *the inequality*

$$
\left[
\begin{array}{l}
\parallel \mathbf{Y}_1 - \mathbf{Y}_2\mathbf{X}_{3l}^*x_{1l} - \mathbf{1}\mathbf{x}_{2l}^* \parallel_{\mathbf{W}}^2 \\[2mm]
:= tr((\mathbf{Y}_1 - \mathbf{Y}_2\mathbf{X}_{3l}^*x_{1l} - \mathbf{1}\mathbf{x}_{2l}^*)^*\mathbf{W}(\mathbf{Y}_1 - \mathbf{Y}_2\mathbf{X}_{3l}^*x_{1l} - \mathbf{1}\mathbf{x}_{2l}^*) \\[2mm]
\leq tr((\mathbf{Y}_1 - \mathbf{Y}_2\mathbf{X}_3^*x_1 - \mathbf{1}\mathbf{x}_2^*)^*\mathbf{W}(\mathbf{Y}_1 - \mathbf{Y}_2\mathbf{X}_3^*x_1 - \mathbf{1}\mathbf{x}_2^*) \\[2mm]
=: \parallel \mathbf{Y}_1 - \mathbf{Y}_2\mathbf{X}_3^*x_1 - \mathbf{1}\mathbf{x}_2^* \parallel_{\mathbf{W}}^2
\end{array}
\right. \tag{15.25}
$$

holds, in particular if $\mathbf{E}_l := \mathbf{Y}_1 - \mathbf{Y}_2\mathbf{X}_{3l}^*x_{1l} - \mathbf{1}\mathbf{x}_{2l}^*$ *has the minimal Frobenius matrix* $\mathbf{W}$*–semi-norm such that* $\mathbf{W} \in \mathbb{R}^{n\times n}$ *is positive semi-definite.*

Note that $\parallel \mathbf{E} \parallel_{\mathbf{W}}^2 := tr(\mathbf{E}^*\mathbf{W}\mathbf{E})$ characterizes the method of least squares tuned to an error matrix $\mathbf{E} \in \mathbb{R}^{n\times 3}$ and a positive semi-definite weight matrix $\mathbf{W}$. Indeed a positive semi-definite weight matrix $\mathbf{W}$ of weights is chosen in-order to have the option to exclude by means of zero weight a particular pseudo-observation, say a particular coordinate row vector $[x_i, y_i, z_i], i \in \mathbb{N}$ arbitrary, but fixed by $w_{ii} = w_i = 0$, which may be an *outlier*. Example 15.4 illustrates details of Definition 15.1.

In-order to construct W-LESS of the inconsistent nonlinear matrix-valued system of equations (15.23) subject to (15.24) we introduce the *Procrustes algorithm II*. The *first* algorithmic step is constituted by the *forward* computation of the transformation parameters $\mathbf{x}_{2l}$ from the unconstraint *Lagrangean* $L(x_1, \mathbf{x}_2, \mathbf{X}_3)$ which is twice the value of the Frobenius error matrix $\mathbf{W}$–semi-norm. As soon as the translation parameters $\mathbf{x}_{2l}$ are *backward* substituted we gain a Lagrangean $L(x_1, \mathbf{X}_3)$ which is centralized with respect to the observation matrix $\mathbf{Y}_1 - \mathbf{Y}_2\mathbf{X}_3^*x_1$. In the *second* algorithmic step the scale parameter x_{1l} is *forward* computed from the centralized Lagrangean $L(x_1, \mathbf{X}_3)$. Its *backward* substitution leads to the Lagrangean $L(\mathbf{X}_3)$ which is only dependent on the rotation matrix $\mathbf{X}_3$. Finally the optimization problem $L(\mathbf{X}_3) = min$ subject to $\mathbf{X}_3^*\mathbf{X}_3 = \mathbf{I}_3, \mid \mathbf{X}_3 \mid = +1$ generates the *third* algorithmic step. This computational step is similar to that of partial Procrustes algorithm of Sect. 8.3. By means of *singular value* *decomposition SVD* the rotation matrix $\mathbf{X}_{3l}$ is *forward* computed and *backward* substituted to gain $(x_1, \mathbf{x}_2, \mathbf{X}_3)$ at the end. The results are collected in Corollaries 15.1 to 15.3.

Corollary 15.1 (Partial W-LESS for the translation vector $\mathbf{x}_{2l}$). *A 3×1 vector $\mathbf{x}_{2l}$ is **partial** W-LESS of (15.23) subject to (15.24) if and only if $\mathbf{x}_{2l}$ fulfills the system of normal equations*

$$\mathbf{1}^*\mathbf{W}\mathbf{1}\mathbf{x}_{2l} = (\mathbf{Y}_1 - \mathbf{Y}_2\mathbf{X}_3^*x_1)^*\mathbf{W}\mathbf{1}. \tag{15.26}$$

The translation vector $\mathbf{x}_{2l}$ always exist and is represented by

$$\mathbf{x}_{2l} = (\mathbf{1}^*\mathbf{W}\mathbf{1})^{-1}(\mathbf{Y}_1 - \mathbf{Y}_2\mathbf{X}_3^*x_1)^*\mathbf{W}\mathbf{1}. \tag{15.27}$$

For the special case $\mathbf{W} = \mathbf{I}_n$, i.e., the weight matrix is unit, the translational parameter vector $\mathbf{x}_{2l}$ is given by

$$\mathbf{x}_{2l} = \frac{1}{n}(\mathbf{Y}_1 - \mathbf{Y}_2\mathbf{X}_3^*x_1)^*\mathbf{1}.$$

Proof. W-LESS is constructed by *unconstraint Lagrangean*

$$\left[\begin{array}{l} L(x_1, \mathbf{x}_2, \mathbf{X}_3) := \dfrac{1}{2} \parallel \mathbf{E} \parallel_\mathbf{W}^2 = \parallel \mathbf{Y}_1 - \mathbf{Y}_2\mathbf{X}_3^*x_1 - \mathbf{1}\mathbf{x}_2^* \parallel_\mathbf{W}^2 \\[3mm] = \dfrac{1}{2}tr(\mathbf{Y}_1 - \mathbf{Y}_2\mathbf{X}_3^*x_1 - \mathbf{1}\mathbf{x}_2^*)^*\mathbf{W}(\mathbf{Y}_1 - \mathbf{Y}_2\mathbf{X}_3^*x_1 - \mathbf{1}\mathbf{x}_2^*) = min, \\[3mm] subject\ to\ \{x_1 \geq 0, \mathbf{x}_2 \in R^{3\times1}, \mathbf{X}_3^*\mathbf{X}_3 = \mathbf{I}_3\} \\[1mm] \dfrac{\partial L}{\partial \mathbf{x}_2^*}(\mathbf{x}_{2l}) = (\mathbf{1}^*\mathbf{W}\mathbf{1})\mathbf{x}_2 - (\mathbf{Y}_1 - \mathbf{Y}_2\mathbf{X}_3^*x_1)^*\mathbf{W}\mathbf{1} = \mathbf{0}, \end{array}\right. \tag{15.28}$$

constitutes the *first necessary condition*. Basics of vector-valued differentials are as given in Table 8.1, p. 96. For more details on matrix properties and manipulations, we refer to [149, pp. 439–451]. As soon as we back-substitute the translation parameter $\mathbf{x}_{2l}$, we are led to the *centralized Lagrangean*

$$\left[\begin{array}{c} L(x_1, \mathbf{X}_3) = \\[2mm] \dfrac{1}{2}tr\{\left[\mathbf{Y}_1 - (\mathbf{Y}_2\mathbf{X}_3^*x_1 + (\mathbf{1}^*\mathbf{W}\mathbf{1})^{-1}\mathbf{1}\mathbf{1}^*\mathbf{W}(\mathbf{Y}_1 - \mathbf{Y}_2\mathbf{X}_3^*x_1))\right]^* \mathbf{W} \\[2mm] * \left[\mathbf{Y}_1 - (\mathbf{Y}_2\mathbf{X}_3^*x_1 + (\mathbf{1}^*\mathbf{W}\mathbf{1})^{-1}\mathbf{1}\mathbf{1}^*\mathbf{W}(\mathbf{Y}_1 - \mathbf{Y}_2\mathbf{X}_3^*x_1))\right]\} \end{array}\right. \tag{15.29}$$

$$
\begin{bmatrix}
L(x_1, \mathbf{X}_3) = \\
\frac{1}{2} tr\{\left[(\mathbf{I} - (\mathbf{1}^*\mathbf{W}\mathbf{1})^{-1}\mathbf{1}\mathbf{1}^*)\mathbf{W}(\mathbf{Y}_1 - \mathbf{Y}_2\mathbf{X}_3^*x_1)\right]^*\mathbf{W} \\
* \left[(\mathbf{I} - (\mathbf{1}^*\mathbf{W}\mathbf{1})^{-1}\mathbf{1}\mathbf{1}^*)\mathbf{W}(\mathbf{Y}_1 - \mathbf{Y}_2\mathbf{X}_3^*x_1)\right]\}
\end{bmatrix} \tag{15.30}
$$

$$
\mathbf{C} := \mathbf{I}_n - (\mathbf{1}^*\mathbf{W}\mathbf{1})^{-1}\mathbf{1}\mathbf{1}^*\mathbf{W} \tag{15.31}
$$

is a definition of the *centering matrix*, namely for $\mathbf{W} = \mathbf{I}_n$

$$
\mathbf{C} := \mathbf{I}_n - \frac{1}{n}\mathbf{1}\mathbf{1},^* \tag{15.32}
$$

being symmetric, in general. Substituting the centering matrix into the *reduced Lagrangean* $L(x_1, \mathbf{X}_3)$, we gain the *centralized Lagrangean*

$$
\begin{matrix}
L(x_1, \mathbf{X}_3) = \\
\frac{1}{2} tr\{[\mathbf{Y}_1 - \mathbf{Y}_2\mathbf{X}_3^*x_1]^* \mathbf{C}^*\mathbf{W}\mathbf{C}[\mathbf{Y}_1 - \mathbf{Y}_2\mathbf{X}_3^*x_1]\}
\end{matrix} \tag{15.33}
$$

□

Corollary 15.2 (Partial W-LESS for the scale factor x_{1l}). *A scalar x_{1l} is **partial** W-LESS of (15.23) subject to (15.24) if and only if*

$$
x_{1l} = \frac{tr\mathbf{Y}_1^*\mathbf{C}^*\mathbf{W}\mathbf{C}\mathbf{Y}_2\mathbf{X}_3^*}{tr\mathbf{Y}_2^*\mathbf{C}^*\mathbf{W}\mathbf{C}\mathbf{Y}_2} \tag{15.34}
$$

holds. For special case $\mathbf{W} = \mathbf{I}_n$ the scale parameter vector x_{1l} is given by

$$
x_{1l} = \frac{tr\mathbf{Y}_1^*\mathbf{C}^*\mathbf{C}\mathbf{Y}_2\mathbf{X}_3^*}{tr\mathbf{Y}_2^*\mathbf{C}^*\mathbf{C}\mathbf{Y}_2} \tag{15.35}
$$

Proof. Partial W-LESS is constructed by the unconstraint centralized Lagrangean

$$
\begin{bmatrix}
L(x_1, \mathbf{X}_3) = \\
= \frac{1}{2} tr\{[(\mathbf{Y}_1 - \mathbf{Y}_2\mathbf{X}_3^*x_1)]^*\mathbf{C}^*\mathbf{W}\mathbf{C}[\mathbf{Y}_1 - \mathbf{Y}_2\mathbf{X}_3^*x_1]\} = min_{x_1, \mathbf{X}_3}, \\
subject \ to \ \{x_1 \geq 0, \mathbf{X}_3^*\mathbf{X}_3 = \mathbf{I}_3\}.
\end{bmatrix} \tag{15.36}
$$

$$
\frac{\partial L}{\partial x_1}(x_{1l}) = x_{1l}tr\mathbf{X}_3\mathbf{Y}_2^*\mathbf{C}^*\mathbf{W}\mathbf{C}\mathbf{Y}_2\mathbf{X}_3^* - tr\mathbf{Y}_1^*\mathbf{C}^*\mathbf{W}\mathbf{C}\mathbf{Y}_2\mathbf{X}_3^* = 0 \tag{15.37}
$$

constitutes the *second necessary condition*. Due to (e.g., cyclic property in Table 8.1, p. 96)

$$tr\mathbf{X}_3\mathbf{Y}_2^*\mathbf{C}^*\mathbf{WCY}_2\mathbf{X}_3^* = tr\mathbf{Y}_2^*\mathbf{C}^*\mathbf{WCY}_2\mathbf{X}_3^*\mathbf{X}_3 = \mathbf{Y}_2^*\mathbf{C}^*\mathbf{WCY}_2,$$

(15.37) leads to (15.34). While the forward computation of $\dfrac{\partial L}{\partial x_1}(x_{1l}) = \mathbf{0}$ enjoyed a representation of the optimal scale parameter x_{1l}, its *backward* substitution into the Lagrangean $L(x_1, \mathbf{X}_3)$ amounts to

$$L(\mathbf{X}_3) =$$
$$tr\left\{\left[\mathbf{Y}_1 - \mathbf{Y}_2\mathbf{X}_3^*\frac{tr\mathbf{Y}_1^*\mathbf{C}^*\mathbf{WCY}_2\mathbf{X}_3^*}{tr\mathbf{Y}_2^*\mathbf{C}^*\mathbf{WCY}_2}\right]\mathbf{C}^*\mathbf{WC}\right.$$
$$\left.*\left[\mathbf{Y}_1 - \mathbf{Y}_2\mathbf{X}_3^*\frac{tr\mathbf{Y}_1^*\mathbf{C}^*\mathbf{WCY}_2\mathbf{X}_3^*}{tr\mathbf{Y}_2^*\mathbf{C}^*\mathbf{WCY}_2}\right]\right\}$$

(15.38)

$$L(\mathbf{X}_3) =$$
$$\frac{1}{2}tr\left\{(\mathbf{Y}_1^*\mathbf{C}^*\mathbf{WCY}_1) - tr(\mathbf{Y}_1^*\mathbf{C}^*\mathbf{WCY}_2\mathbf{X}_3^*)\frac{tr\mathbf{Y}_1^*\mathbf{C}^*\mathbf{WCY}_2\mathbf{X}_3^*}{tr\mathbf{Y}_2^*\mathbf{C}^*\mathbf{WCY}_2}\right.$$
$$-tr(\mathbf{X}_3\mathbf{Y}_2^*\mathbf{C}^*\mathbf{WCY}_1)\frac{tr\mathbf{Y}_1^*\mathbf{C}^*\mathbf{WCY}_2\mathbf{X}_3^*}{tr\mathbf{Y}_2^*\mathbf{C}^*\mathbf{WCY}_2}$$
$$\left.+tr(\mathbf{X}_3\mathbf{Y}_2^*\mathbf{C}^*\mathbf{WCY}_2\mathbf{X}_3^*)\frac{[tr\mathbf{Y}_1^*\mathbf{C}^*\mathbf{WCY}_2\mathbf{X}_3^*]^2}{[tr\mathbf{Y}_2^*\mathbf{C}^*\mathbf{WCY}_2]^2}\right\}$$

(15.39)

$$L(\mathbf{X}_3) =$$
$$= \frac{1}{2}tr(\mathbf{Y}_1^*\mathbf{C}^*\mathbf{WCY}_1) - \frac{[tr\mathbf{Y}_1^*\mathbf{C}^*\mathbf{WCY}_2\mathbf{X}_3^*]^2}{[tr\mathbf{Y}_2^*\mathbf{C}^*\mathbf{WCY}_2]}$$
$$+\frac{1}{2}\frac{[tr\mathbf{Y}_1^*\mathbf{C}^*\mathbf{WCY}_2\mathbf{X}_3^*]^2}{[tr\mathbf{Y}_2^*\mathbf{C}^*\mathbf{WCY}_2]}$$

(15.40)

$$L(\mathbf{X}_3) =$$
$$= \frac{1}{2}tr(\mathbf{Y}_1^*\mathbf{C}^*\mathbf{WCY}_1) - \frac{1}{2}\frac{[tr\mathbf{Y}_1^*\mathbf{C}^*\mathbf{WCY}_2\mathbf{X}_3^*]^2}{[tr\mathbf{Y}_2^*\mathbf{C}^*\mathbf{WCY}_2]} = min,$$

(15.41)

$$subject\ to\ \{\mathbf{X}_3^*\mathbf{X}_3 = \mathbf{I}_3\}$$

Corollary 15.3 (Partial W-LESS for the rotation matrix X_{3l}).
A 3×3 orthonormal matrix X_3 is partial W-LESS of (15.41) if and only if

$$X_{3l} = UV^* \tag{15.42}$$

holds, where $A := Y_1^* C^* WCY_2 = U\Sigma_s V^*$ is a singular value decomposition with respect to a left orthonormal matrix $U, U^* U = I_3$, a right orthonormal matrix $V, VV^* = I_3$, and $\Sigma_s = Diag(\sigma_1, \sigma_2, \sigma_3)$ a diagonal matrix of singular values $(\sigma_1, \sigma_2, \sigma_3)$. The singular values are the canonical coordinates of the right eigenspace $(A^* A - \Sigma_s^2 I)V = 0$. *The* left eigenspace *is based upon* $U = AV\Sigma_s^{-1}$.

Proof. In (15.41) $L(X_3)$ subject to $X_3^* X_3 = I_3$ is minimal if

$$tr(Y_1^* C^* WCY_2 X_3^*) = min, \quad subject\ to\ \{x_1 \geq 0, X_3^* X_3 = I_3\}. \tag{15.43}$$

Let $A := Y_1^* C^* WCY_2 = U\Sigma_s V^*$ be a singular value decomposition with respect to a left orthonormal matrix $U, U^* U = I_3$, a right orthonormal matrix $V, VV^* = I_3$, and $\Sigma_s = Diag(\sigma_1, \sigma_2, \sigma_3)$ a diagonal matrix of singular values $(\sigma_1, \sigma_2, \sigma_3)$. Then

$$\left[\begin{array}{c} tr(AX_3^*) = tr(U\Sigma_s V^* X_3^*) \\ = tr(\Sigma_s V^* X_3^* U) = \sum_{i=1}^3 \sigma_i r_{ii} \leq \sum_{i=1}^3 \sigma_i, \end{array} \right. \tag{15.44}$$

holds since $R = V^* X_3^* U = [r_{ij}] \in \mathbb{R}^{3\times3}$ is orthonormal with $\mid r_{ii} \mid \leq 1$. The identity $tr(AX_3^*) = \sum_{i=1}^3 \sigma_i$ applies if $V^* X_3^* U = I_3$, that is $X_3^* = VU^*$, $X_3 = UV^*$, namely if $tr(AX_3^*)$ is maximal:

$$\left[\begin{array}{c} tr(AX_3^*) = max_{\{X_3^* X_3 = I_3\}} \\ \Leftrightarrow R = V^* X_3^* U = I_3. \end{array} \right. \Leftrightarrow trAX_3^* = \sum_{i=1}^3 \sigma_{\cdot i} \tag{15.45}$$

An alternative proof of Corollary 15.3 based on formal differentiation of traces and determinants has been given in Sect. 8.3.2 of Chap. 8. Finally we collect our sequential results in Theorem 15.1 identifying the *stationary point* of W-LESS of (15.23) specialized for $W = I$, i.e., matrix of unit weight in Corollary 15.4. The highlight is the *Procrustes algorithm II* we have developed in Solution 15.5.

Theorem 15.1 (W-LESS of $\mathbf{Y}_1 = \mathbf{Y}_2\mathbf{X}_3^*x_1 + \mathbf{1}\mathbf{x}_2^* + \mathbf{E}$).

(i)   *The parameter array $\{x_1, \mathbf{x}_2, \mathbf{X}_3\}$ is W-LESS of (15.23) if*

$$x_{1l} = \frac{tr\,\mathbf{Y}_1^*\mathbf{C}^*\mathbf{W}\mathbf{C}\mathbf{Y}_2\mathbf{X}_{3l}^*}{tr\,\mathbf{Y}_2^*\mathbf{C}^*\mathbf{W}\mathbf{C}\mathbf{Y}_2} \tag{15.46}$$

$$\mathbf{x}_{2l} = (\mathbf{1}^*\mathbf{W}\mathbf{1})^{-1}(\mathbf{Y}_1 - \mathbf{Y}_2\mathbf{X}_{3l}^*x_{1l})^*\mathbf{W}\mathbf{1} \tag{15.47}$$

$$\mathbf{X}_{3l} = \mathbf{U}\mathbf{V}^*, \tag{15.48}$$

subject to the singular value decomposition of the general 3×3 matrix

$$\mathbf{Y}_1^*\mathbf{C}^*\mathbf{W}\mathbf{C}\mathbf{Y}_2 = \mathbf{U}Diag(\sigma_1, \sigma_2, \sigma_3)\mathbf{V}^*, \tag{15.49}$$

namely

$$\left[\begin{array}{c} [(\mathbf{Y}_1^*\mathbf{C}^*\mathbf{W}\mathbf{C}\mathbf{Y}_2)^*(\mathbf{Y}_1^*\mathbf{C}^*\mathbf{W}\mathbf{C}\mathbf{Y}_2) - \sigma_i^2\mathbf{I}]\mathbf{v}_i = \mathbf{0}\ \forall i \in \{1, 2, 3\} \\ \mathbf{V} = [\mathbf{v}_1, \mathbf{v}_2, \mathbf{v}_3]\,, \mathbf{V}\mathbf{V}^* = \mathbf{I}_3 \end{array}\right.$$

$$\tag{15.50}$$

$$\left[\begin{array}{c} \mathbf{U} = \mathbf{Y}_1^*\mathbf{C}^*\mathbf{W}\mathbf{C}\mathbf{Y}_2\mathbf{V}Diag(\sigma_1^{-1}, \sigma_2^{-1}, \sigma_3^{-1}) \\ \mathbf{U}^*\mathbf{U} = \mathbf{I}_3 \end{array}\right.$$

$$\tag{15.51}$$

and as well as to the centering matrix

$$\mathbf{C} := \mathbf{I}_n - (\mathbf{1}^*\mathbf{W}\mathbf{1})^{-1}\mathbf{1}\mathbf{1}^*\mathbf{W}. \tag{15.52}$$

(ii)   *The empirical error matrix of type W-LESS accounts for*

$$\mathbf{E}_l = [\mathbf{I}_n - \mathbf{1}\mathbf{1}^*\mathbf{W}(\mathbf{1}^*\mathbf{W}\mathbf{1})^{-1}]\left\{\mathbf{Y}_1 - \mathbf{Y}_2\mathbf{V}\mathbf{U}^*\frac{tr\,\mathbf{Y}_1^*\mathbf{C}^*\mathbf{W}\mathbf{C}\mathbf{Y}_2\mathbf{V}\mathbf{U}^*}{tr\,\mathbf{Y}_2^*\mathbf{C}^*\mathbf{W}\mathbf{C}\mathbf{Y}_2}\right\},$$

$$\tag{15.53}$$

with the related Frobenius matrix $\mathbf{W}$–semi-norm

$$\left[\begin{array}{c} \| \mathbf{E} \|_{\mathbf{W}}^2 := tr(\mathbf{E}_l^*\mathbf{W}\mathbf{E}_l) = \\[2mm] tr\{(\mathbf{Y}_1 - \mathbf{Y}_2\mathbf{V}\mathbf{U}^*\frac{tr\,\mathbf{Y}_1^*\mathbf{C}^*\mathbf{W}\mathbf{C}\mathbf{Y}_2\mathbf{V}\mathbf{U}^*}{tr\,\mathbf{Y}_2^*\mathbf{C}^*\mathbf{W}\mathbf{C}\mathbf{Y}_2})^*. \\[2mm] .[\mathbf{I}_n - \mathbf{1}\mathbf{1}^*\mathbf{W}(\mathbf{1}^*\mathbf{W}\mathbf{1})^{-1}]^*\mathbf{W}[\mathbf{I}_n - \mathbf{1}\mathbf{1}^*\mathbf{W}(\mathbf{1}^*\mathbf{W}\mathbf{1})^{-1}]. \\[2mm] .(\mathbf{Y}_1 - \mathbf{Y}_2\mathbf{V}\mathbf{U}^*\frac{tr\,\mathbf{Y}_1^*\mathbf{C}^*\mathbf{W}\mathbf{C}\mathbf{Y}_2\mathbf{V}\mathbf{U}^*}{tr\,\mathbf{Y}_2^*\mathbf{C}^*\mathbf{W}\mathbf{C}\mathbf{Y}_2})\}, \end{array}\right.$$

$$\tag{15.54}$$

and the representative scalar measure of the error of type W-LESS given by

$$\||\; E_l \;\||_{\mathbf{W}} := \sqrt{tr(\mathbf{E}_l^* \mathbf{W} \mathbf{E}_l)/3n}. \tag{15.55}$$

Corollary 15.4 (I-LESS of $\mathbf{Y}_1 = \mathbf{Y}_2 \mathbf{X}_3^* x_1 + \mathbf{1} \mathbf{x}_2^* + \mathbf{E}$).

(i)   The parameter array $\{x_1, \mathbf{x}_2, \mathbf{X}_3\}$ is I-LESS of (15.23) if

$$x_{1l} = \frac{tr\,\mathbf{Y}_1^* \mathbf{C} \mathbf{Y}_2 \mathbf{X}_{3l}^*}{tr\,\mathbf{Y}_2^* \mathbf{C} \mathbf{Y}_2} \tag{15.56}$$

$$\mathbf{x}_{2l} = \frac{1}{n}(\mathbf{Y}_1 - \mathbf{Y}_2 \mathbf{X}_{3l}^* x_{1l})^* \mathbf{1} \tag{15.57}$$

$$\mathbf{X}_{3l} = \mathbf{U} \mathbf{V}^*, \tag{15.58}$$

subject to the singular value decomposition of the general 3×3 matrix

$$\mathbf{Y}_1^* \mathbf{C} \mathbf{Y}_2 = \mathbf{U} Diag(\sigma_1, \sigma_2, \sigma_3) \mathbf{V}^*, \tag{15.59}$$

namely

$$\left[\begin{array}{c} [(\mathbf{Y}_1^* \mathbf{C} \mathbf{Y}_2)^* (\mathbf{Y}_1^* \mathbf{C} \mathbf{Y}_2) - \sigma_i^2 \mathbf{I}] \mathbf{v}_i = \mathbf{0}\, \forall i \in \{1,2,3\} \\ \mathbf{V} = [\mathbf{v}_1, \mathbf{v}_2, \mathbf{v}_3], \mathbf{V} \mathbf{V}^* = \mathbf{I}_3 \end{array} \right. \tag{15.60}$$

$$\mathbf{U} = \mathbf{Y}_1^* \mathbf{C} \mathbf{Y}_2 \mathbf{V} Diag(\sigma_1^{-1}, \sigma_2^{-1}, \sigma_3^{-1}) \\ \mathbf{U}^* \mathbf{U} = \mathbf{I}_3 \tag{15.61}$$

and as well as to the centering matrix

$$\mathbf{C} := \mathbf{I}_n - \frac{1}{n} \mathbf{1} \mathbf{1}^*. \tag{15.62}$$

(ii)   The empirical error matrix of type I-LESS accounts for

$$\mathbf{E}_l = [\mathbf{I}_n - \frac{1}{n} \mathbf{1} \mathbf{1}^*] \left\{ \mathbf{Y}_1 - \mathbf{Y}_2 \mathbf{V} \mathbf{U}^* \frac{tr\,\mathbf{Y}_1^* \mathbf{C} \mathbf{Y}_2 \mathbf{V} \mathbf{U}^*}{tr\,\mathbf{Y}_2^* \mathbf{C} \mathbf{Y}_2} \right\}, \tag{15.63}$$

with the related Frobenius matrix $\mathbf{W}$–semi-norm

$$
\left[
\begin{array}{c}
\| \mathbf{E} \|_{\mathbf{I}}^2 := tr(\mathbf{E}_l^* \mathbf{E}_l) = \\[2mm]
tr\{(\mathbf{Y}_1 - \mathbf{Y}_2\mathbf{V}\mathbf{U}^* \dfrac{tr\mathbf{Y}_1^*\mathbf{C}\mathbf{Y}_2\mathbf{V}\mathbf{U}^*}{tr\mathbf{Y}_2^*\mathbf{C}\mathbf{Y}_2})^* \cdot \\[2mm]
\cdot [\mathbf{I}_n - \frac{1}{n}\mathbf{1}\mathbf{1}^*] \cdot \\[2mm]
\cdot (\mathbf{Y}_1 - \mathbf{Y}_2\mathbf{V}\mathbf{U}^* \dfrac{tr\mathbf{Y}_1^*\mathbf{C}\mathbf{Y}_2\mathbf{V}\mathbf{U}^*}{tr\mathbf{Y}_2^*\mathbf{C}\mathbf{Y}_2})\}
\end{array}
\right.
\tag{15.64}
$$

and the representative scalar measure of the error of type I-LESS

$$
\||\ \mathbf{E}_l\ \||_{\mathbf{I}} := \sqrt{tr(\mathbf{E}_l^* \mathbf{E}_l)/3n}.
$$

In the proof of Corollary 15.4, we only sketch the results that the matrix $\mathbf{I}_n - \frac{1}{n}\mathbf{1}\mathbf{1}^*$ *is idempotent:*

$$
\left[
\begin{array}{c}
(\mathbf{I}_n - \frac{1}{n}\mathbf{1}\mathbf{1}^*)(\mathbf{I}_n - \frac{1}{n}\mathbf{1}\mathbf{1}^*) = \\[2mm]
= \mathbf{I}_n - \frac{2}{n}\mathbf{1}\mathbf{1}^* + \frac{1}{n^2}(\mathbf{1}\mathbf{1}^*)^2 = \\[2mm]
= \mathbf{I}_n - \frac{2}{n}\mathbf{1}\mathbf{1}^* + \frac{1}{n^2}n\mathbf{1}\mathbf{1}^* = \mathbf{I}_n - \frac{1}{n}\mathbf{1}\mathbf{1}^*.
\end{array}
\right.
$$

Solution 15.5 (Procrustes algorithm II).

$$\boxed{Step\quad 1}$$

<u>Read</u> : $\mathbf{Y}_1 = \begin{bmatrix} x_1\ y_1\ z_1 \\ .\quad .\quad . \\ x_n\ y_n\ z_n \end{bmatrix}$ and $\begin{bmatrix} X_1\ Y_1\ Z_1 \\ .\quad .\quad . \\ X_n\ Y_n\ Z_n \end{bmatrix} = \mathbf{Y}_2$

$$\boxed{Step\quad 2}$$

<u>Compute</u> : $\mathbf{Y}_1^*\mathbf{C}\mathbf{Y}_2$ subject to $\mathbf{C} := \mathbf{I}_n - \frac{1}{n}\mathbf{1}\mathbf{1}^*$

$$\boxed{Step\quad 3}$$

<u>Compute</u> : SVD $\mathbf{Y}_1^*\mathbf{C}\mathbf{Y}_2 = \mathbf{U}Diag(\sigma_1, \sigma_2, \sigma_3)\mathbf{V}^*$

<u>3-1</u> $|\ (\mathbf{Y}_1^*\mathbf{C}\mathbf{Y}_2)^*(\mathbf{Y}_1^*\mathbf{C}\mathbf{Y}_2) - \sigma_i^2\mathbf{I}\ | = 0 \Rightarrow (\sigma_1, \sigma_2, \sigma_3)$

$$((\mathbf{Y}_1^*\mathbf{C}\mathbf{Y}_2)^*(\mathbf{Y}_1^*\mathbf{C}\mathbf{Y}_2) - \sigma_i^2\mathbf{I})\mathbf{v}_i = \mathbf{0} \,\forall i \in \{1, 2, 3\}$$

3-2
$$\Rightarrow \mathbf{V} = \begin{bmatrix} \mathbf{v}_1, \mathbf{v}_2, \mathbf{v}_3 \end{bmatrix} \quad right\,eigenvectors$$
$$(right\,eigencoloumns)$$

3-3
$$\mathbf{U} = \mathbf{Y}_1^*\mathbf{C}\mathbf{Y}_2\mathbf{V}Diag(\frac{1}{\sigma_1}, \frac{1}{\sigma_2}, \frac{1}{\sigma_3}) \; left\,eigenvectors$$
$$(left\,eigencoloumns)$$

$$\boxed{Step \quad 4}$$

Compute : $\mathbf{X}_{3l} = \mathbf{U}\mathbf{V}^*$ (rotation)

$$\boxed{Step \quad 5}$$

Compute : $x_{1l} = \dfrac{tr\,\mathbf{Y}_1^*\mathbf{C}\mathbf{Y}_2\mathbf{X}_3^*}{tr\,\mathbf{Y}_2^*\mathbf{C}\mathbf{Y}_2}$ (dilatation)

$$\boxed{Step \quad 6}$$

Compute : $\mathbf{x}_{2l} = \frac{1}{n}(\mathbf{Y}_1 - \mathbf{Y}_2\mathbf{X}_3^*x_1)^*\mathbf{1}$ (translation)

$$\boxed{Step \quad 7}$$

Compute : $\mathbf{E}_l = \mathbf{C}\left\{\mathbf{Y}_1 - \mathbf{Y}_2\mathbf{V}\mathbf{U}^*\dfrac{tr\,\mathbf{Y}_1^*\mathbf{C}\mathbf{Y}_2\mathbf{V}\mathbf{U}^*}{tr\,\mathbf{Y}_2^*\mathbf{C}\mathbf{Y}_2}\right\}$ (error

matrix)

<u>*"optional control"*</u>

$$\mathbf{E}_l := \mathbf{Y}_1 - (\mathbf{Y}_2\mathbf{X}_{3l}^*x_{1l} + \mathbf{1}\mathbf{x}_{2l}^*)$$

$$\boxed{Step \quad 8}$$

Compute : $\| \mathbf{E}_l \|_{\mathbf{I}} := \sqrt{tr(\mathbf{E}_l^*\mathbf{E}_l)}$ (error matrix)

$$\boxed{Step \quad 9}$$

Compute : $\|| \mathbf{E}_l \||_{\mathbf{I}} := \sqrt{tr(\mathbf{E}_l^*\mathbf{E}_l)/3n}$ (mean error matrix)

15.2.4 Weighted Procrustes Transformation

As already stated in Sect. 15.1, other than the problems associated with linearization and iterations, the 7-datum transformation problem (conformal group $\mathbb{C}_7(3)$) is compounded with the problem of incorporating the weights of the systems involved. This section presents Procrustes algorithm II; a reliable means of solving the problem. We have already seen that the problem could be solved using Gauss-Jacobi combinatorial algorithm in Sect. 15.2.2. Procrustes algorithm II presented in the preceding section offers therefore an alternative that is not computationally intensive as the combinatorial method.

To obtain the weight matrix used in Corollaries 15.1, 15.2 and 15.3 in the *weighted Procrustes problem*, we proceed via the variance-covariance matrix of Theorem 15.2, whose proof is given in Solution 15.6 where the dispersion matrices of two sets of coordinates in $\mathbb{R}^3$ are presented in (15.66) and (15.68). They are used in (15.70) and (15.71) to obtain the dispersion of the error matrix $\mathbf{E}$ in (15.72). In-order to simplify (15.72), we make use of Corollary 15.5 adopted from [149, Appendix A, p. 419] to express $vec\, x_1\mathbf{X}_3\mathbf{Y}_2^*$ as in (15.73) and substitute it in (15.72) to obtain (15.74). We provide as a summary the following:

Theorem 15.2 (variance-covariance matrix). *Let* $vec\, \mathbf{E}^*$ *denote the vector valued form of the transposed error matrix* $\mathbf{E} := \mathbf{Y}_1 - \mathbf{Y}_2\mathbf{X}_3^*x_1 - \mathbf{1x}_2^*$. *Then*

$$
\left[\begin{array}{c}
\boldsymbol{\Sigma}_{vec\mathbf{E}^*} = \boldsymbol{\Sigma}_{vec\,\mathbf{Y}_1^*} + (\mathbf{I}_n \otimes x_1\mathbf{X}_3)\boldsymbol{\Sigma}_{vec\,\mathbf{Y}_2^*}(\mathbf{I}_n \otimes x_1\mathbf{X}_3)^* \\
-2\boldsymbol{\Sigma}_{vec\mathbf{Y}_1^*,(\mathbf{I}_n\otimes x_1\mathbf{X}_3)vec\mathbf{Y}_2^*}
\end{array}\right. \tag{15.65}
$$

is the exact representation of the dispersion matrix (variance-covariance matrix) $\Sigma_{vec\,\mathbf{E}^*}$ *of* $vec\, \mathbf{E}^*$ *in terms of dispersion matrices (variance-covariance matrices)* $\boldsymbol{\Sigma}_{vec\,\mathbf{Y}_1^*}$ *and* $\boldsymbol{\Sigma}_{vec\,\mathbf{Y}_2^*}$ *of the two coordinates sets* $vec\,\mathbf{Y}_1^*$ *and* $vec\,\mathbf{Y}_2^*$ *as well as of their covariance matrix*

$$
\boldsymbol{\Sigma}_{vec\mathbf{Y}_1^*,(\mathbf{I}_n\otimes x_1\mathbf{X}_3)vec\mathbf{Y}_2^*}.
$$

Proof. By means of Solution 15.6 we define the dispersion matrices, also called variance-covariance matrices, of $vec\,\mathbf{Y}_1^*$ and $vec\,\mathbf{Y}_2^*$ of the two sets of coordinates.

Solution 15.6 (Dispersion matrices of two sets of coordinates in $\mathbb{R}^3$).

$$\boldsymbol{\Sigma}_{vec\,\mathbf{Y}_1^*} = E\{ \begin{bmatrix} x_1 - E\{x_1\} \\ x_2 - E\{x_2\} \\ ... \\ x_n - E\{x_n\} \end{bmatrix} \left[(x_1 - E\{x_1\})^* ... (x_n - E\{x_n\})^* \right] \}$$

(15.66)

$$E\{(vec\mathbf{Y}_1^* - E\{\mathbf{Y}_1^*\})(vec\mathbf{Y}_1^* - E\{\mathbf{Y}_1^*\})^*\}$$

(15.67)

$$\boldsymbol{\Sigma}_{vec\,\mathbf{Y}_2^*} = E\{ \begin{bmatrix} X_1 - E\{X_1\} \\ X_2 - E\{X_2\} \\ ... \\ X_n - E\{X_n\} \end{bmatrix} \left[(X_1 - E\{X_1\})^* ... (X_n - E\{X_n\})^* \right] \}$$

(15.68)

Next from the transposed error matrix

$$\mathbf{E}^* := \mathbf{Y}_1^* - (x_1\mathbf{X}_3\mathbf{Y}_2^* + \mathbf{x}_2\mathbf{1}\mathbf{x}_2^*)$$

(15.69)

we compute the dispersion matrix (variance-covariance matrix) $\boldsymbol{\Sigma}_{vec\,\mathbf{E}^*}$

$$\begin{bmatrix} \boldsymbol{\Sigma}_{vec\,\mathbf{E}^*} := E\{[vec\,\mathbf{E}^* - E\{vec\,\mathbf{E}^*\}][vec\,\mathbf{E}^* - E\{vec\,\mathbf{E}^*\}]^*\} \\ = E\{[vec\,\mathbf{Y}_1^* - E\{vec\,\mathbf{Y}_1^*\} - x_1(vec\,\mathbf{X}_3\mathbf{Y}_2^* - E\{vec\,\mathbf{X}_3\mathbf{Y}_2^*\})] \\ \times [vec\,\mathbf{Y}_1^* - E\{vec\,\mathbf{Y}_1^*\} - x_1(vec\,\mathbf{X}_3\mathbf{Y}_2^* - E\{vec\,\mathbf{X}_3\mathbf{Y}_2^*\})]^*\} \end{bmatrix}$$

(15.70)

$$\begin{bmatrix} \boldsymbol{\Sigma}_{vec\,\mathbf{E}^*} = E\{[vec\,\mathbf{Y}_1^* - E\{vec\,\mathbf{Y}_1^*\}][vec\,\mathbf{Y}_1^* - E\{vec\,\mathbf{Y}_1^*\}]^*\} \\ + x_1^2 E\{[vec\,\mathbf{X}_3\mathbf{Y}_2^* - E\{vec\,\mathbf{X}_3\mathbf{Y}_2^*\}][vec\,\mathbf{X}_3\mathbf{Y}_2^* - E\{vec\,\mathbf{X}_3\mathbf{Y}_2^*\}]^* \\ - 2x_1 E\{[vec\,\mathbf{Y}_1^* - E\{vec\,\mathbf{Y}_1^*\}][vec\,\mathbf{X}_3\mathbf{Y}_2^* - E\{vec\,\mathbf{X}_3\mathbf{Y}_2^*\}]^*\} \end{bmatrix}$$

(15.71)

$$\{ \boldsymbol{\Sigma}_{vec\,\mathbf{E}^*} = \boldsymbol{\Sigma}_{vec\,\mathbf{Y}_1^*} + \boldsymbol{\Sigma}_{vec\,x_1\mathbf{X}_3\mathbf{Y}_2^*} - 2\boldsymbol{\Sigma}_{vec\,\mathbf{Y}_1^*,vec\,x_1\mathbf{X}_3\mathbf{Y}_2^*}$$

(15.72)

Corollary 15.5.

$$\begin{bmatrix} vec\,\mathbf{AB} = (\mathbf{I}_q \otimes \mathbf{A})vec\,\mathbf{B} \, for \, all \, \mathbf{A} \in \mathbb{R}^{n \times m}, \mathbf{B} \in \mathbb{R}^{m \times q} \\ vec\,\mathbf{X}_3\mathbf{Y}_2^* = (\mathbf{I}_n \otimes x_1\mathbf{X}_3)vec\,\mathbf{Y}_2^* \, for \, all \, \mathbf{X}_3 \in \mathbb{R}^{3 \times 3}, \mathbf{Y}_2^* \in \mathbb{R}^{3 \times n} \end{bmatrix}$$

(15.73)

As soon as we implement the *Kronecker-Zehfu decomposition* of the $vec\,\mathbf{AB}$ we arrive at the general representation of the dispersion matrix $\mathbf{\Sigma}_{vec\,\mathbf{E}^*}$, namely

$$\left[\begin{array}{c} \mathbf{\Sigma}_{vec\,\mathbf{E}^*} = \mathbf{\Sigma}_{vec\,\mathbf{Y}_1^*} + (\mathbf{I}_n \otimes x_1\mathbf{X}_3)\mathbf{\Sigma}_{vec\,\mathbf{Y}_2^*}(\mathbf{I}_n \otimes x_1\mathbf{X}_3)^* - \\ -2\mathbf{\Sigma}_{vec\mathbf{Y}_1^*,(\mathbf{I}_n\otimes x_1\mathbf{X}_3)vec\mathbf{Y}_2^*} \end{array}\right. \qquad (15.74)$$

□

The results of Theorem 15.2 are interpreted in more detail as follows: The variance-covariance matrix of the vectorized error matrix $\mathbf{E}^*$ depends on;

(i) the variance-covariance matrix $\mathbf{\Sigma}_{vec\,\mathbf{Y}_1^*}$ of the local coordinate set $(x_1, y_1, z_1, \ldots, x_n, y_n, z_n)$,

(ii) the variance-covariance matrix $\mathbf{\Sigma}_{vec\,\mathbf{Y}_2^*}$ of the global coordinate set $(X_1, Y_1, Z_1, \ldots, X_n, Y_n, Z_n)$,

(iii) the covariance matrix between $vec\,\mathbf{Y}_1^*$ and $(\mathbf{I}_n \otimes x_1\mathbf{X}_3)vec\mathbf{Y}_2^*$ of the global coordinate set $vec\,\mathbf{Y}_2^*$ as well as

(iv) the nonlinearity of the parameter model on the unknowns x_1, $\mathbf{X}_3$ of type *"scale factor"* and *"rotation matrix"* coupled to $(\mathbf{I}_n\otimes x_1\mathbf{X}_3)$.

So as to take advantage of the equivalence theorem between least squares approximation and best linear uniformly unbiased estimation, e.g., [149, §3, pp. 339–340], which holds for linear Gauss-Markov model, it is tempting to identify the weight matrix $\mathbf{W}$ of W-LESS with $\mathbf{\Sigma}_{vec\,\mathbf{E}^*}^{-1}$ shrunk to a *locally isotropic error situation*. Such a shrinking procedure is outlined in Example 15.5, namely by taking in account *isotropic, but inhomogeneous criterion matrices*.

Example 15.5 (Computation of transformation parameters incorporating weights). We consider Cartesian coordinates of seven stations given in the local and global system (WGS-84) as in Tables 15.1 and 15.2 on pp. 267 and 267 respectively. Desired are the 7-datum transformation parameters; scale x_1, the translation vector $\mathbf{x}_2 \in \mathbb{R}^{3\times1}$ and the rotation matrix $\mathbf{R} \in \mathbb{R}^{3\times3}$. In addition to these 7 datum transformation parameters, we compute for control purposes the residual (error matrix) $\mathbf{E}$ upon which the mean error norm (15.55) is determined as a scalar measure of error of types W-LESS. A two step procedure is carried out as follows:

In the first step, we computed the 7 transformation parameters using I-LESS (with weight matrix as identity) from Corollary 15.4 on p. 283. The computed values of scale x_1 and the rotation matrix $\mathbf{R} \in \mathbb{R}^{3\times3}$

are used in (15.74) to obtain the dispersion of the error matrix $\mathbf{E}$. In-order to obtain the dispersions $\Sigma_{vec\mathbf{Y}_1^*}$ and $\Sigma_{vec\mathbf{Y}_1^*}$ of the pseudo-observations in the local and global systems respectively, we make use of "positional error sphere" for each point (position) in both systems. Here, the positional error sphere refers to the average of the variances $(\sigma_i^2 = \sqrt{(\sigma_x^2 + \sigma_y^2 + \sigma_z^2)/3})$ for the $i = 7$ points involved so as to achieve the isotropic condition.

The identity matrices are multiplied by these positional error spheres so as to obtain the dispersion matrices $\Sigma_{vec\mathbf{Y}_1^*}$ and $\Sigma_{vec\mathbf{Y}_2^*}$ which fulfill the isotropic condition. One obtains therefore the dispersion matrices $\Sigma_{vec\mathbf{Y}_1^*}$ and $\Sigma_{vec\mathbf{Y}_2^*}$ as being diagonal block matrices with each block corresponding to the variance-covariance matrices of the respective position. For points 1 and 2 in the local system for instance, assuming no correlation between the two points, one obtains

$$
\Sigma_{vec\mathbf{Y}_1^*} = \begin{bmatrix} \sigma_1^2 & & & & & \\ & \sigma_1^2 & & & & \\ & & \sigma_1^2 & & & \\ & & & \sigma_2^2 & & \\ & & & & \sigma_2^2 & \\ & & & & & \sigma_2^2 \end{bmatrix}, \tag{15.75}
$$

where $\{\sigma_1^2, \sigma_2^2\}$ are positional error spheres for points 1 and 2 respectively. This is also performed for $\Sigma_{vec\mathbf{Y}_2^*}$ and the resulting dispersion matrices used in (15.74) to obtain the dispersion matrix of the error matrix $\mathbf{E}$.

Since the obtained block diagonal error matrix $\mathbf{E}$ is a $3n \times 3n$ matrix, the $n \times n$ matrix is extracted by taking the trace of the block diagonal matrices of $\mathbf{E}$. Adopting such a matrix from [138, Table 7] as

$$
\mathbf{W} = \begin{bmatrix} 1.8110817 & 0 & 0 & 0 & 0 & 0 & 0 \\ 0 & 2.1843373 & 0 & 0 & 0 & 0 & 0 \\ 0 & 0 & 2.1145291 & 0 & 0 & 0 & 0 \\ 0 & 0 & 0 & 1.9918578 & 0 & 0 & 0 \\ 0 & 0 & 0 & 0 & 2.6288452 & 0 & 0 \\ 0 & 0 & 0 & 0 & 0 & 2.1642460 & 0 \\ 0 & 0 & 0 & 0 & 0 & 0 & 2.359370 \end{bmatrix},
$$
$$
\tag{15.76}
$$

In Tables. 15.9 and 15.10, the results of I-LESS (step 1) and W-LESS (step 2) Procrustes transformation are presented. Presented are the 7-datum transformation parameters namely; the scale, rotation matrix and the translation parameters. We also present the residual (error) matrix and the norms of the error matrices. The computed residuals can be compared with those of linearized least squares procedure in Table 15.7 on p. 271.

Table 15.9. Results of the I-LESS Procrustes transformation

	Values			
Rotation Matrix $\mathbf{X}_3 \in \mathbb{R}^{3 \times 3}$	1.00000 $\quad$ -4.33276e-6 $\quad$ 4.81463e-6 $\quad$ -4.81465e-6 $\quad$ 1.00000 $\quad$ -4.84085e-6 $\quad$ 4.33274e-6 $\quad$ 4.84087e-6 $\quad$ 1.00000			
Translation $\mathbf{x}_2 \in \mathbb{R}^{3 \times 1} (m)$	641.8804 $\quad$ 68.6553 $\quad$ 416.3982			
Scale $x_1 \in \mathbb{R}$	1.00000558251985			
Residual matrix $\mathbf{E}(m)$	Site	$X(m)$	$Y(m)$	$Z(m)$
	Solitude	0.0940	0.1351	0.1402
	Buoch Zeil	0.0588	-0.0497	0.0137
	Hohenneuffen	-0.0399	-0.0879	-0.0081
	Kuelenberg	0.0202	-0.0220	-0.0874
	Ex Mergelaec	-0.0919	0.0139	-0.0055
	Ex Hof Asperg	-0.0118	0.0065	-0.0546
	Ex Keisersbach	-0.0294	0.0041	0.0017
Error matrix norm (m) $\||\ E_l\ \||_{\mathbf{W}} := \sqrt{tr(\mathbf{E}_l^* \mathbf{E}_l)}$	0.2890			
Mean error matrix norm (m) $\||\ E_l\ \||_{\mathbf{W}} := \sqrt{tr(\mathbf{E}_l^* \mathbf{E}_l)/3n}$	0.0631			

Table 15.10. Results of the W-LESS Procrustes transformation

	Values			
Rotation Matrix $\mathbf{X}_3 \in \mathbb{R}^{3 \times 3}$	1.00000 4.77976e-6 -4.34410e-6 -4.77978e-6 1.00000 -4.83730e-6 4.34408e-6 4.83731e-6 1.00000			
Translation $\mathbf{x}_2 \in \mathbb{R}^{3 \times 1}(m)$	641.8377 68.4743 416.2159			
Scale $x_1 \in \mathbb{R}$	1.00000561120732			
Residual matrix $\mathbf{E}(m)$	Site	$X(m)$	$Y(m)$	$Z(m)$
	Solitude	0.0948	0.1352	0.1407
	Buoch Zeil	0.0608	-0.0500	0.0143
	Hohenneuffen	-0.0388	-0.0891	-0.0072
	Kuelenberg	0.0195	-0.0219	-0.0868
	Ex Mergelaec	-0.0900	0.0144	-0.0052
	Ex Hof Asperg	-0.0105	0.0067	-0.0542
	Ex Keisersbach	-0.0266	0.0036	0.0022
Error matrix norm *(m)* $\lVert\lvert E_l \rvert\rVert_{\mathbf{w}} := \sqrt{tr(\mathbf{E}_l^* \mathbf{W} \mathbf{E}_l)}$	0.4268			
Mean error matrix norm *(m)* $\lVert\lvert E_l \rvert\rVert_{\mathbf{w}} := \sqrt{tr(\mathbf{E}_l^* \mathbf{W} \mathbf{E}_l)/3n}$	0.0930			

15.3 Concluding Remark

The chapter has illustrated how the algebraic technique of Groebner basis explicitly solves the nonlinear 7-parameter datum transformation equations once they have been converted into algebraic (polynomial) form. In particular, the algebraic tool of Groebner basis provides symbolic solutions; showing the scale parameter to fulfill a quartic polynomial and the rotation parameters are given by linear functions in scale. It has also been demonstrated how overdetermined version of the problem can be solved using the Gauss-Jacobi combinatorial algorithm and the general Procrustes algorithm. Both approaches incorporate the stochasticity of both systems involved in the transformation problem.

Although computationally intensive, the Gauss-Jacobi combinatorial algorithm solves the weighted transformation problem without any assumption. The general Procrustes algorithm functions well with the isotropic assumptions, i.e., all the three coordinates $\{X_i, Y_i, Z_i\}$ of a point i are given the same weight. The weights are further assumed to be inhomogeneous, i.e., the weights of a point i differ from those of the point j. Both these assumptions are ideal and may not necessarily hold in practice. The subject of transformation in general is still an active area of research as evidenced in the works of [2, 3, 6, 11, 13, 20, 23, 34,

35, 45, 50, 52, 62, 66, 106, 137, 138, 143, 153, 154, 155, 158, 159, 169, 179, 196, 204, 206, 258, 259, 260, 309, 319, 329, 339, 347].

Computer Algebra Systems (CAS)

16.1 General and Special Purpose CAS

In high school algebra, one learnt how to perform long division of univariate polynomials such as the one we illustrated in Example 4.7 on p. 36. This is just but one example of operations that do not involve numerical manipulation. Expanding and factoring of polynomials, solution of ordinary differential equations and integration without limits are some non numeric problems that require hand computation; which had often required a paper and a pen. Some derivations and computations were however long and labour intensive such that one gave up mid way. With the advent of computers, the paper and pen approach is slowly being replaced by software developed to undertake these tasks. Specifically, where symbolic rather than numerical solutions are desired, these software normally come in handy. Software which perform symbolic computations are called *Computer Algebra System* (CAS). Computer algebra system made its appearance in early 1970s as a product of research in artificial intelligence.

The symbolic mathematical software that form computer algebra system are normally divided roughly into two: These are the general and special purpose computer algebra systems. The general purpose computer algebra systems consists of software such as MATHEMATICA, MAPLE, AXIOM, MACSYMA, REDUCE and MAGMA etc. They often solve broad range of problems and contain in-built functions to solve several tasks such as trigonometry, square roots etc. They are widely used by scientists and engineers and offer nice user interface. They have the capability to plot graphs (see e.g., Fig. 16.2) and manipulate data. Special purpose computer algebra system on the contrary are tailored towards specific areas of computer algebra. In most cases, they are developed by specific research groups with the users being limited.

In this chapter, the general purpose computer algebra systems are discussed. We examine possible areas where they can be of use in geodesy and geoinformatics and presents references where further materials could be obtained. Indeed, computer algebra system has widely been used for:

(a) Manipulating symbolic and numerical computations. In this regard for instance, one would be interested in explicit solutions of functions. In the preceding chapters, we have actually employed the computer algebra system (e.g., MATHEMATICA) to provide explicit solutions to various problems. Besides the provision of explicit formulae, CAS also find use in expanding and factoring algebraic expressions.

(b) Provision of graphical visualization. In vehicle production industry for example, computer algebra systems are used to design vehicles. Various graphs of vehicles are drawn from which experts analyze shapes in order to obtain best possible products.

(c) Modelling engineering curves during constructions of roads, rails etc.

16.2 Some CAS Software Useful in Geodesy and Geoinformatics

Several general purpose computer algebra software exist that maybe of use. In this section, four software are discussed namely: MATLAB, MAPLE, MATHEMATICA and REDUCE.

16.2.1 MATLAB

Perhaps in the list of the four above, MATLAB appears to be the odd one out. This is because in most computer algebra textbooks, it is not listed as one of the computer algebra software. It however performs functions such as factoring of polynomials, symbolic and numerical operations. Since these operations are performed also by computer algebra software such as MATHEMATICA and MAPLE, we find it useful to include MATLAB in this group.

MATLAB is essentially based on linear algebra, operating best with vectors and matrices. Whereas in Fortran programming, one has to define the sizes of vectors and matrices, and declare whether variables are real or integers, MATLAB does not have such requirements. Most functions and subroutines are in-built. One therefore does not need to call subroutines during execution of files as does Fortran. It provides nice user interface and has the capability to plot and manipulate graphs (see e.g., Fig. 16.1). Besides a nice user interface, it also has an editor where the user can write programmes. Both online as well as textbook help are provided to the user. There exists commercial as well as students' version of MATLAB, e.g., [174]. It can manipulate both numerical as well as symbolic computations. MATLAB has several tool boxes for different tasks. Some of them, which may be of use, include:

- Curve Fitting Toolbox.
- Data Acquisition Toolbox.
- Filter Design Toolbox.
- Image Processing Toolbox.
- Mapping Toolbox.
- Optimization Toolbox.
- Differential Equation Toolbox.
- Signal Processing Toolbox.
- Spline Toolbox.
- Statistics Toolbox.
- Symbolic Math Toolbox.
- Wavelet Toolbox.

Several versions of MATLAB exist, with the latest (2004) being version 6.5. Most examples presented in this book and figures were prepared using versions 5.2 and 6.5. Some of the toolboxes listed above could be used for:

1. Mapping. The mapping toolbox offers nice facilities for plotting in different coordinate systems and different map projections. Figure 16.1 for example is a MATLAB graphic interface viewmaps used to display various map projections. In the figure, we have directed MATLAB to display equal area cylindrical map projection on the second box, top right corner. One can further constrain what one wants MATLAB to perform. For example, if one is interested in displaying an equal area cylindrical map projection with only topography and continents, the boxes for these features are checked.

The resulting figure, when the apply button is pressed, is displayed on the left of Fig. 16.1. One can compare several types of map projections and play around with the origin and orientation. Several other tasks can be performed using the mapping toolbox.

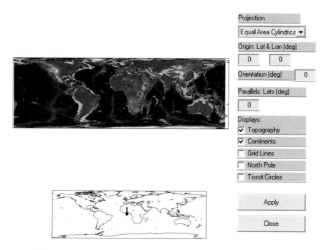

Fig. 16.1. Interactive graphic windows in matlab

2. Symbolic computations of problems in calculus and algebra. In the Examples that follow, we demonstrate how MATLAB 6.5 solves the problems of numeric as well as symbolic nature.

Example 16.1 (MATLAB's symbolic computation). Compute the symbolic solution of the following function

$$f(x) := \int x^6 dx. \tag{16.1}$$

One proceeds as follows: In the MATLAB windows, the symbolic toolbox is activated by declaring the variables x, y as symbols using the command
$>>$ syms x y.
The solution of (16.1) then proceeds by first entering the function to be solved in MATLAB's user interface as

$$>> f = x^\wedge 6,$$

then using "int" function as
$>>$ solution1=int(f,x).

The "int" function is a MATLAB in-built function for integration. In the execution above, MATLAB is told to integrate the function f with respect to x. One can learn more about the command "int" by accessing MATLAB's online help or by simply typing $>> help$ *int* on MATLAB window. The resulting solution is
solution1 $= 1/7 * x^{\wedge}7$,
which does not look good. The "pretty" command is used to make the solution more presentable, e.g.,
$>>$ pretty(solution1)
leading to $1/7x^7$

Example 16.2 (MATLAB's numerical computation). Let us consider a case where the limits to Example 16.1 are given as

$$g(x) = \int_1^5 x^6 dx. \tag{16.2}$$

One proceeds to solve 16.2 by writing
$>>$ solution2=int(x,1,5),
which tells MATLAB to integrate the function g with respect to x from a lower limit of 1 to an upper limit of 5. On pressing the enter/return key, the answer is given as 12.

Example 16.3 (Interaction with MATLAB). In this example, we illustrate how MATLAB responds when there is an error in the entered expressions. Consider a case where we have requested MATLAB to find the partial derivative of the function

$$h(x, y) = x^2 + 6xy + x^3 y^3, \tag{16.3}$$

with respect to x. In trying to enter (16.3) into MATLAB, let us assume that one typed the expression erroneously in MATLAB as

$$>> h = (x^{\wedge}2 + 6x * y + x^{\wedge}3 * y^{\wedge}3).$$

After pressing the return key, MATLAB will return the following error message

$$???f2 = (x^{\wedge}2 + 6x * y + x^{\wedge}3 * y^{\wedge}3).$$

Error: ")" expected, "identifier" found.
In the expression above MATLAB noticed that between 6 and x, a

multiplication sign {*} was missing and reported an error message. The correct expression is then entered and MATLAB returns

$$>> h = (x^\wedge 2 + 6 * x * y + x^\wedge 3 * y^\wedge 3).$$

MATLAB is then instructed to find the partial derivative of h with respect to x using "diff" command as
$>>$ solution3=diff(h,x)
which leads to
$>>$ solution3 $= 2 * x + 6 * y + 3 * x^\wedge 2 * y^\wedge 3.$

3. Faster computation of problems that are not labour intensive such as those illustrated in the preceding chapters. Although MATLAB can be used for geoid computations, other users have complained that it is too slow compared to Fortran.
4. Plotting and manipulating graphs. It permits graphs to be produced in encapsulated postscript (eps) format that can easily be placed in text editors such as Microsoft word or latex.
5. Statistical analysis of data. In Chap. 14 for example, the median used was obtained using MATLAB's "median" command. For a vector **X**, median(X) is the median value of the elements in X.

More information on MATLAB can be found online from its web page[1] and also in [174].

16.2.2 MAPLE

There are several versions of MAPLE with the latest (2004) being MAPLE 9.5. MAPLE software can be run on desktop windows' systems and also SUN systems. All the versions of MAPLE software use the same kernel for computing. The difference between the versions appear only in the user interface. Information on MAPLE software can be obtained from its web page[2]. With this software, users have an interactive system that displays input, output, text and graphics on the same worksheet. MAPLE further offers the possibilities to:

- Solve symbolic computations. Problems that require expansion, factorization and solving of systems of equations can be undertaken using MAPLE.

---

[1]http://www.mathworks.com/
[2]http://www.maplesoft.com/

- Manipulate calculus. Problems requiring differentiation, integration, series expansion and limits can be performed.
- Solve differential equations.

16.2.3 MATHEMATICA

MATHEMATICA software exist in several versions with the latest (2004) being MATHEMATICA 5. These versions have the same computing kernel and differ only in the user interface. Both symbolic and numerical computations can be performed using this software. When it is started, an empty notebook appears which allows the user to enter input to be evaluated. When the input is typed, a blue bracket at the right-hand-side appears. This is the cell to which the input is placed. By pressing shift-enter keys, the input is executed. Similar to MATLAB and MAPLE, it also offers an online help. MATHEMATICA information can be obtained from its web page[3]. The following capabilities are offered to users:

- Integration of numerical and symbolic problems.
- Graphics system (e.g., Example 16.4).

  Example 16.4 (Using MATHEMATICA 3.0 interface). Let as consider a simple case where we are required to plot a graph of $Cos(xy)$ for $x = 0, \pi$ and $y = 0, \pi$. This is entered in MATHEMATICA interface as:
  in[1]:=Plot3D[Cos[x y],{x,0,Pi},{y,0,Pi}],
  which leads to Fig. 16.2.

- Programming language. The software allows users to write their own programs.
- It can be used for calculus computations as is the case with MAPLE.
- Documentation system.
- Advanced connectivity to other applications. For instance, it permits one to save documents in latex format. Figure (16.2) for example was plotted in MATHEMATICA 3.0 and saved as a latex file which was easily integrated to this text.

16.2.4 REDUCE

This is an interactive computer algebra software that manipulates problems of algebraic nature. There are several versions of this software with

---

[3]http://www.wolfram.com/

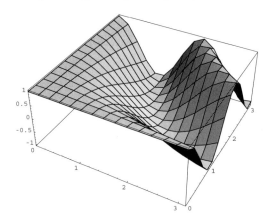

Fig. 16.2. MATHEMATICA plot of $Cos(xy)$

the latest (2004) being REDUCE 3.8, released on 15th April 2004. Its Information can be obtained online from the web[4]. Similar to modern calculators which have replaced hand calculations, the software can be seen as an algebraic hand calculator which could be used for:

- Expanding, factorizing and ordering polynomials and rational functions.
- Substitutions and pattern matching in a wide variety of forms
- Simplification of expressions.
- Symbolic calculations, e.g., solving algebraic equations.
- Arbitrary precision integer and real arithmetic.
- Defining new functions and extending program syntax.
- Solving analytically differentiation and integration problems.
- Manipulating expressions in variety of formats.
- Optimizing numerical programs from symbolic input.

16.3 Concluding Remarks

The chapter has only presented an overview of computer algebra systems. More details can be obtained e.g., in [178, 292, 324, 334]: In the book of [334] for instance, it is pointed out that some systems could give

---

[4]http://www.uni-koeln.de/REDUCE/

wrong answers. This finding is thus an eye-opener to users who tend to believe that every computer algebra result is truth. The book further discusses a list of computer algebra systems e.g., MathCAD and their resources. Harper [178] also provides a thorough comparison of five computer algebra systems namely: REDUCE, MACSYMA, MAPLE, MATHEMATICA and DRIVE. Their advantages and disadvantages are documented therein.

We conclude this chapter by advocating the use of computer algebra systems. They may make a difference between geodesists and geoinformatists sticking to the paper and pencil snail phase approach or replacing it!

Appendix

Appendix A-1: Definitions

To enhance the understanding of the theory of Groebner bases presented in Chap. 4, the following definitions supplemented with examples are presented.

Three commonly used monomial ordering systems are; *lexicographic ordering, graded lexicographic ordering* and *graded reverse lexicographic ordering*. First we define the monomial ordering before considering the three types.

Definition .1 (Monomial ordering). *A monomial ordering on $k[x_1, \ldots, x_n]$ is any relation $>$ on $\mathbb{Z}_{\geq 0}^n$ or equivalently any relation on the set x^α, $\alpha \in \mathbb{Z}_{\geq 0}^n$ satisfying the following conditions:*
(a) is total (or linear) ordering on $\mathbb{Z}_{\geq 0}^n$
(b) If $\alpha > \beta$ and $\gamma \in \mathbb{Z}_{\geq 0}^n$, then $\alpha + \gamma > \beta + \gamma$
(c) $<$ is a well ordering on $\mathbb{Z}_{\geq 0}^n$.
This condition is satisfied if and only if every strictly decreasing sequence in $\mathbb{Z}_{\geq 0}^n$ eventually terminates.

Definition .2 (Lexicographic ordering). *This is akin to the ordering of words used in dictionaries. If we define a polynomial in three variables as $P = k[x, y, z]$ and specify an ordering $x > y > z$, i.e., x comes before y and y comes before z, then any term with x will supersede that of y which in tern supersedes that of z. If the powers of the variables for respective monomials are given as $\alpha = (\alpha_1, \ldots, \alpha_n)$ and $\beta = (\beta_1, \ldots, \beta_n), \alpha, \beta \in \mathbb{Z}_{\geq 0}^n$, then $\alpha >_{lex} \beta$ if in the vector difference $\alpha - \beta \in \mathbb{Z}^n$, the most left non-zero entry is positive. For the same variable (e.g., x) this subsequently means $x^\alpha >_{lex} x^\beta$.*

Example .1. $x > y^5z^9$ is an example of lexicographic ordering. As a second example, consider the polynomial $f = 2x^2y^8 - 3x^5yz^4 + xyz^3 - xy^4$, we have the *lexicographic order*; $f = -3x^5yz^4 + 2x^2y^8 - xy^4 + xyz^3 \,|x > y > z$.

Definition .3 (Graded lexicographic ordering). *In this case, the total degree of the monomials is taken into account. First, one considers which monomial has the highest total degree before looking at the lexicographic ordering. This ordering looks at the left most (or largest) variable of a monomial and favours the largest power. Let* $\alpha, \beta \in \mathbb{Z}_{\geq 0}^n$,

then $\alpha >_{grlex} \beta$ *if* $|\alpha| = \sum\limits_{i=1}^n \alpha_i > |\beta| = \sum\limits_{i=1}^n \beta_i$ *or* $|\alpha| = |\beta|$, *and* $\alpha >_{lex} \beta$,

in $\alpha - \beta \in \mathbb{Z}^n$, *the most left non zero entry is positive.*

Example .2. $x^8y^3z^2 >_{grlex} x^6y^2z^3 \,|(8,3,2) >_{grlex} (6,2,3)$, since $|(8,3,2)| = 13 > |(6,2,3)| = 11$ and $\alpha - \beta = (2,1,-1)$. Since the left most term of the difference (2) is positive, the ordering is graded lexicographic. As a second example, consider the polynomial $f = 2x^2y^8 - 3x^5yz^4 + xyz^3 - xy^4$, we have the graded lexicographic order; $f = -3x^5yz^4 + 2x^2y^8 - xy^4 + xyz^3 \,|x > y > z.$

Definition .4 (Graded reverse lexicographic ordering). *In this case, the total degree of the monomials is taken into account as in the case of graded lexicographic ordering. First, one considers which monomial has the highest total degree before looking at the lexicographic ordering. In contrast to the graded lexicographic ordering, one looks at the right most (or largest) variable of a monomial and favours the smallest power. Let* $\alpha, \beta \in \mathbb{Z}_{\geq 0}^n$, *then* $\alpha >_{grevlex} \beta$ *if* $|\alpha| = \sum\limits_{i=1}^n \alpha_i >$

$|\beta| = \sum\limits_{i=1}^n \beta_i$ *or* $|\alpha| = |\beta|$, *and* $\alpha >_{grevlex} \beta$, *and in* $\alpha - \beta \in \mathbb{Z}^n$ *the right most non zero entry is negative.*

Example .3. $x^8y^3z^2 >_{grevlex} x^6y^2z^3 \,|(8,3,2) >_{grevlex} (6,2,3)$ since $|(8,3,2)| = 13 > |(6,2,3)| = 11$ and $\alpha - \beta = (2,1,-1)$. Since the right most term of the difference (-1) is negative, the ordering is *graded reverse lexicographic*. As a second example, consider the polynomial $f = 2x^2y^8 - 3x^5yz^4 + xyz^3 - xy^4$, we have the *graded reverse lexicographic order*: $f = 2x^2y^8 - 3x^5yz^4 - xy^4 + xyz^3 \,|x > y > z.$

If we consider a non-zero polynomial $f = \sum_\alpha a_\alpha x^\alpha$ in $k\,[x_1, \ldots, x_n]$ and fix the monomial order, the following additional terms can be defined:

Definition .5.

Multidegree of f : Multideg (f)=$\max(\alpha \in \mathbb{Z}_{\geq 0}^n \,|\, a_\alpha \neq 0)$

Leading Coefficient of f : LC (f)=$a_{\text{multideg}(f)} \in k$

Leading Monomial of f : LM (f)=$x^{\text{multideg}(f)}$ *(with coefficient 1)*

Leading Term of f : LT (f)=LC (f) LM (f)

Example .4. Consider the polynomial $f = 2x^2y^8 - 3x^5yz^4 + xyz^3 - xy^4$
with respect to lexicographic order $\{x > y > z\}$, we have

Multideg $(f)=(5,1,4)$

LC $(f)=$ -3

LM $(f)= x^5yz^4$

LT $(f)= -3x^5yz^4$

The definitions of polynomial ordering above have been adopted from [94, pp. 52–58].

Appendix A-2: C. F. Gauss Combinatorial Formulation

CARL FRIEDRICH GAUSS

WERKE

NEUNTER BAND.

HERAUSGEGEBEN

VON DER

KÖNIGLICHEN GESELLSCHAFT DER WISSENSCHAFTEN

ZU

GÖTTINGEN.

IN COMMISSION BEI B. G. TEUBNER IN LEIPZIG.

1903.

BESTIMMUNG

DES

BREITENUNTERSCHIEDES

ZWISCHEN DEN

STERNWARTEN VON GÖTTINGEN UND ALTONA

DURCH

BEOBACHTUNGEN AM RAMSDENSCHEN ZENITHSECTOR

VON

CARL FRIEDRICH GAUSS,

RITTER DES GUELPHEN- UND DANNEBROG-ORDENS; K. GROSSBR. HANNOVERSCHEM HOFRATH;
PROFESSOR DER ASTRONOMIE UND DIRECTOR DER STERNWARTE IN GÖTTINGEN;
MITGLIED DER AKADEMIEN UND SOCIETÄTEN VON BERLIN, COPENHAGEN, EDINBURG, GÖTTINGEN,
LONDON, MÜNCHEN, NEAPEL, PARIS, PETERSBURG, STOCKHOLM,
DER AMERIKANISCHEN, ITALIENISCHEN, KURLÄNDISCHEN, LONDONER ASTRONOMISCHEN U. A.

GÖTTINGEN,

BEI VANDENHOECK UND RUPRECHT.

1828.

NACHLASS.

[1.]

Endresultat für den Ort eines Punktes in einer Ebene, der von drei bekannten aus angeschnitten ist.

Es bedeuten 10, 20, 30 die drei beobachteten Richtungen [nach P] und a, β, γ die entsprechenden Entfernungen.

Die drei einzelnen Resultate aus den Combinationen 2—3, 1—3, 1—2 seien A, B, C, zugleich die Winkel des durch jene gebildeten Dreiecks; die ihnen gegenüber stehenden Seiten a, b, c.

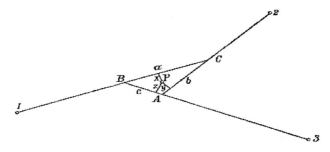

Perpendikel von dem gesuchten Orte auf a, b, c seien x, y, z. S doppelter Flächeninhalt des Dreiecks.

Es sind dann

$$\frac{x}{\alpha}, \quad \frac{y}{\beta}, \quad \frac{z}{\gamma}$$

die übrig bleibenden Fehler, also

$$\frac{xx}{\alpha\alpha} + \frac{yy}{\beta\beta} + \frac{zz}{\gamma\gamma} \quad \text{Minimum}$$

und

$$ax + by + cz = S.$$

Also werden x, y, z proportional den Grössen $\alpha\alpha a$, $\beta\beta b$, $\gamma\gamma c$;

$$x = \frac{\alpha\alpha a S}{\alpha\alpha a a + \beta\beta b b + \gamma\gamma c c},$$

etc.

[Bezeichnet (ABC) die Fläche des Dreiecks ABC, u. s. f., so ist

$$S = 2(ABC) = (\alpha\alpha a a + \beta\beta b b + \gamma\gamma c c)k$$
$$2(BPC) = \alpha\alpha a a k$$
$$2(APC) = \beta\beta b b k$$
$$2(APB) = \gamma\gamma c c k,$$

wo k die Correlate der Bedingungsgleichung ist. P ist der durch die Perpendikel x, y, z bestimmte Punkt.

Folglich wird, wenn A, B, C, P die complexen Grössen bedeuten, denen die Eckpunkte des Dreiecks ABC und der Punkt P entsprechen:

$$(\alpha\alpha a a + \beta\beta b b + \gamma\gamma c c)P = \alpha\alpha a a A + \beta\beta b b B + \gamma\gamma c c C.]$$

Es folgt hieraus, dass das Endresultat*)

$$\frac{\alpha\alpha a a A + \beta\beta b b B + \gamma\gamma c c C}{\alpha\alpha a a + \beta\beta b b + \gamma\gamma c c}$$

also ein Mittel aus den drei particllen Resultaten A, B, C ist, indem man diesen die Gewichte

$$\alpha\alpha a a, \qquad \beta\beta b b, \qquad \gamma\gamma c c$$

beilegt, oder

$$\alpha\alpha \sin A^2, \qquad \beta\beta \sin B^2, \qquad \gamma\gamma \sin C^2.$$

Offenbar ist hier A zugleich der Winkel zwischen 20 und 30, u. s. f.

307

References

1. Abel JS, Chaffee JW (1991) Existence and uniqueness of GPS solutions. IEEE Transactions on Aerospace and Electronic Systems 27: 952–956
2. Abd-Elmotaal H, El-Tokhey M (1995) Effect of spherical approximation on datum transformation. Manuscripta Geodaetica 20: 469–474
3. Abusali PAM, Schutz BE, Tapley BD (1994) Transformation between SLR/VLBI and WGS-84 frames. Bulletin Geodesique 69: 61–72
4. Aduol FWO (1987) Detection of outliers in geodetic networks using principal component analysis and bias parameter estimation. Institute of Geodesy, University of Stuttgart, Technical Report No. 2, Stuttgart
5. Aduol FWO (1994) Robust geodetic parameter estimation through iterative weighting. Survey Review 32: 359–367
6. Aduol FWO, Gacoki TG (2002) Transformation between GPS coordinates and local plane UTM coordinates using the excel spreadsheet. Survey Review 36: 449–462
7. Aduol FWO, Schaffrin B (1986) On outlier identification in geodetic networks using principal component analysis. Conference on Influential Data Analysis, University of Sheffield
8. Ansermet A (1910) Eine Auflösung des Rückwärtseinschneidens. Zeitschrift des Vereins Schweiz. Konkordatsgeometer, Jahrgang 8, pp. 88–91
9. Anthes R (2003): The Constellation Observing system for Meteorology Ionosphere and Climate (COSMIC). International Workshop on GPS Meteorology, 14th-17th January 2003, Tsukuba, Japan
10. Anthes RA (2004) Application of GPS Remote Sensing to Meteorology and Related Fields, Journal of Meteorological Society of Japan, Vol. 82, No. 1B
11. Awange JL (2002a) Groebner bases, multipolynomial resultants and the Gauss-Jacobi combinatorial algorithms-adjustment of nonlinear GPS/LPS observations. Ph.D. thesis, Department of Geodesy and GeoInformatics, Stuttgart University, Germany. Technical Reports, Report Nr. 2002 (1)
12. Awange JL (2002b) Groebner basis solution of planar resection. Survey Review 36: 528–543
13. Awange JL (2003a): Partial procrustes solution of the threedimensional orientation problem from GPS/LPS observations. In: Grafarend EW, Krumm FW, Schwarze VS (eds) Geodesy - the Challenge of the 3rd Millennium. Springer, Heidelberg pp.277–286

14. Awange JL (2003b) Buchberger algorithm applied to planar lateration and intersection problems. Survey Review 37: 319–329

15. Awange JL (2004): Diagnosis of Outlier of type Multipath in GPS Pseudo-range observations. Survey Review in press

16. Awange JL, Aduol FWO (1999) An evaluation of some robust estimation techniques in the estimation of geodetic parameters. Survey Review 35: 146–162

17. Awange JL, Aduol FWO (2002) An evaluation of some robust estimation techniques in the estimation of geodetic parameters-part II. Survey Review 36: 380–389

18. Awange JL, Fukuda Y (2003) On possible use of GPS-LEO satellite for flood forecasting. Accepted to the International Civil Engineering Conference on Sustainable Development in the 21st Century "The Civil Engineer in Development" 12 - 16 August 2003 Nairobi, Kenya

19. Awange JL, Grafarend EW (2002a) Sylvester resultant solution of planar ranging problem. Allgemeine Vermessungs-Nachrichten 108: 143–146

20. Awange JL, Grafarend EW (2002b) Linearized least squares and nonlinear Gauss-Jacobi combinatorial algorithm applied to 7-parameter datum transformation. Zeitschrift für Vermessungswessen 127: 109–117

21. Awange JL, Grafarend EW (2002c) Algebraic solution of GPS pseudo-ranging equations. Journal of GPS Solutions 4: 20–32

22. Awange JL, Grafarend EW (2002d) Nonlinear adjustment of GPS observations of type pseudo-range. Journal of GPS Solutions 4: 80–93

23. Awange JL, Grafarend EW (2003a) Closed form solution of the overdetermined nonlinear 7-parameter datum transformation. Allgemeine Vermessungs Nachrichtern 109: 130–148

24. Awange JL, Grafarend EW (2003b) Groebner basis solution of the three-dimensional resection problem (P4P). Journal of Geodesy, 77: 327–337

25. Awange JL, Grafarend EW (2003c) Multipolynomial resultant solution of the threedimensional resection problem (P4P). Bollettino di Geodesia e Science Affini 62: 79–102

26. Awange JL, Grafarend EW (2003d) Explicit solution of the overdetermined three-dimensional resection problem. Journal of Geodesy 76: 605–616

27. Awange JL, Grafarend EW (in press) From space angles to point position using Sylvester resultant. Allgemeine Vermessungs-Nachrichten, in press

28. Awange JL, Fukuda Y, Takemoto S, Grafarend EW (2003a) Direct Polynomial approach to nonlinear distance (ranging) problems. Earth, Planets and Space 55: 231–241

29. Awange JL, Fukuda Y, Takemoto S, Ateya I, Grafarend EW (2003b): Ranging algebraically with more observations than unknowns. Earth, Planets and Space 55 (2003) 387-394.

30. Awange JL, Grafarend EW, Fukuda Y (2003c) Closed form solution of the triple three-dimensional intersection problem. Zeitschrift für Geodaesie, Geoinfornation und Landmanagement 128: 395–402

31. Awange JL, Fukuda Y, Takemoto S, Grafarend EW (2003d) Resultants approach to the triple three-dimensional intersection problem. Journal of Geodetic Society of Japan 49: 243–256

32. Awange JL, Fukuda Y, Takemoto S (2004a) B. Strumfel's resultant solution of planar resection problem. Allgemeine Vermessungs-Nachrichten 111(6): 214–219

310

33. Awange JL, Fukuda Y, Takemoto S, Wickert J, Aoyama Y (2004) Analytic solution of GPS atmospheric sounding refraction angles. Earth, Planets and Space 56: 573–587

34. Awange JL, Grafarend EW, Fukuda Y (in press a) Exact solution of the nonlinear 7-parameter datum transformation by Groebner basis. Bollettino di Geodesia e Scienze Affini, to appear on the issue no. 2/2004

35. Awange JL, Grafarend EW, Fukuda Y, Takemoto S (in press b) The role of algebra in Modern day Geodesy. In press Springer

36. Awange JL, Grafarend EW, Fukuda Y, Takemoto S (in press d) Application of commutative algebra to Geodesy. Journal of Geodesy, in press

37. Awange JL, Grafarend EW, Fukuda Y (in press e): A combinatorial scatter approach to the overdetermined three-dimensional intersection problem. Bollettino di Geodesia e Scienze Affini, in press to appear on the issue no.3/2004 or no.4/2004

38. Baarda W (1967a) A generalization of the concept strength of the figure. Publications on Geodesy, New Series 2(4), Delft

39. Baarda W (1967b) Statistical concepts in geodesy. The Netherlands geodetic commission, Publication in geodesy, New Series 2, No. 4 , Delft

40. Baarda W (1968a) Statistics - A compass for the land surveyor. Computing centre of the Delft Geodetic Institute, Delft

41. Baarda W (1968b) A testing procedure for use in geodetic networks. Publication in geodesy, New Series 2, No. 5 , Delft

42. Baarda W (1973) S-transformation and criterion matrices. Netherlands Geodetic Commission. Publications on Geodesy, New Series 5(1), Delft

43. Bajaj C, Garity T, Waren J (1988) On the applications of multi-equational resultants. Department of Computer Science, Purdue University, Technical Report CSD-TR-826, pp. 1-22

44. Baker HC, Dodson AH, Penna NT, Higgins M, Offiler D (2001) Ground-based GPS water vapour estimation: Potential for meteorological forecasting, Journal of Atmospheric and Solar-Terrestrial Physics 63(12): 1305–1314.

45. Baki IzH, Chen YQ (1999) Tailored datum transformation model for locally distributed data. Journal of Surveying Engineering 125: 25–35

46. Balodimos DD, Korakitis R, Lambrou E, Pantazis G. (2003) Fast and accurate determination of astronomical coordinates Θ, Λ and azimuth using total station and GPS receiver. Survey Review 37: 269–275

47. Bancroft S (1985) An algebraic solution of the GPS equations. IEEE Transaction on Aerospace and Electronic Systems AES-21: 56–59.

48. Bancroft S (1985) An algebraic solution of the GPS equations. IEEE Transaction on Aerospace and Electronic Systems AES-21: 56–59

49. Barbeau EJ (2003) Polynomials. Problem books in mathematics. Springer, New York, Berlin

50. Barsi A (2001) Performing coordinate transformation by artificial neural network. Allgemeine Vermessungs-Nachrichten 108: 134–137

51. Bartelme N, Meissl P (1975) Ein einfaches, rasches und numerisch stabiles Verfahren zur Bestimmung des kürzesten Abstandes eines Punktes von einem sphäroidischen Rotationsellipsoid. Allgemeine Vermessungs-Nachrichten 82: 436–439

52. Bazlov YA, Galazin VF, Kaplan BL, Maksimov VG, Rogozin VP (1999) Propagating PZ 90 and WGS 94 transformation parameters. GPS Solutions 3: 13–16

53. Bähr HG (1988) A quadratic approach to the non-linear treatment of non-redundant observations. Manuscripta Geodaetica 13: 191–197

54. Bähr HG (1991) Einfach überbestimmtes ebenes Einschneiden, differentialgeometrisch analysiert. Zeitschrift für Vermessungswesen 116: 545–552

55. Becker T, Weispfenning V (1993) Gröbner bases. A computational approach to commutative algebra. Graduate Text in Mathematics 141, Springer, New York

56. Becker T, Weispfenning V (1998) Gröbner bases. A computational approach to commutative algebra. Graduate Text in Mathematics 141, 2nd Edition, Springer, New York

57. Beinat A, Crosilla F (2001) Generalised Procrustes analysis for size and shape 3-D object reconstructions. Optical 3-D measurement techniques, Wien 1-4 October, V, pp. 345–353

58. Beinat A, Crosilla F (2003) Generalised Procrustes algorithm for the conformal updating of a cadastral map. Zeitschrift für Geodasie, Geoinformation und Landmanagement 128: 341-349

59. Bevis M, Businger S, Chiswell S, Herring TA, Anthes RA, Rocken C, Ware RH (1994) GPS Meteorology: Mapping zenith wet delays onto precipitable water. Journal of Applied Meteorology. 33: 379–386

60. Benning W (1974) Der kürzeste Abstand eines in rechtwinkligen Koordinaten gegebenen Außenpunktes vom Ellipsoid. Allgemeine Vermessungs-Nachrichten 81: 429–433

61. Benning W (1987) Iterative ellipsoidische Lotfußpunktberechung. Allgemeine Vermessungs-Nachrichten 94: 256–260

62. Beranek M (1997) Uberlegungen zur Normalverteilung und theoretische Analyse der 3- und 4-Parameter-Transformation. Allgemeine Vermessungs-Nachrichten 104: 137–141

63. Biagi L, Sanso F (2004) Sistemi di riferimento in geodesia: algebra e geometria dei minimi quadrati per un modello con deficienza di rango (parte seconda). Bollettino di Geodesia e Science Affini 63: 29–52

64. Bil WL (1992) Sectie en Projectie. NGT (Dutch Geodetic Magazine) Geodesia 92-10: 405–411

65. Bingham C, Chang T, Richards D (1992) Approximating the matrix fischer and Bingham distributions: Applications to spherical regression and procrustes analysis. Journal of Multivariate Analysis 41:314–337

66. Birardi G (1996) The future global geodetic network and the transformation of local onto WGS coordinates. Bullettino di Geodesia e Scienze Affini 55: 49–56

67. Blaha G, Besette RP (1989) Nonlinear least squares method via an isomorphic geometrical setup. Bulletin Geodesique 63: 115–138

68. Bock W (1959) Mathematische und geschichtliche Betrachtungen zum Einschneiden. Schriftenreihe Niedersaechsisches Landesvermessungsamt. Report 9, Hannover

69. Bojanczyk AW, Lutoborski A (2003) The procrustes problem for orthogonal Kronecker products. SIAM Journal of Scientific Computation 25: 148–163

70. Borg I, Groenen P (1997) Modern multidimensional scaling. Springer, New York

71. Borkowski KM (1987) Transformation of geocentric to geodetic coordinates without approximation. Astrophys. Space. Sci. 139: 1–4

72. Borkowski KM (1989) Accurate algorithm to transform geocentric to geodetic coordinates. Bull. Geod. 63: 50–56

73. Bowring BR (1976) Transformation from spatial to geographical coordinates. Survey Review 23: 323–327

74. Bowring BR (1985) The accuracy of geodetic latitude and height equations. Survey Review 28: 202–206

75. Brandstätter G (1974) Notiz zur analytischen Lösung des ebenen Rückwärtsschnittes. Österreichische Zeitschrift für Vermessungswesen 61: 34–136

76. Brokken FB (1983) Orthogonal Procrustes rotation maximizing congruence. Psychometrika 48: 343–352

77. Brunner FK (1979) On the analysis of geodetic networks for the determination of the incremental strain tensor. Survey Review 25: 146–162

78. Buchberger B (1965) An algorithm for finding a basis for the residue class ring of a zero dimensional polynomial ideal (German). Ph.D. thesis, Institute of Mathematics, University of Innsbruck

79. Buchberger B (1970) Ein algorithmisches Kriterium für die Lösbarkeit eines algebraischen Gleichungsystems. Aequationes Mathematicae 4: 374–383

80. Buchberger B (1979) A criterion for detecting unnecessary reductions in the construction of Groebner bases. Proceedings of the 1979 European Symposium on Symbolic and Algebraic computation, Springer lecture notes in Computer Science 72, pp. 3-21, Springer, Berlin-Heidelberg-New York

81. Buchberger B (2001) Gröbner bases. A short introduction for system theorists. In: Moreno-Diaz R et al. (eds): EUROCAST 2001, LNCS 2178, pp. 1–19

82. Buchberger B, Winkler F (1998) Groebner bases and applications. London mathematical society lecture note series 251, Cambridge university press

83. Canny JF (1988) The complexity of robot motion planning. ACM doctoral dissertation award, MIT Press

84. Canny JF, Kaltofen E, Yagati L (1989) Solving systems of nonlinear polynomial equations faster. Proceedings of the International Symposium on Symbolic and Algebraic Computations ISSAC, July 17-19, Portland, Oregon, pp.121–128

85. Cattani E, Dickenstein A, Sturmfels B (1998) Residues and resultants. J. Math. Sci. University of Tokyo 5: 119–148

86. Chaffee JW, Abel JS (1994) On the exact solutions of the pseudorange equations. IEEE Transactions on Aerospace and Electronic Systems 30: 1021–1030

87. Chen G, Herring TA (1997) Effects of atmospheric azimuthal asymmetry on the analysis of apace geodetic data. Journal of Geophysical Research 102(B9): 20489–20502

88. Cheng C-L, Van Ness JW (1999) Statistical regression with measurement error. Oxford University Press, 198 Madison Avenue, New York

89. Chu MT, Driessel R (1990) The projected gradient method for least squares matrix approximations with spectral constraints. SIAM J. Numer. Anal. 27(4): 1050–1060

90. Chu MT, Trendafilov NT (1998) Orthomax rotation problem. A differential equation approach. Behaviormetrika 25(1): 13–23

91. Chrystal G (1964) Textbook of Algebra (Vol. 1). Chelsea, New York

92. Cox DA (1998) Introduction to Gröbner bases. Proceedings of Symposia in Applied Mathematics 53: 1–24

93. Cox TF, Cox MA (1994) Multidimensional scaling. St. Edmundsbury Press, St. Edmunds, Suffolk

94. Cox D, Little J, O'Shea D (1997) Ideals, Varieties, and Algorithms. An introduction to computational algebraic geometry and commutative algebra, Springer, New York

95. Cox D, Little J, O'Shea D (1998) Using algebraic geometry. Graduate Text in Mathematics 185. Springer, New York

96. Croceto N (1993) Point projection of topographic surface onto the reference ellipsoid of revolution in geocentric Cartesian coordinates. Survey Review 32: 233–238

97. Crosilla F (1983a) A criterion matrix for a second order design of control networks. Bull. Geod. 57: 226–239

98. Crosilla F (1983b) Procrustean transformation as a tool for the construction of a criterion matrix for control networks. Manuscripta Geodetica 8: 343–370

99. Crosilla F, Beinat A (2003) Procrustes analysis and geodetic sciences. In: Grafarend EW, Krumm FW, Schwarze VS (eds) Geodesy - the Challenge of the 3rd Millennium. Springer, Heidelberg pp. 277–286

100. Crosilla F (2003) Use of generalised Procrustes analysis for the photogrammetric block adjustment by independent models. Journal of Photogrammetric & Remote sensing 56: 195–209

101. Dach R (2000) Einfluß von Auflasteffekten auf Präzise GPS-Messungen, DGK, Reihe C, Heft Nr. 519

102. Davenport JH, Siret Y, Tournier E (1988) Computer algebra. Systems and algorithms for algebraic computation. Academic Press Ltd., St. Edmundsbury, London

103. Davis JL, Herring TA, Shapiro II, Rogers AE, Elgered G (1985) Geodesy by radio interferometry: Effects of atmospheric modeling errors on estimates of baseline length. Radio Sci 20: 1593–1607

104. Dixon AL (1908) The elimination of three quantics in two independent variables. Proc. London Mathematical Society series 2 6: 468–478

105. Dryden IL (1998) General shape and registration analysis. In: Barndorff-Nielsen O, Kendall WS, van Lieshout MNM (eds) Stochastic Geometry: likelihood and computation. Chapman and Hall, London pp: 333–364

106. Featherstone WE (1997) A comparison of existing co-ordinate transformation models and parameters in Australia. Cartography 26: 13–25

107. Finsterwalder S, Scheufele W (1937) Das Rückwartseinschneiden im Raum. Sebastian Finsterwalder zum 75 Geburtstage, pp. 86–100, Verlag Hebert Wichmann, Berlin

108. Fischbach FF (1965) A satellite method for pressure and temperature below 24km. Bull. Am. Meteorol. 46: 528–532

109. Fischler MA, Bolles RC (1981) Random sample consensus: A paradigm for modell fitting with application to image analysis and automated cartography. Communications of the ACM 24: 381–395

110. Fitzgibbon A, Pilu M, Fisher RB (1999) Direct least squares fitting of ellipses. IEEE Transactions on Pattern Analysis and Machine Intelligence 21: 476–480

111. Flores A, Ruffini G, Rius A (2000) 4D Tropospheric Tomography Using GPS Slant Wet Delay. Ann. Geophys. 18: 223–234

112. Fotiou A (1998) A pair of closed expressions to transform geocentric to geodetic coordinates. Zeitschrift für Vermessungswesen 123: 133–135

113. Foulds LR (1984) Combinatorial optimization for undergraduates. Springer, New York

114. Francesco D, Mathien PP, Senechal D (1997) Conformal field theory. Springer, Heidelberg, New York
115. Fröhlich H, Hansen HH (1976) Zur Lotfußpunktrechnung bei rotationsellipsoidischer Bezugsfläche. Allgemeine Vermessungs-Nachrichten 83: 175–179
116. Fukushima T (1999) Fast transform from geocentric to geodetic coordinates. Journal of Geodesy 73: 603–610
117. Gander W, Golub GH, Strebel R (1994) Least-Squares fitting of circles and ellipses. BIT No. 43: 558–578
118. Gelfand IM, Kapranov MM, Zelevinsky AV (1990) Generalized Euler Integrals and A-Hypergeometry Functions. Advances in Mathematics 84: 255–271
119. Gelfand IM, Kapranov MM, Zelevinsky AV (1994) Discriminants, resultants and multidimensional determinants. Birkhäuser, Boston
120. Gelfand MS, Mironor AA, Perzner PA (1996) Gene recognition via spliced sequence alignment. Proc. Natl. Aead. Sci. USA 93: 9061–9066
121. Golub GH (1987) Least squares, singular values and matrix approximation. Aplikace matematiky 13: 44–51
122. Goodall C (1991) Procrustes methods in statistical analysis of shape. J. Royal Statistical Soc. B53: 285–339
123. Gordon SJ, Lichti DD (2004) Terrestrial laser scanners with a narrow field of view: the effect on 3D resection solutions. Survey Review 37:448–468
124. Gotthardt E (1940) Zur Unbestimmtheit des räumlichen Rückwärtseinschnittes, Mitteilungen der Ges. f. Photogrammetry e.V., Jänner 1940, Heft 5
125. Gotthardt E (1974) Ein neuer gefährlicher Ort zum räumlichen Rückwärtseinschneiden, Bildm. u. Luftbildw.
126. Gower JC (1975) Generalized procrustes analysis. Psychometrika 40(1): 33–51
127. Grafarend EW (1975) Three dimensional Geodesy 1. The holonomity problem. Zeitschrift für Vermessungswesen 100: 269–280
128. Grafarend EW (1981) Die Beobachtungsgleichungen der dreidimensionalen Geodäsie im Geometrie- und Schwereraum. Ein Beitrag zur operationellen Geodäsie. Zeitschrift für Vermessungswesen 106: 411–429
129. Grafarend EW (1985) Variance-Covariance component estimation; theoretical results and geodetic applications. Statistics & Decisions, Supplement Issue No. 2, 407–441
130. Grafarend EW (1988) Azimuth transport and the problem of orientation within geodetic traverses and geodetic networks. Vermessung, Photogrammetrie, Kulturtechnik 86: 132–150
131. Grafarend EW (1989) Photogrammetrische Positionierung. Festschrift für Prof. Dr.-Ing. Dr. h.c Friedrich Ackermann zum 60. Geburtstag, Institut für Photogrammetrie, Univerität Stuttgart, Heft 14, pp.44-55, Stuttgart
132. Grafarend EW (1990) Dreidimensionaler Vorwaertschnitt. Zeitschrift für Vermessungswesen 115: 414–419
133. Grafarend EW (1991) Application of Geodesy to Engineering. In: Eds. Linkwitz K, Eisele V, Mönicke HJ, IAG-Symposium No. 108, Springer, Berlin-Heidelberg-New York
134. Grafarend EW (2000) Gaußsche flächennormale Koordinaten im Geometrie- und Schwereraum. Erste Teil: Flächennormale Ellipsoidkoordinaten. Zeitschrift für Vermessungswesen 125: 136–139

135. Grafarend EW (2000) Gaußsche flächennormale Koordinaten im Geometrie- und Schwereraum. Erste Teil: Flächennormale Ellipsoidkoordinaten. Zeitschrift für Vermessungswesen 125: 136–139

136. Grafarend EW, Ardalan A (1999) World geodetic datum 2000. Journal of Geodesy 73: 611–623

137. Grafarend EW, Awange JL (2000) Determination of vertical deflections by GPS/LPS measurements. Zeitschrift für Vermessungswesen 125: 279–288

138. Grafarend EW, Awange JL (2003) Nonlinear analysis of the three-dimensional datum transformation (conformal group $\mathbb{C}_7(3)$. Journal of Geodesy 77: 66–76

139. Grafarend EW, Keller W (1995) Setup of observational functionals in gravity space as well as in geometry space. Manuscripta Geodetica 20: 301–325

140. Grafarend EW, Kunz J (1965) Der Rückwärtseinschnitt mit dem Vermessungskreisel. Bergbauwissenschaften 12: 285–297

141. Grafarend EW, Lohse P (1991) The minimal distance mapping of the topographic surface onto the (reference) ellipsoid of revolution. Manuscripta Geodaetica 16: 92–110

142. Grafarend EW, Mader A (1993) Robot vision based on an exact solution of the threedimensional resection-intersection. Applications of Geodesy to Engineering. In K. Linkwitz, V. Eisele and H-J Moenicke, Symposium No. 108, Spinger Berlin-Heidelberg-Newyork-London-Paris-Tokyo-HongKong-Barcelona-Budapest.

143. Grafarend EW, Okeke F (1998) Transformation of conformal coordinates of type Mercator from global datum (WGS 84) to local datum (regional, national). Marine Geodesy 21: 169–180

144. Grafarend EW, Richter B (1977) Generalized Laplace condition. Bull. Geod. 51: 287–293

145. Grafarend EW, Sanso F (1985) Optimization and design of geodetic networks. Springer, Berlin-Heidelberg-New York-Tokyo

146. Grafarend EW, Schaffrin B (1974) Unbiased Freenet adjustment. Survey Review 22: 200–218

147. Grafarend EW, Schaffrin B (1989) The geometry of nonlinear adjustment- the planar trisection problem-. In: Kejlso E, Poder K, Tscherning CC (eds) Festschrift to T. Krarup, pp. 149-172, Denmark

148. Grafarend EW, Schaffrin B (1991) The planar trisection problem and the impact of curvature on non-linear least -squares estimation. Computational statistics & data analysis 12: 187–199

149. Grafarend EW, Schaffrin B (1993) Ausgleichungsrechnung in Linearen Modellen. B. I. Wissenschaftsverlag, Mannheim

150. Grafarend EW, Shan J (1996) Closed-form solution of the nonlinear pseudoranging equations (GPS). ARTIFICIAL SATELLITES, Planetary Geodesy 31: 133–147

151. Grafarend EW, Shan J (1997a) Closed-form solution of P4P or the three-dimensional resection problem in terms of Möbius barycentric coordinates. Journal of Geodesy 71: 217–231

152. Grafarend EW, Shan J (1997b) Closed form solution of the twin P4P or the combined three dimensional resection-intersection problem in terms of Möbius barycentric coordinates. Journal of Geodesy 71: 232–239

153. Grafarend EW, Syffus R (1997) Strip transformation of conformal coordinates of type Gauss-Kruger and UTM. Allgemeine Vermessungs-Nachrichten 104: 184–190

154. Grafarend EW, Syffus R (1998) Transformation of conformal coordinates of type Gauss-Krüger or UTM from local datum (regional, national, European) to global datum (WGS 84) part 1: The transformation equations. Allgemeine Vermessungs-Nachrichten 105: 134–141

155. Grafarend EW, Hendricks A, Gilbert A (2000) Transformation of conformal coordinates of type Gauss-Kruger or UTM from a local datum (Regional, National, European) to a global datum (WGS 84, ITRF 96). Allgemeine Vermessungs-Nachrichten 107: 218–222

156. Grafarend EW, Lohse P, Schaffrin B (1989) Dreidimensionaler Rückwärtsschnitt. Zeitschrift für Vermessungswesen 114: 61–67,127–137,172–175,225–234,278–287

157. Grafarend EW, Knickmeyer EH, Schaffrin B (1982) Geodätische Datumtransformationen. Zeitschrift für Vermessungswesen 107: 15–25

158. Grafarend EW, Krumm F, Okeke F (1995) Curvilinear geodetic datum transformation. Zeitschrift für Vermessungswesen 120: 334–350

159. Grafarend EW, Syffus R, You RJ (1995) Projective heights in geometry and gravity space. Allgemeine Vermessungs-Nachrichten 102: 382–402

160. Green B (1952) The orthogonal approximation of an oblique stracture in factor analysis. Psychometrika 17: 429–440

161. Grewal MS, Weill LR, Andrews AP (2001) Global Positioning Systems, Inertial Navigation and Integration, John Wiley & Sons, New York

162. Grunert JA (1841) Das Pothenotsche Problem in erweiterter Gestalt; nebst Bemerkungen über seine Anwendungen in der Geodäsie. Grunerts Archiv für Mathematik und Phsyik 1 pp. 238–241

163. Guckenheimer J, Myers M, Sturmfels B (1997) Computing Hopf birfucations. SIAM J. Numerical Analysis 34: 1–21

164. Gui Q, Zhang J (1998) Robust biased estimation and its applications in geodetic adjustments. Journal of Geodesy 72: 430–435

165. Gulliksson M (1995a) The partial Procrustes problem - A first look. Department of Computing Science, Umea University, Report UMINF-95.11, Sweden

166. Gulliksson M (1995b) Algorithms for the partial Procrustes problem. Department of Industrial Technology, Mid Sweden University s-891 18, Report 1995:27, Ornskoldsvik, Sweden

167. Gullikson M, Söderkvist I (1995) Surface fitting and parameter estimation with nonlinear least squares. Zeitschrift für Vermessungswesen 25: 611–636

168. Guolin L (2000) Nonlinear curvature measures of strength and nonlinear diagnosis. Allgemein Vermessungs-Nachrichten 107: 109–111

169. Guo J, Jin F (2001) A new model of digitizing coordinate transformation and its nonlinear solution. Allgemeine Vermessungs-Nachrichten 108: 311–317

170. Gurbunov ME, Gurvich AS, Bengtsson L (1996) Advanced algorithms of inversion of GPS/MET satellite data and their application to the reconstruction of temperature and humidity, Rep. No. 211, Max-Plunk-Institut für Meteorologie, Hamburg, Germany

171. Han SC, Kwon JH, Jekeli C (2001) Accurate absolute GPS positioning through satellite clock error estimation. Journal of Geodesy 75: 33–43

172. Hampel FR, Ronchetti EM, Rousseeuw P, Stahel WA (1986) Robust Statistic - the approach based non influence Functions. John Wiley & Sons, New York

173. Hammer E (1896) Zur graphischen Ausgleichung beim trigonometrischen Einschneiden von Punkten. Optimization methods and softwares 5: 247–269

174. Hanselman D, Littlefield B (1997) The student edition of Matlab. Prentice-Hall, New Jersey
175. Hanssen RF, Weckwerth TM, Zebker HA, Klees R (1999) High-Resolution Water Vapor Mapping from Interferometric Radar Measurements. Science 283: 1297–1299
176. Haralick RM, Lee C, Ottenberg K, Nölle M (1991) Analysis and solution of the three point perspective pose estimation problem. Proc. IEEE Org. on Computer Vision and Pattern Recognition, pp. 592–598
177. Haralick RM, Lee C, Ottenberg K, Nölle M (1994) Review and analysis of solution of the three point perspective pose estimation problem. International Journal of Computer Vision 13 3: 331–356
178. Harper D, Wooff C, Hodgkinson D (1991) A guide to computer algebra system. John Wiley & Sons, New York
179. Harvey BR (1986) Transformation of 3D coordinates. The Australian Surveyor 33: 105–125
180. Healey S, Jupp A, Offiler D, Eyre J (2003) The assimilation of radio occultation measurements. In Reigber C, Lühr H, Schwintzer P (eds), First CHAMP mission results for gravity, magnetic and atmospheric studies, Springer, Heidelberg
181. Heck B (1987) Rechenverfahren und Auswertemodelle der Landesvermessung. Wichmann Verlag, Karlsruhe, Germany
182. Heikkinen M (1982) Geschlossene Formeln zur Berechnung räumlicher geodätischer Koordinaten aus rechtwinkligen Koordinaten. Zeitschrift für Vermessungswesen 107: 207–211
183. Heindl G (1982) Experiences with non-statistical method of detecting outliers. International symposium on geodetic network and computations of the I. A. G. Munich, Aug. 30th to Sept. 5, 5: 19–28
184. Heiskanen WA, Moritz H (1967) Physical Geodesy, Freeman and Company, London
185. Hirvonen R, Moritz H (1963) Practical computation of gravity at high altitudes, Institute of Geodesy, Photogrammetry and Cartography. Ohio State University, Report No. 27, Ohio
186. Hofman-Wellenhof B, Lichtenegger H, Collins J (2001) Global Positioning System: Theory and practice, 5th Edition, Springer, Wien
187. Horaud R, Conio B, Leboulleux O (1989) An analytical solution for the perspective 4-point problem. Computer Vision, Graphics and Image Processing 47: 33–44
188. Hornoch AT (1950) Über die Zurückführung der Methode der kleinsten Quadrate auf das Prinzip des arithmetischen Mittels. Zeitschrift für Vermessungswesen 38: 13–18
189. Huber PJ (1964) Robust estimation of a location parameter. Annals of Mathematical Statistics 35: 73–101
190. Huber PJ (1972) Robust Statistics; A review. Annals of Mathematical Statistics 43: 1041–1067
191. Huber PJ (1981) Robust Statistics. John Wiley & Sons, New York
192. Ireland K, Rosen M (1990) A classical introduction to modern number theory. Springer, New York
193. Irving RS (2004) Integers, polynomials, and rings. Springer, New York
194. Jacobi CGI (1841) Deformatione et proprietatibus determinantum, Crelle's Journal für die reine und angewandte Mathematik, Bd. 22

318

195. Kahmen H, Faig W (1988) Surveying. Walter de Gruyter, Berlin
196. Kampmann G (1996) New adjustment techniques for the determination of transformation parameters for Cadastral and Engineering purposes. Geomatica 50: 27–34
197. Killian K (1990) Der gefährliche Ort des überbestimmten räumlichen Rückwärtseinschneidens. Öst. Zeitschrift für Vermessungswesen und Photogrammetry 78: 1–12
198. Kleusberg A (1994) Die direkte Lösung des räumlichen Hyperbelschnitts. Zeitschrift für Vermessungswesen 119: 188-192
199. Kleusberg A (2003) Analytical GPS navigation solution. In: Grafarend EW, Krumm FW, Schwarze VS (eds) Geodesy - the Challenge of the 3rd Millennium. Springer, Heidelberg pp.93–96
200. Koch KR (1999) Parameter estimation and hypothesis testing in linear models. Springer, Berlin, Heidelberg
201. Koch KR (2001) Bermekung zu der Veröffentlichung "Zur Bestimmung eindeutiger transformationparameter". Zeitschrift für Vermessungswesen 126: 297
202. Koch KR, Yang Y (1998a) Konfidenzbereiche und Hypothesenteste für robuste Parameterschätzungen. ZfV 123: 20–26
203. Koch KR, Yang Y (1998b) Robust Kalman filter for rank deficient observation models. Journal of Geodesy 72: 436–441
204. Koch KR, Fröhlich H, Bröker G (2000) Transformation rumlicher variabler Koordinaten. Allgemeine Vermessungs-Nachrichten 107: 293–295
205. Kuo Y.-H, Sokolovski SV, Anthens RA, Vandenberghe F (2000) Assimilation of the GPS radio occultation data for numerical weather prediction. Terrestrial, Atmospheric and Oceanic Science, 11: 157–186
206. Krarup T (1979) S transformation or how to live without the generalized inverse - almost. Geodetisk Institut, Charlottenlund, Denmark
207. Krarup T (1982) Nonlinear adjustment and curvature. In: Forty years of thought, Delft, pp. 145–159
208. Krause LO (1987) A direct solution of GPS-type navigation equations. IEEE Transactions on Aerospace and Electronic Systems 23: 225–232
209. Krishna S, Manocha D (1995) Numerical algorithms for evaluating one-dimensional algebraic sets. Proceedings of the International Symposium on Symbolic and Algebraic Computation ISSAC, July 10-12, pp. 59–67, Montreal, Canada
210. Kubik KK (1967) Iterative Methoden zur Lösunge des nichtlinearen Ausgleichungsproblemes. Zeitschrift für Vermessungswesen 91: 145–159
211. Kursinski ER, Hajj GA, Schofield JT, Linfield RP, Hardy KR (1997) Observing Earth's atmosphere with radio occultation measurements using Global Positioning System. J. Geophy. Res. 102: 23429–23465
212. Kurz S (1996) Positionierung mittels Rückwartsschnitt in drei Dimensionen. Studienarbeit, Geodätisches Institut, University of Stuttgart, Stuttgart
213. Lam TY (2003) Exercises in classical ring theory. Springer, New York, Tokyo
214. Lannes A, Durand S (2003) Dual algebraic formulation of differential GPS. Journal of Geodesy 77: 22–29
215. Lapaine M (1990) A new direct solution of the transformation problem of Cartesian into ellipsoidal coordinates. In: Rapp RH. Sanso F, Determination of the geoid: Present and future. pp. 395-404. Springer, New York
216. Larson LW (1996) Destructive water: Water-caused natural disasters, their abetment and control. IAHS conference, Anaheim California, June 24-28

217. Lauritzen N (2003) Concrete abstract algebra. From numbers to Gröbner bases. Cambridge University Press, UK

218. Leick A (2003) GPS satellite surveying, 3rd Edition, John Wiley & Sons, New York

219. Lenzmann E, Lenzmann L (2001a) Zur Bestimmung eindeutiger transforma-tionparameter. Zeitschrift für Vermessungswesen 126: 138–142

220. Lenzmann E, Lenzmann L (2001b) Erwiderung auf die Anmerkung von Jörg Reinking und die Bermekungen von Karl-Rudolf Koch zu unserem Meitrag "Zur Bestimmung eindeutiger transformationparameter". Zeitschrift für Ver-messungswesen 126: 298–299

221. Lichtenegger H (1995) Eine direkte Lösung des räumlichen Bogenschnitts. Österreichische Zeitschrift für Vermessung und Geoinformation 83: 224–226

222. Lidl R, Pilz G (1998) Applied abstract algebra. 2nd edition, Springer, New York

223. Lin KC, Wang J (1995) Transformation from geocentric to geodetic coordinates using Newton's iteration. Bulletin Geodesique 69: 300–303

224. Linnainmaa S, Harwood D, Davis LS (1988) Pose determination of a three-dimensional object using triangle pairs. IEEE transaction on pattern analysis and Machine intelligence 105: 634–647

225. Lohse P (1990) Dreidimensionaler Rückwärtsschnitt. Ein Algorithmus zur Streckenberechnung ohne Hauptachsentransformation. Zeitschrift für Vermes-sungswesen 115: 162–167

226. Lohse P (1994) Ausgleichungsrechnung in nichtlinearen Modellen. DGK, Reihe C, Heft Nr. 429

227. Loskowski P (1991) Is Newton's iteration faster than simple iteration for trans-formation between geocentric and geodetic coordinates? Bulletin Geodesique 65: 14–17

228. Lyubeznik G (1995) Minimal resultant system. Journal of Algebra 177: 612–616

229. Macaulay F (1902) On some formulae in elimination. Proceeding in London Mathematical Society, pp. 3-27

230. Macaulay F (1916) The algebraic theory of modular systems. Cambridge Tracts in Mathematics 19, Cambridge University Press, Cambridge

231. Macaulay F (1921) Note on the resultant of a number of polynomials of the same degree. Proceeding in London Mathematical Society 21: 14–21

232. Mackenzie FT (2003) Our changing planet; an introduction to Earth system science and global environmental change. 3rd edition, Prentice Hall, New Jersey

233. Manocha D (1992) Algebraic and numeric techniques for modeling and robotics. Ph.D. thesis, Computer Science Division, Department of Electrical Engineering and Computer Science, University of California, Berkeley

234. Manocha D (1993) Efficient algorithms for multipolynomial resultant. The Computer Journal 36: 485–496

235. Manocha D (1994a) Algorithms for computing selected solutions of polynomial equations. Extended abstract appearing in the proceedings of the ACM ISSAC 94.

236. Manocha D (1994b) Computing selected solutions of polynomial equations. Proceedings of the International Sypmosium on Symbolic and Algebraic Com-putations ISSAC, July 20-22, pp.1-8, Oxford

237. Manocha D (1994c) Solving systems of polynomial equations. IEEE Computer Graphics and application 14: 46–55

320

238. Manocha D (1998) Numerical methods for solving polynomial equations. Proceedings of Symposia in Applied Mathematics 53: 41–66

239. Manocha D, Canny J (1991) Efficient techniques for multipolynomial resultant algorithms. Proceedings of the International Symposium on Symbolic Computations, July 15-17, 1991, pp. 86–95, Bonn

240. Manocha D, Canny J (1992) Multipolynomial resultant and linear algebra. Proceedings of the International Symposium on Symbolic and Algebraic Computations ISSAC, July 27-29, pp. 158-167, Berkeley

241. Manocha D, Canny J (1993) Multipolynomial resultant algorithms. Journal of Symbolic Computations 15: 99–122

242. Mardia K (1978) Some properties of classical multidimensional scaling. Commun. Statist.-Theory Meth. A7(13): 1233–1241

243. Mathar R (1997) Multidimensionale Skalierung. B. G. Teubner Verlag, Stuttgart

244. Mathes A (1998) GPS und GLONASS als Teil eines hybrid Meßsystems in der Geodäsie am Beispiel des Systems HIGGINS, Dissertationen, DGK, Reihe C, Nr. 500

245. Mautz R (2001) Zur Lösung nichtlinearer Ausgleichsprobleme bei der Bestimmung von Frequenzen in Zeitreihen. DGK, Reihe C, Nr. 532

246. McCoy NH, Janusz GJ (2001) Introduction to abstract algebra. Harcout Academic Press, San Diego

247. Meissl P (1982) Least squares adjustment. A modern approach, Mitteilungen der geodätischen Institut der Technischen Universität Craz, Folge 43. 17

248. Melbourne WG, Davis ES, Duncan CB, Hajj GA, Hardy K, Kursinski R, Mechan TK, Young LE, Yunck TP (1994) The application of spaceborne GPS to atmospheric limb sounding and global change monitoring. JPL Publication 94-18

249. Merritt EL (1949) Explicit Three-point resection in space. Phot. Eng. 15: 649–665

250. Mittermayer E (1972) A generalization of least squares adjustment of free networks. Bull. Geod. 104: 139–155

251. Monhor D (2001) The real linear algebra and linear programming. Müszaki Könyvkiadó, Budapest

252. Monhor D (2002) Clarification of and complements to the concept of outlier. Geodezia es Kartografia 12: 21–27

253. Morgan AP (1992) Polynomial continuation and its relationship to the symbolic reduction of polynomial systems. In Symbolic and Numerical Computations for Artificial Intelligence, pp. 23–45

254. Mukherjee K(1996) Robust estimation in nonlinear regression via minimum distance method. Mathematical methods of statistics, Vol 5, No. 1, Allerton Press. Inc., New York

255. Müller FJ (1925) Direkte (Exakte) Lösungen des einfachen Rückwarscnittseinschneidens im Raum. 1 Teil. Zeitschrift für Vermessungswesen 37: 249–255, 265–272, 349–353, 365–370, 569–580

256. Nicholson WK (1999) Introduction to abstract algebra. Second Edition, John Wiley & Sons, New York-Chichester-weinheim-Brisbane-Singapore

257. Niell AE (1996) Global Mapping Functions for the Atmosphere Delay at Radio Wavelengths. Journal of Geophysical Research 101(B2): 3227–3246

258. Newsome G, Harvey BR (2003) GPS coordinate transformation parameters for Jamaica. Survey Review 37: 218–233

321

259. Nitschke M, Knickmeyer EH (2000): Rotation parameters - a survey of techniques. Journal of Surveying Engineering 126: 83–105

260. Okeke FI (1998) The curvilinear datum transformation model, DGK, Reihe C, Heft Nr. 481

261. Ozone MI (1985) Non-iterative solution of the ϕ equations. Surveying and Mapping 45: 169–171

262. Paul MK (1973) A note on computation of geodetic coordinates from geocentric (Cartesian) coordinates. Bull. Geod. No. 108: 135–139

263. Penev P (1978) The transformation of rectangular coordinates into geographical by closed formulas. Geo. Map. Photo 20: 175–177

264. Perelmuter A (1979) Adjustment of free networks. Bull. Geod. 53: 291–295

265. Pick M (1985) Closed formulae for transformation of Cartesian coordinates into a system of geodetic coordinates. Studia geoph. et geod. 29: 653–666

266. Pistone G, Wynn HP (1996) Generalized confounding with Gröbner bases. Biometrika 83: 112–119

267. Pope A (1982) Two approaches to non-linear least squares adjustments. The Canadian Surveyor 28: 663–669

268. Preparata FP, Shamos MI (1985) Computational geometry. An Introduction. Springer, New York-Berlin-Heidelberg-London-Paris-Tokyo-Hong Kong-Barcelona-Budapest

269. Press WH, Teukolsky SA, Vetterling WT, Flannery BP (1992) Numerical recipes in Fortran 77: The art of scientific computing, 2nd edition, Cambridge University Press

270. Prestel A, Delzell CN (2001) Positive polynomials: from Hilbert's 17th problem to real algebra. Springer, Berlin

271. Rao CR (1967) Least squares theory using an estimated dispersion matrix and its application to measurement of signals. Procedure of the Fifth Barkeley Symposium, Barkeley

272. Rao CR (1971) Estimation of variance and covariance components - MINQUE theory. Journal of Multivariate Analysis 1: 257–275

273. Rao CR (1973) Representation of the best linear unbiased estimators in the Gauss-Markov model with singular dispersion matrix. Journal of multivariate analysis 3: 276–292

274. Rao CR (1978) Choice of the best linear estimators in the Gauss-Markov model with singular dispersion matrix. Comm. Stat. Theory Meth. A7 (13): 1199–1208

275. Rao CR, Kleffe J (1979) Variance and covariance components estimation and applications. Technical Report No. 181, Ohio State University, Dept. of Statistics, Columbus, Ohio

276. Reigber C, Lühr H, Schwintzer P (2003) First CHAMP mission results for gravity, magnetic and atmospheric studies, Springer, Heidelberg

277. Reinking J (2001) Anmerkung zu "Zur Bestimmung eindeutiger transformationparameter". Zeitschrift für Vermessungswesen 126: 295–296

278. Richter B (1986) Entwurf eines nichtrelativistischen geodätisch-astronomischen Bezugssystems, DGK, Reihe C, Heft Nr. 322

279. Rinner K (1962) Über die Genauigkeit des räumlichen Bogenschnittes. Zeitschrift für Vermessungswesen 87: 361–374

280. Ritt JF (1950) Differential algebra. AMS colloquium publications 18

281. Rocken C, Anthes R, Exner M, Hunt D, Sokolovski S, Ware R, Gorbunov M, Schreiner S, Feng D, Hermann B, Kuo Y.-H, Zou X (1997) Analysis and

validation of GPS/MET data in the neutral atmosphere. J. Geophys. Res. 102: 29849–29860

282. Runge C (1900) Graphische Ausgleichung beim Rückwätseinchneiden. Zeitschrift für Vermessungswesen 29: 581–588

283. Saito T (1973) The non-linear least squares of condition equations. Bull. Geod. 110: 367–395

284. Saleh J (2000) Robust estimation based on energy minimization principles. Journal of Geodesy 74: 291–305

285. Salmon G (1876) Lessons Introductory to modern higher algebra. Hodges, Foster and Co., Dublin

286. Schaffrin B (1983) Varianz-Kovarianz-Komponenten-Schätzung bei der Ausgleichung heterogener Wiederholungsmessungen, DGK, Reihe C, Heft Nr.282

287. Schek HJ, Maier P (1976) Nichtlineare Normalgleichungen zur Bestimmung der Unbekannten und deren Kovarianzmatrix. Zeitschrift für Vermessungswesen 101: 140–159

288. Schönemann PH (1966) Generalized solution of the orthogonal Procrustes problem. Psychometrika 31: 1–10

289. Schönemann PH, Caroll RM (1970) Fitting one matrix to another under choice of a certain dilatation and rigid motion. Psychometrika 35(2): 245–255

290. Schottenloher M (1997) A mathematical introduction to conformal field theory. Springer, Berlin, Heidelberg, New York

291. Schram TG (1988) Properties of gravitational lens mapping. In: Kaiser N, Lasenby AN (eds) The post-recombination universe. Kluwer Academic Publishers pp. 319–321

292. Schram TG (1998) Computer algebra system in engineering education. Global Journal of Engng. Educ. 2: 187–194

293. Shut GH (1958/59) Construction of orthogonal matrices and their application in analytical Photogrammetrie. Photogrammetria XV: 149–162

294. Schwarze VS (1995) Satellitengeodätische Positionierung in der relativistischen Raum-Zeit, DGK, Reihe C, Heft Nr.449

295. Shut GH (1958/59) Construction of orthogonal matrices and their application in analytical Photogrammetrie. Photogrammetria XV: 149-162

296. Singer P, Ströbel D, Hördt R, Bahndorf J, Linkwitz K (1993) Direkte Lösung des räumlichen Bogenschnitts. Zeitschrift für Vermessungswesen 118: 20–24

297. Sjöberg LE (1999) An efficient iterative solution to transform rectangular geocentric coordinates to geodetic coordinates. Zeitschrift für Vermessungswesen 124: 295–297

298. Soler T, Hothem LD (1989) Important parameters used in geodetic transformations. Journal of Surveying Engineering 115: 414–417

299. Steiner AK (1998) High resolution sounding of key climate variables using the radio occultation technique. Dissertation, Institute for Meteorology and Geophysics, University of Graz, No. 3

300. Steiner AK, Kirchengast G, Foelsche U, Kornblueh L, Manzini E, Bengtsson L (2001) GNSS occultation sounding for climate monitoring. Phys. Chem. Earth (A) 26: 113–124

301. Stillwell J (2003) Elements of number theory, Springer, New York

302. Strang G, Borre K (1997) Linear Algebra, Geodesy and GPS, Wellesley Cambridge Press, Wellesley

303. Sturmfels B (1994) Multigraded resultant of Sylvester type. Journal of Algebra 163: 115–127

304. Sturmfels B (1996) Gröbner bases and convex polytopes. American Mathematical Society, Providence

305. Sturmfels B (1998) Introduction to resultants. Proceedings of Symposia in Applied Mathematics 53: 25–39

306. Sünkel H (1999) Ein nicht-iteratives Verfahren zur Transformation geodätischer Koordinaten. Öster. Zeitschrift für Vermessungswesen 64: 29–33

307. Ten Berge, JMF (1977) Orthogonal procrustes rotation for two or more matrices. Psychometrika 42: 267–276

308. Teunissen PJG (1990) Nonlinear least squares. Manuscripta Geodaetica 15: 137–150

309. Teunissen PJG (1988) The non-linear 2d symmetric Helmert transformation: an exact nonlinear least squares solution. Bull. Geod. 62: 1–15

310. Teunissen PJG, Knickmeyer EH (1988) Non-linearity and least squares. CISM Journal ASCGC 42: 321–330

311. Thompson EH (1959a) A method for the construction of orthogonal matrices. Photogrammetria III: 55–59

312. Thompson EH (1959b) An exact linear solution of the absolute orientation. Photogrammetria XV: 163–179

313. Torge W (1991) Geodesy. 2nd Edition. Walter de Gruyter, Berlin

314. Trefethen LN, Bau D (1997) Numerical linear algebra. SIAM, Philadelphia

315. Tsuda T, Heki K, Miyazaki S, Aonashi K, Hirahara K, Tobita M, Kimata F, Tabei T, Matsushima T, Kimura F, Satomura M, Kato T, Naito I (1998) GPS meteorology project of Japan-Exploring frontiers of geodesy-. Earth Planets Space, 50(10): i–v

316. Tsuda T, Hocke K (2004) Application of GPS occultation for studies of atmospheric waves in the Middle Atmosphere and Ionosphere. In Anthens et al. (eds). Application of GPS Remote Sensing to Meteorology and Related Fields, Journal of Meteorological Society of Japan, Vol. 82, No. 1B, pp. 419–426

317. Van Mierlo J (1988) Rückwärtschnitt mit Streckenverhältnissen. Algemain Vermessungs Nachrichten 95: 310–314

318. Vanicek P, Krakiwski EJ (1982) Geodesy: The concepts. North-Holland Publishing Company. Amsterdam. New York-Oxford

319. Vanicek P, Steeves RR (1996) Transformation of coordinates between two horizontal geodetic datums. Journal of Geodesy 70: 740–745

320. Vasconcelos WV (1998) Computational methods in commutative algebra and algebraic geometry, Springer, Berlin-Heidelberg

321. Vincenty T (1978) Vergleich zweier Verfahren zur Berechnung der geodätischen Breite und Höhe aus rechtwinkligen koorninaten. Allgemeine Vermessungs-Nachrichten 85: 269–270

322. Vincenty T (1980) Zur räumlich-ellipsoidischen Koordinaten-Transformation. Zeitschrift für Vermessungswesen 105: 519–521

323. Voigt C (1998) Prokrustes Transformationen. Geodätisches Institut, Stuttgart

324. Von zur Gathen J, Gerhard J (2003) Modern computer algebra. 2nd edition, Cambridge University Press, UK

325. Vorob'ev, VV, Krasil'nikova TG (1994) Estimation of the accuracy of atmospheric refractive index recovery from Doppler shift measurements at frequencies used in the NAVSTAR system. Phys. of Atmos. and Oceans 29: 602–609

326. Van Der Waerden BL (1950) Modern Algebra. 3rd Edition, F. Ungar Publishing Co., New York

327. Weiss J (1993) Resultant methods for the inverse kinematics problem. In: Angeles et al. (eds.) Computational Kinematics, Kluwer Academic Publishers, Netherlands

328. Wellisch S (1910) Theorie und Praxis der Ausgleichsrechnung. Bd. II: Probleme der Ausgleichsrechnung

329. Welsch WM (1993) A general 7-parameter transformation for the combination, comparison and accuracy control of the terrestrial and satellite network observations. Manuscripta Geodaetica 18: 295–305

330. Ware H, Fulker D, Stein S, Anderson D, Avery S, Clerk R, Droegmeier K, Kuettner J, Minster B, Sorooshian S (2000) SuomiNet: A real time national GPS network for atmospheric research and education. Bull. Am. Meteorol. Soc. 81: 677–694

331. Werkmeister P (1916) Trigonometrische Punktbestimmung durch einfaches Einschneiden mit Hilfe von Vertikalwinkeln. Zeitschrift für Vermessungswesen 45: 248–251

332. Werkmeister P (1920) Über die Genauigkeit trigonometrischer Punktbestimmungen. Zeitschrift für Vermessungswesen 49: 401–412,433–456

333. Werner D (1913) Punktbestimmung durch Vertikalwinkelmessung. Zeitschrift für Vermessungswesen 42: 241–253

334. Wester MJ (1999) Computer algebra system- a practical guide. John Wiley & Sons, Chichester, United Kingdom

335. Wickert J (2002) Das CHAMP-Radiookkultationsexperiment: Algorithmen, Prozessierungssystem und erste Ergebnisse. Dissertation. Scientific Technical Report STR02/07, GFZ Potsdam

336. Wild F (2001) Test an der Geschlossenen Lösung des "Twin P4P-Problems": Dreidimensionaler Vorwärts- und Rückwärtsschnitt. Studienarbeit, Geodetic Institute, Stuttgart University

337. Winkler F (1996) A polynomial algorithm in computer algebra. Springer, Wien

338. Wieser A, Brunner FK, 2002 Short static GPS sessions: Robust estimation results. Journal of GPS Solutions 5: 70–79

339. Wolfrum O (1992) Merging terrestrial and satellite networks by a ten-parameter transformation model. Manuscripta Geodaetica 17: 210–214

340. Wu W (1984) On the decision problem and mechanization of the theorem proving elementary geometry. Scientia Sinica 21: 150–172

341. Xu G (2003) GPS. Theory, algorithms and applications, Springer, Berlin Heidelberg

342. Xu P (1987) A test method for many outliers. I. T. C. Journal 4: 314–317

343. Xu P (1989a) Statistical criteria for robust methods. I. T. C. Journal 1: 37–40

344. Xu P (1989b) On robust estimation with correlated observations. Bull. Geod. 63: 237–252

345. Xu P (2002) A hybrid global optimization method: the one-dimensional case. Journal of Computation and Applied mathematics 147: 301–314

346. Xu P (2003) A hybrid global optimization method: the multi-dimensional case. Journal of Computation and Applied mathematics 155: 423–446

347. Yang Y (1999) Robust estimation of geodetic datum transformation. Journal of Geodesy 73: 268–274

348. Yang Y, Cheng MK, Shum CK, Tapley BD (1999) Robust estimation of systematic errors of satellite laser range. Journal of Geodesy 73: 345–349

349. You RJ (2000) Transformation of Cartesian to geodetic coordinates without iterations. Journal of Surveying Engineering 126: 1–7

350. Youcai H, Mertikas SP (1995) On the design of robust regression estimators. Man. Geod. 20: 145–160

351. Yunck TP (2003): The promise of spaceborne GPS for Earth remote sensing. International Workshop on GPS Meteorology, 14th-17th January 2003, Tsukuba, Japan

352. Zhang S (1994) Anwendung der Drehmatrix in Hamilton normierten Quaternionenen bei der Bündelblock Ausgleichung. Zeitschrift für Vermessungswesen 119: 203-211

353. Zippel R (1993) Effective polynomial computation, Kluwer Academic Publications, Boston 1993

Index